Charlotte

Enhancing Biological Control

Enhancing Biological Control

Habitat Management to Promote Natural Enemies of Agricultural Pests

CHARLES H. PICKETT
ROBERT L. BUGG

EDITORS

UNIVERSITY OF CALIFORNIA PRESS
Berkeley   Los Angeles   London

The editors thank the following individuals for facilitating the development of this book:
Larry Bezark, Ray Gill and Fred Hrusa, California Department of Food and Agriculture;
Lyndon Hawkins and Kathy Brunetti, California Department of Pesticide Regulation;
M. Christine Sparks and Darle E. Tilly, The Publications Department; and Timothy A. Rice.

University of California
Berkeley and Los Angeles, California

University of California
London, England

Enhancing biological control: habitat management to promote natural enemies of agricultural pests /
   Charles H. Pickett & Robert L. Bugg, editors.
      p.   cm.
Includes bibliographical references and index
ISBN 0-520-21362-9 (alk. paper)
1. Arthropod pests—Biological control. 2. Biological pest-control agents. I. Pickett, C.H. (Charles
   Hammond), 1952–   . II. Bugg, Robert Lyman.
   SB975.E55   1998
   632'.72—DC20

                                                                                      95-48418
                                                                                         CIP

Printed in the United States of America
9 8 7 6 5 4 3 2 1

The paper used in this publication meets the minimum requirements of American National Standard for
Information Sciences—Permanence of Paper for Printed Library Materials, ANSI Z39.48-1984.

Preface

Refuges have long been used to maintain flora and fauna that could not exist otherwise. Today, the preservation of wetlands, rainforests and other sensitive habitats has become a major priority of environmentalists throughout the world. Field biologists know that without certain habitat, migratory waterfowl and other threatened species will disappear. Farmers in the United States have planted hedges and maintained ponds for the purpose of attracting game birds. Rarely, however, do farmers consider setting aside land for aiding in pest management, although certain kinds of habitat can increase the diversity and abundance of natural enemies.

Over the last ten years an increasing number of field entomologists and farmers have recognized that conservation of natural enemies is important to effective biological control in many agricultural systems. Researchers and extension entomologists in the United States and Europe are studying the roles of non-crop vegetation and natural enemies on organic farms. Farmers in these areas grow a diversity of crops within a small area and often maintain a mix of non-crop vegetation to increase the diversity of natural enemies. Growers on small farms in developing countries have traditionally intercropped when growing food crops. Both types of growers know through practical experience that diversity within their farming system reduces pest problems and provides other benefits, e.g. enhanced soil quality. However, conventional agriculture in the United States has never relied on such practices because of the widespread use of pesticides and fertilizers and the lack of well-understood alternatives.

Biological control through importation of new natural enemies has been used to reduce arthropod pest problems for 100 years. However, the full potential of introduced and native natural enemies has rarely been realized in conventional agricultural because most crops are treated with broad-spectrum pesticides. Furthermore, annual crops are continually rotated or replanted, thus disrupting predator–prey associations. In these systems, lack of habitat, hosts, or prey may prevent natural enemies from persisting. Permanent strips of vegetation within a field may allow for the year-round presence of important natural enemies. These plantings can provide the continuity seen in forests and orchard systems where "successes" in classical biological control have been most frequent.

Most research efforts in biological control concern the importation, establishment and evaluation of new biological control agents, not the enhancement or maintenance of resident natural enemies. Only recently have there been sufficient, especially ongoing, field studies using habitat modification to warrant a book on the subject. This book, however, does not present habitat management as a panacea for pest control. Non-crop vegetation and intercrops may compete or otherwise interfere with economic crops. The same vegetation may support key pests or serve as a source of plant pathogens.

We include contributions from the United States, Finland, Germany, Great Britain, New Zealand, People's Republic of China, and Switzerland. Chapters summarize recent findings on aspects of habitat management, with an emphasis on regional perspectives. The introductory chapter by Bugg and Pickett provides a historical overview of the field of habitat modification to enhance natural enemies. That field is discussed in relation to other aspects of biological control, conventional pest control, integrated pest management, and integrated farming. Subsequent chapter topics include habitat modification in the following settings: (1) within fields (Chaney; Coll; Helenius; Riechert; Schoenig et al.), orchards (Häni and Keller), or vineyards (Häni and Keller; Roltsch et al.); (2) along or near the perimeters of fields (Bugg et al.; Murphy et al.; Nentwig; Wratten et al.); (3) distant from fields, including hedges or other non-cultivated areas (Beane and Bugg; Corbett; Häni et al.). Generalist and specialist natural enemies are discussed, as are both theoretical and practical issues. Whereas most of the chapters emphasize vegetational diversification, those by Beane and Bugg and by Olkowski and Zhang also detail the use of human-constructed arthropod nesting and overwintering sites. Schoenig et al. discuss issues of experimental design, analysis, and interpretation. Corbett presents a computer analysis of natural enemy movement that has important implications for both experimental and practical enhancement schemes. Häni et al. discuss "The Third Way," a constantly evolving technique that is intermediate between conventional and organic farming, compatible with biological control, and driven in part by public policy.

At a time when agricultural practices are changing rapidly, we hope that this book helps researchers, agricultural advisers, and progressive farmers alike to design and implement appropriate systems for enhancing biological control of agricultural pests.

We thank the following individuals for assisting in the development of this book: Larry Bezark of the California Department of Food and Agriculture; Ray Gill and Fred Hrusa of the California Department of Food and Agriculture for providing valuable taxonomical expertise; Lyndon Hawkins and Kathy Brunetti of the California Department of Pesticide Regulation; Darle E. Tilly and M. Christine Sparks of The Publications Department for their dedication to finalizing this book; Timothy A. Rice for page design and typesetting.

Charles H. Pickett and Robert L. Bugg
Davis, California, March, 1998

The editors dedicate this book to their parents:
Marion N. and Morris J. Pickett
Elizabeth B. and Sterling L. Bugg

CONTENTS

Introduction:
Enhancing Biological Control—
Habitat Management to Promote Natural
Enemies of Agricultural Pests

ROBERT L. BUGG—University of California Sustainable Agriculture Research and Education Program, University of California, Davis, CA 95616-8716

CHARLES H. PICKETT—Biological Control Program, California Department of Food and Agriculture, 3288 Meadowview Road, Sacramento, CA 95832

Definitions

DeBach (1964) narrowly defined biological control as "the action of parasites, predators, or pathogens in maintaining another organism's population density at a lower average than would occur in their absence." DeBach outlined three categories: importation, augmentation and conservation of natural enemies (Table 1). In our usage, "habitat management to enhance biological control" means the provision of resources to natural enemies to improve their effectiveness at controlling pests. This represents a subcategory for conservation of natural enemies. The latter term includes both modification of the environment and careful use of pesticides to protect natural enemies. Our concept includes the former but not the latter aspect. Here we consider this concept in some of its complexity, including the basis and origin of the practice; complications in research and implementation; integration with other aspects of biological control and with cultural control; the importance of vegetational diversification, diets and movement patterns of biological control agents; and reducing pesticide dependency through integrated farming.

Basis and Origin of the Practice

Habitat management can enhance biological control of arthropod pests by providing various environmental requisites to natural enemies, including: (1) supplementary foods (alternate hosts or prey, or in some cases pollen); (2) complementary foods (honeydew, pollen, nectar); (3) modified climate (e.g., windbreaks, Reed et al. 1970; Pickett et al. 1990); or (4) overwintering or nesting habitat (Janvier 1956; Huang & Yang 1987; Olkowski & Zhang this volume). This general approach is not new. Chinese citrus

Table 1. Components of biological control according to DeBach 1964.

Term	Original Definition
Biological Control	The action of predators, parasites, or pathogens in maintaining the densities of pests at a level lower than would occur in their absence.
Augmentation of Natural Enemies	Manipulation of natural enemies in order to make them more efficient in regulating pests: achieved through either inundative or inoculative releases of mass reared natural enemies.
Biological Control	The use of predators, parasites, and pathogens to reduce a pests density to a level lower than would occur in their absence.
Classical Biological Control	The importation and establishment of exotic natural enemies for control of pests (usually exotic).
Conservation of Natural Enemies	Modification of environment and judicious use of pesticides in order to conserve natural enemies.
Naturally-occurring Biological Control	The maintenance of a population density of an organism within certain upper and lower limits by action of both biotic and abiotic environmental factors; permanent control.
Inoculative Release of Natural Enemies	Release of mass reared enemies to control a pest by action of released individuals and progeny.
Inundative Release of Natural Enemies	Releases of mass reared enemies to control a pest, primarily by the released individuals, not their progeny.

growers have for centuries promoted the activity of the predaceous ant *Oecophylla smaragdina subnitida* Emery (Hymenoptera: Formicidae) by placing bamboo poles between trees previously inoculated with ant colonies (McCook 1882; Yang 1982; Huang & Yang 1987; Way & Khoo 1992). *Oecophylla smaragdina subnitida* Emery is an important natural enemy of a coreid pest of coconuts in the Solomon Islands (O'Connor 1950). Despite its demonstrated value, the theory and practice of enhancing natural enemies has been relatively neglected. This neglect is understandable, because biological control is seldom simple to document, and habitat management to enhance ongoing biological control leads to further complications.

Complications in Research and Implementation

Biological control in its simplest manifestations is nonetheless complex. Natural enemy effectiveness has been evaluated by many workers. For example, the late George Tamaki and co-workers explored qualitative and quantitative dimensions of "predator power and efficacy." In Tamaki's view, predator effectiveness depended on numerous interactive factors, including issues of temporal and spatial synchrony and colonization/dispersal dynamics of populations, as well as thermal activity patterns, voracity, relative reproductive rates, and availability of alternative foods (Tamaki et al. 1974; Tamaki & Long 1978; Tamaki 1981; Tamaki et al. 1981). In assessing successful classical biological programs, Huffaker (1974) found that effective imported parasites and predators shared the following traits:

1. High searching capacity (ability to find host or prey when these exist at low densities);
2. Host specificity;
3. High potential rate of increase (high fecundity, short developmental time);
4. Ability to occupy the same niche as the host or prey.

In evaluating the role of habitat modification, we must add the following complicating issues:

1. Inadvertent direct or indirect effects of the habitat management scheme on crop plants;
2. Variable effects of scale (including plot size and proximity) (Corbett and Plant 1993); distance that a natural enemy must travel between overwintering and summer habitat may affect its abundance and impact (Kido et al. 1984; Pickett et al. 1990; Thomas et al. 1991, Nentwig this volume);
3. Year-to-year variation in populations of pests and natural enemies;
4. Intra-guild interactions among natural enemies (Rosenheim & Wilhoit 1993; Rosenheim et al. 1993);
5. Differential performance of pests or natural enemies in various farming systems and plot configurations (Coll & Bottrell 1994; Kruess & Tscharntke 1994), crops (Braimah & Van Emden 1994; Coll & Ridgway 1995), cultivars (Obrycki & Tauber 1985; Barbour et al. 1993, 1997; Stoner 1996), or microhabitats (Wilson and Gutierrez 1980; Coll & Bottrell 1992).

From the above, it may be obvious that evaluation of enhancement of biological control is a daunting problem, far different from "spray-and-count" entomology or even from the incremental advance to the "count-and-spray" approach of conventional "integrated pest management." Such complexity is likely to deter many scientists. Nonetheless, habitat management to enhance biological control has proven valuable in many settings, and it may be more expensive to ignore the concept than to fund the relevant research and development.

Integration with Other Aspects of Biological Control

Although conceptually distinct from various other categories of biological control (Table 1), habitat management may in practice interrelate with these. Practitioners of classical biological control, the introduction of natural enemies against pests of exotic origin, have long been aware of the importance of physical environmental requisites. Typically, the first consideration in foreign exploration for new natural enemies is the degree of climatic similarity between the native and target regions. Even if physical environmental requisites are met, however, biological requisites may remain unfulfilled, leading to failed importations (Drea & Hendrickson 1986; Rose & DeBach 1990). Habitat management could be used in conjunction with classical biological control, to enhance the establishment of introduced natural enemies (De Charmoy 1917; Shahjahan 1968; Pickett et al. 1996).

Habitat management may also be used to support "augmentative biological control," which involves repeated releases of natural enemies, usually purchased from a commercial insectary. In fact, some habitat-management schemes effectively create field insectaries that could enable released natural enemies to better control pests. In California, related scenarios are currently being explored with the use of cover crops or "nursery plants" to promote build-up following inoculative release of the generalist predatory mite *Euseius tularensis* (Congdon) (Acari: Phytoseiidae) in California citrus orchards (Ouyang et al. 1992; E.E. Grafton-Cardwell & Y. Ouyang personal communication 1997) and *Anaphes iole* Girault (Hymenoptera: Mymaridae), an egg parasite of *Lygus hesperus* Knight (Hemiptera: Miridae) in strawberries (*Fragaria* x*annanassa*) (Norton et al. 1992; S. Udayagiri & S.C. Welter personal communication1997).

Integration with Cultural Control

Habitat management to enhance biological control may also interrelate with aspects of cultural control, which involves farming practices that make the environment less favorable to pests. In some cases, the same practices that favor natural enemies may interfere directly with pest survival, reproduction, dispersal, or colonization. In such instances, habitat management can represent the dovetailing of biological and cultural controls. This could occur in the use of trap crops (Hokkanen 1991) that arrest movement by the pest *Lygus hesperus* into California strawberry fields (S. Udayagiri & S.C. Welter personal communication1997). Experimental trap crops include such nectar-bearing plants as buckwheat (*Fagopyrum esculentum*) and sweet alyssum (*Lobularia maritima*), which may also attract and sustain beneficial arthropods, including *Anaphes iole*, which parasitizes *Lygus* eggs.

In some cases, however, cultural controls and biological controls may be at cross purposes. For example, in California almond production, orchard sanitation procedures that remove "mummy nuts" reduce overwintering both by the pest navel orangeworm (*Amyelois transitella* [Walker], Lepidoptera: Pyralidae) (Barnett et al. 1989), and by its parasite *Goniozus legneri* Gordh (Hymenoptera: Bethylidae) (Legner & Warkentin 1988; Legner & Gordh 1992; W. Bentley personal communication1997).

Habitat modification is likely to affect a broad spectrum of resident indigenous or naturalized and insectary-reared natural enemies. This may introduce complications in that some generalist predators may interfere with biological control by key natural enemies, as Rosenheim & Wilhoit (1993) and Rosehneim et al. (1993) showed for control of cotton aphid (*Aphis gossypii* Glover, Homoptera: Aphididae) by common green lacewing (*Chrysoperla carnea* [Stephens], Neuroptera: Chrysopidae).

Vegetational Diversification

A key issue in enhancement is vegetational diversification. The concept of biodiversity does not merely concern the number of species present, but also includes aspects of composition, structure, and function. In turn, each of these aspects includes multiple levels. As discussed by Franklin (1988) and Noss (1990), the three hierarchies of biodiversity can be envisioned as:

I. Composition
    A. Landscape types
        1. Communities, ecosystems
            a. Species, populations
                i. Genes
II. Structure
    A. Landscape patterns
        1. Physiognomy, habitat structure
            a. Population structure
                ii. Genetic structure
III. Function
    A. Landscape processes, disturbances; land-use trends
        1. Interspecific interactions, ecosystem processes
            a. Demographic processes, life histories
                i. Genetic processes

These hierarchies are actually not distinct, but interrelated, and have been represented graphically as interconnected spheres (Noss 1990).

Evidence from native grassland ecosystems shows that increased plant species richness (number of species) leads to enhanced total plant canopy cover, biomass production, and plant survival through droughts, as well as to reduced nitrate leaching (Tilman & Downing 1994; Tilman 1996; and Tilman et al. 1996), and that, in general, community parameters are stabilized by diversity, but that population parameters are not. This suggests that artful management of biodiversified systems may be needed to favor the principal economic "target species," and that multiple economically useful species should be employed in such systems, to ensure overall success. Economically useful species could include those providing multiple benefits. For example, plants that serve as food crops could also promote beneficial insects, improve soil quality, etc.

Farm functions may be influenced by numerous aspects of biodiversity. In particular, ecologists have long held that vegetational diversity can affect densities of arthropods. In some well-documented cases, diversification may lead to reduced incidence of phytophagous arthropods (Costello 1994; Costello & Altieri 1995). Vegetational diversification can affect pests in various ways, with or without the mediation of natural enemies. Diversification may exacerbate pests (notably, *Heliothis* spp. [Lepidoptera: Noctuidae], *Lygus* spp. [Hemiptera: Miridae], stink bugs [Hemiptera: Pentatomidae]) if these build up on one crop and then disperse to another while the latter is vulnerable. Diversification could also provide pests with nectar, overwintering habitat, etc. (Kennedy & Margolies 1985; Andow 1988; Bugg 1991; Zhao et al. 1992). By contrast, diversity may reduce pests if it interferes with pest movement, colonization, and reproduction, or otherwise interrupts pest life cycles. If diversification is used to reduce pest dispersal, colonization, or reproduction on target crops, this is usually termed cultural control. This can involve: (1) maintenance of trap or diversionary crops (Hokkanen 1991; Fleischer & Gaylor 1987), (2) confusing pests visually or olfactorally thus reducing their colonization of target crops (Kareiva 1983), (3) host plant nutritional changes that reduce pest success (Hauptli et al. 1990; Robinson 1996), and (4) microclimatic changes that reduce pest success (Pickett et al. 1990; Solbreck & Sillén-Tullberg, 1990; Volkl et al. 1993; Coll this volume). By contrast, if the aim is to enhance performance of natural enemies, this is considered by strict constructionists to be an aspect of biological control. At times, both cultural and biological controls may operate simultaneously (see reviews by Altieri & Letourneau [1982] and Andow [1988]).

When diversifying agroecoystems, there are dangers in applying simplistic thinking, as illustrated in the case of California grape vineyards. Spiders are important generalist predators in vineyards (Costello et al. 1995; Roltsch et al. this volume). Cover crops may enhance densities of certain spiders and thereby reduce incidence of grape leafhopper (Settle et al. 1986; R. Hanna personal communication 1997), or may in some cases exacerbate problems with orange tortrix moth (*Argyrotaenia citrana* [Fernald], Lepidoptera: Tortricidae) or false chinchbug (*Nysius* spp., Hemiptera: Lygaeidae) (C. Shannon personal communication 1995; R. Bartolucci personal communication 1993). Various wild and domestic plants outside the vineyard can be important sources of beneficial insects, such as the wasp *Anagrus epos* (Girault) (Hymenoptera: Mymaridae) which parasitizes eggs of grape leafhopper (*Erythroneura elegantula* Osborn, Homoptera: Cicadellidae) (Murphy et al. 1996; Murphy et al. this volume). By contrast, important pests may be harbored on various wild plants. Vectors of Pierce's disease, e.g., bluegreen sharpshooter (*Graphocephala atropunctata* [Signoret], Homoptera: Cicadellidae), are harbored on various riparian trees and shrubs. In some cases, the same plant harbors both the beneficial (*A. epos*) and pest (sharpshooter) insects (e.g., wild blackberry, wild grape) (Doutt & Nakata 1973; A. Purcell personal communication 1997).

Root (1973) proposed two alternative hypotheses to explain observations of reduced densities of pests in some vegetationally diverse systems. By the "resource concentration hypothesis," specialist herbivores are more likely to find and remain in pure stands of plants, resulting in higher populations as compared to diverse stands. By contrast, according to the "enemies hypothesis," predators and parasitoids cause greater mortality to herbivores in diverse versus pure stands of plants. Regardless of which mechanism is more important, most studies show that fewer crop pests are found as the diversity of an agroecosystem increases (Risch et al. 1983; Russell 1989; Andow 1991). To date, no comprehensive theory has been proposed to predict under what conditions and to what extent vegetational diversity may increase the effectiveness of natural enemies.

The term "farmscaping" has been coined to describe the conscious and integrated modification of agricultural settings (Bugg et al. this volume). The objectives of farmscaping are to provide habitat for wildlife, enhance the aesthetics of the farm, provide natural weed control through competition, and establish insect control through enhancement of natural enemy activity. Habitat management for enhancement of natural enemies can be considered a component of farmscaping. Plants can be employed that have special entomological associations relevant to fields, orchards, and vineyards (Nentwig this volume). Within fields, such plants could include cash crops, cover crops, and resident vegetation (Chaney this volume; Wratten et al. this volume). Outside of fields, natural enemy activity could be affected through management of field margins, hedgerows, windbreaks, and specific vegetation planted along roadsides, ditches, and adjoining wildlands (Häni et al. this volume; Murphy et al. this volume; Olkowski & Zhang this volume). Habitat modification can affect crops, pests, and beneficial arthropods through transforming the physical environment, inducing microclimatic and local climatic changes. For example, windbreak trees may affect wind, insolation, insulation, temperature, and humidity. Preferred overwintering niches and nesting sites could also be provided (Beane & Bugg this volume). The biotic environment could be modified through provision of alternate prey or hosts, pollen, nectar, and honeydew.

Diets of Biological Control Agents

The host or prey ranges of biological control agents may vary from a few species to many. The diets may also include plant materials or products. If restricted to one species of host or prey, a natural enemy is termed monophagous; if it attacks a narrow range, oligophagous; if a broad range, polyphagous. Specialist and generalist natural enemies are at opposite extremes of this continuum. Specialists and generalists may be affected differently by vegetational diversification. Increased plant diversity may increase the effectiveness of generalist natural enemies but decrease that of specialists (Sheehan 1986). The temporal stability of a system may also affect natural enemies differently. Based on studies in northern Europe, Helenius (this volume) suggests that generalists can withstand repeated disruptions of their local habitat and are thus likely to be more abundant and play a greater role in pest suppression.

A distinction is made between foods that sustain reproduction (essential foods) and those that do not. For example, in lieu of aphids, convergent lady beetle (*Hippodamia convergens* Guerin-Ménéville, Coleoptera: Coccinellidae) can subsist on pollen or nectar (Hagen 1974) or flower thrips (Thysanoptera: Thripidae) (Bugg et al. 1990); these foods permit fat accumulation, but not egg development (Hagen 1974). There are also foods of intermediate value. For example, the pink lady beetle (*Coleomegilla maculata* DeGeer, Coleoptera: Coccinellidae), an important generalist predator in vegetable crops (Groden et al. 1990), can reproduce on a diet of pollen alone (Smith 1960; Hodek et al. 1978), but the addition of pea aphid to the diet increases the proportion of larvae that reach adulthood (Smith 1960).

It is also important to distinguish between complementary and supplementary foods (Leon & Tumpson 1975; Rapport 1980). Complementary foods are those that taken together provide a balanced diet that can sustain a beneficial arthropod: thus, such foods complement one another. Supplementary foods, by contrast, are those that may substitute for one another. DeLima (1980) showed that western bigeyed bug, (*Geocoris pallens* Stål, Hemiptera: Lygaeidae) attained maximum longevity, fecundity, and per capita prey consumption rates when cotton extrafloral nectar was available. Thus, nectar and various prey species (including agricultural pests) may serve as complementary resources. That is, optimal diets may involve appropriate combinations of the two types of food. Bigeyed bugs (*Geocoris* spp., Hemiptera: Lygaeidae) are hemimetabolous; adults and nymphs have similar diets. Moreover, bigeyed bugs disperse rather slowly while foraging (Corbett & Plant 1993). Therefore, simultaneous presentation of complementary foods is important in sustaining these predators. By contrast, ichneumonid wasps (Hymenoptera: Ichneumonidae) are highly mobile as adults, which may feed on nectar of floral (Maingay et al. 1991; Jervis et al. 1993) or extrafloral origin (Bugg et al. 1989), but as larvae may parasitize caterpillars that infest various vegetable crops. Thus, the caterpillars (protein- and lipid-rich) and nectar (carbohydrate-rich) are dissimilar foods that can be said to complement one another. In some cases, simultaneous provision of dietary complements can increase voracity of generalist natural enemies (DeLima 1980).

By contrast with complementary dietary items, similar foods, such as different kinds of caterpillars serving as alternate hosts for ichneumonid larvae, are said to supplement one another. If supplementary foods (e.g., alternate prey) are presented simultaneously, these could actually detract from biological control of insect pests on crop plants by causing natural enemies to concentrate on non-pest hosts or prey (see Ridgway & Jones 1968; Ables et al. 1978; Bugg et al. 1987; Riechert & Bishop 1990; Bugg 1992). By contrast, enhancement of biological control might be more likely through sequential, rather than simultaneous and continuous, presentation of supplementary foods (Bugg et al. 1991a, b). During late winter and early spring, rye (*Secale cereale*) sustains bird–cherry oat aphid (*Rhopalosiphum padi* [L.], Homoptera: Aphididae), whereas hairy vetch (*Vicia villosa*) later harbors abundant pea aphid (*Acyrthosiphon pisum* [Harris], Homoptera: Aphididae). These aphid species serve as prey to various lady beetles (Coleoptera: Coccinellidae), and the patterns of aphid abundance are reflected in lady beetle attendance and reproduction on these plants, whether grown in monocultures (Bugg et al. 1990) or in mixtures (Bugg et al. 1991a). The sequential supply of prey aphids may enable lady beetles to persist in areas from which they might otherwise disperse, but such patterns have not reliably led to enhanced densities of lady beetles in Georgia pecan trees (Bugg et al. 1991; Bugg & Dutcher 1993).

The issue of sequential vs. simultaneous presentation of alternative supplementary foods may be expected to arise in the case of relay-intercropping vs. simultaneous intercrops. For example, relay intercrops of subterranean clover (*Trifolium subterraneum*) and other late-maturing winter-annual legumes harbored abundant pea aphid and other potential prey, and led to increased densities of the generalist predator bigeyed bug *Geocoris punctipes* (Say) (Hemiptera: Lygaeidae). As these annual crops senesced and died during late spring and early summer, *G. punctipes* dispersed to and accumulated in adjoining cantaloupe (*Cucumis melo*) (Bugg et al. 1991). By contrast, several studies on simultaneous intercropping with annual or perennial forage legumes have not shown distinct enhancement of generalist predator densities or efficiencies amid the associated vegetable crops (e.g., Ryan et al. 1980; Rämert 1996).

Patterns of Natural Enemy Movement

The impact of environmental heterogeneity on the movement of natural enemies is highlighted by Corbett (this volume), whose simulation model concerns the diffusion and density of natural enemies in an agroecosystem involving strips of alfalfa (*Medicago sativa*) planted amid cotton (*Gossypium hirsutum*). If planted at the same time or earlier than cotton, the alfalfa serves as a source of the predatory mite *Metaseiulus occidentalis* (Nesbitt) (Acari: Phytoseiidae). If planted later, the alfalfa functions as a sink, reducing the relative abundance of natural enemies in an adjoining crop. Corbett (this volume) also used a computer simulation model to predict relative enhancement of natural enemies in edges vs. the centers of cotton fields interspersed with alfalfa strips. He determined that distance between strips had varying effects in enhancing natural enemies that disperse slowly, versus those that disperse rapidly. An arthropod may undergo short-range, intermediate, or long-range dispersal at various points in its life cycle (Beane & Bugg this volume). The scale of habitat modification should relate meaningfully to the scale of dispersal by the target arthropods.

As suggested above, spatial issues include the scale at which habitat modification is practiced. Diversification can be implemented using closely to widely interspersed vegetation. Schettini (1992) regarded within-field interspersion as being on the "micro" scale, between-field on the "meso" scale, and between-farm on the "macro" scale. Regional issues might be regarded as operative on a "mega" scale. Recent studies have indicated that pest density increases with increasing patch size of crops (Litsinger et al. 1991; Walde 1991), suggesting that intercropping may become less effective at reducing pest numbers as the area-wide acreage of the target crop increases. This might be expected because perimeter/area ratio decreases as the area of a monocrop increases, so fewer individuals (pests) will be lost by emigration (Kareiva 1985).

Reducing Pesticide Dependency Through Integrated Farming

With declining availability and efficacy of some insecticides, growers and researchers alike are expressing increased interest in non-chemical approaches to pest management. Cultural and biological control of arthropod pests can play important roles in farming systems in which the use of pesticides is minimized (Ellis 1990; Prokopy 1991; Benbrook et al. 1996; Thrupp 1996). Several terms have been coined to describe other types of farming systems that share a common goal, to decrease dependency on synthetic pesticides, fertilizers, and other purchased inputs. These types of farming systems are often pooled under the term 'alternative agriculture' (Table 2), and include various branches of organic farming.

There have been several evaluations of arthropod faunae of various types of "sustainable" agricultural systems, including organic farms (e.g., Dritschilo & Wanner 1980; Kromp 1990; Hesler et al. 1993). Such studies appear to indicate an awareness of the importance of evaluating "baseline" arthropod and other biotic activity in systems that are not disrupted by heavy agrichemical inputs.

In general, practitioners of alternative agriculture share the belief that a single tactic can have multiple benefits. This belief suggests a holistic perspective, and proponents of integrated agriculture typically emphasize soil improvement and provision of habitat for beneficial arthropods as approaches to preventive pest management. These growers may actually regard so-called pests as potential allies or assets that should be managed. For instance, so-called weeds may be managed as soil-improving cover crops (McKee 1910; Kourik 1986). Flower thrips (*Frankliniella* spp., Thysanoptera: Thripidae), regarded by many as pests of crop plants, also serve as predators of spider mites, and as alternate prey to important generalist predators (González et al. 1982; Pickett et al. 1988).

As reported by Stanhill (1990), organic farming systems attain on aver-

Table 2. Systems of agricultural production.

Term	Source	Original Definition
Bio-Dynamic Farming	Whipple 1987 Sattler & vonWistinghausen 1992	Through soil health, seeks to minimize or eliminate the farm's vulnerability to external costs; goal is quality rather than quantity of yield.
Integrated Farming	Vereijken 1989	Integrates economics of farm, well-being of environment and society; biological and physical techniques of pest control are given preference over chemical.
Low-Input Farming	Ikerd 1990	Reduced use of external, off farm purchases, such as nonrenewable energy, inorganic, or synthetic agricultural products; greater reliance on internal biological processes, i.e. cover crops and biological pest control.
Low-Input Sustainable Agriculture	Ingels & Liebhardt 1991	A combination of sustainable agriculture and low-input farming.
Organic Farming	Lockeretz 1989 Scofield 1986	As originally used by Lord Northbourne, the term meant a farm that was a biologically complete, living, integrated whole; now refers to farming systems that avoid use of synthetic pesticides and fertilizers.
Permaculture	Mollison 1988 Dahlberg 1991	Ecologically-based landscape design emphasizing integration of multi-storied plant complexes, aquaculture, and household systems. Seeks to develop new systems rather than rely on traditional agroecosystems.
Regenerative Agriculture	Rodale 1983	Systems designed to regenerate over the long term entire food systems, including the farming community itself and all aspects of food production: distribution, processing, recycling, and disposal.
Sustainable Agriculture	Ingels & Liebhardt 1991	An agricultural system that is economically viable, environmentally sound, and socially responsible.
Third Way	Häni (this volume)	Integrated Farming, per Vereijken (1989), with the addition of a formal certification program to ensure compliance by farmers.

age about 91% of the yields of the corresponding conventional farming systems. It is perhaps a sobering thought for entomologists that arthropod pests are of relatively minor concern to organic growers in the United States and elsewhere; weeds are the primary pest problems (USDA 1980; Peacock & Norton 1990).

The concept of "integrated pest management," or IPM, reflects the idea that control may require the coordinated use of several discrete tactics: biological, cultural, and chemical controls (Table 3) (Frisbie & Smith 1989). By contrast, as mentioned, alternative agriculture stresses the multiple benefits of each tactic (e.g., of cover cropping, which can enhance soil fertility and pest management), as well as integration of tactics. Despite its conceptual emphasis on coordination, the IPM philosophy, with its focus on pest management *per se*, reflects greater reductionism than do the various ideologies of alternative agriculture. Thus, the underlying premises of IPM and alternative agriculture are different; however, the two approaches share a common theme: diversification and integration. An implied tenet of each is that the greater the complexity of the agroecosystem or number of management tactics used in controlling a pest, the lower the probability of pest outbreaks. Thus, although the central themes of alternative agriculture and IPM indicate a difference in perspective, they may be viewed as complementary. For example, a grower may incorporate the use of a specific control strategy, such as confusant pheromone for control of a key pest, thus eliminating the use of a disruptive, broad-spectrum insecticide. This in turn would allow a cover crop, already aiding soil fertility and weed control, to sustain complexes of natural enemies that assist in the control of secondary pests. Thus, an IPM innovation could allow an alternative agricultural tactic to more fully express one of its multiple benefits.

The essential problem with integrated pest management as it has been practiced for the last 30 years is its primary reliance on, and the overriding influence of, synthetic pesticides (Vereijken 1989). These pesticides can neutralize potentially complementary tactics. A heavy emphasis on pesticides has occurred despite the philosophical underpinnings of IPM that stress the harmonious linkage of biological, cultural, and chemical controls. IPM protocols are dictated primarily by short-term economic considerations, and not by any overriding incentives to reduce pesticide use. In some cases, these protocols (e.g., use of economic thresholds, sampling, or simulation models) can result in reductions, but many result in increased use (Allen et al. 1987). Pesticide applications, whether based on calendar dates or action thresholds, can have adverse side effects, including: (1) destruction of natural enemies leading to resurgence of target pests and outbreaks of secondary pests (DeBach 1974; van den Bosch 1978; Luckmann & Metcalf 1982); (2) reduced vegetational diversity where herbicides are heavily used, including impairment of crop rotations and reductions of innocuous plants (Bird et al. 1990; Francis & Clegg 1990); (3) impairment of plant health by insecticides

Table 3. Approaches to pest control.

Term	Source	Original Definition
Biological Control	DeBach 1964	The action of parasites, predators, and pathogens to maintain another organism's population density at a lower average than would occur in its absence.
Biologically-Intensive Integrated Pest Management	Frisbie & Smith 1989	The use of biological control, host plant resistance and cultural management to meet the traditional philosophical tenets of IPM.
Chemical Control	Watson et al. 1976	The use of synthetic pesticides to control arthropods; reduces populations temporarily.
Cultural Control	Watson et al. 1976	The use of farming practices associated with production practices that make the environment less favorable for the survival of pests.
Host Plant Resistance	Watson et al. 1976	Resistance by plants through inherited traits that reduce the damage by insects over that which would occur in their absence.
Integrated Control or Integrated Pest Management	Stern et al. 1959 Smith & Reynolds 1966	A pest management system that utilizes all suitable techniques in a compatible manner to reduce pest populations and maintain them at levels below those causing economic injury.
Regulatory Control	Watson et al. 1976	The use of regulations to impose area-wide cultural practices or quarantines designed to control or exclude pests.
Levels and Stages of Integrated Pest Management	Prokopy et al. (1990)	A hierarchical scheme for categorizing pest management technologies, including levels and stages within levels, with incrementally increasing integration.
Physical Control	Watson et al. 1976	Direct or indirect measures that kill the insect or disrupt its physiology by means other than insecticides.

as well as herbicides (Jones et al. 1986); (4) resistance of pests to pesticides, necessitating higher doses, mixes of pesticides, or new pesticides.

As indicated by Frisbie and Smith (1989), IPM is changing. These authors distinguished between "chemically-dependent" and "biologically-intensive" or "bio-intensive" IPM (Table 3). The mainstay of the former has been synthetic pesticides, whereas the latter would rely on the combined action of biological and cultural controls, and host plant resistance. Bio-intensive IPM as originally defined depends in part on classical plant breeding and genetic engineering of crop plants and microorganisms. Such an approach would, of course, still require special off-farm purchases (e.g., special crop cultivars, microbial pesticides), but these tools will probably prove less disruptive of natural pest controls than broad-spectrum pesticides have been. Growers using such "soft" tools will be able to benefit from biological interactions that may otherwise be attenuated or distorted under an umbrella of pesticide use.

We (R.L.B. & C.H.P.) are encountering increasing numbers of growers interested in diversifying farming systems in order to reduce the need for pesticides. Diversification may take several forms: increased crop rotation, intercropping, use of cover crops, and refuges for natural enemies. We now see the development of teams of farmers, scientists, educators, and consultants who work to develop and refine such tools and their integration in developing productive whole farming systems (McGrath et al. 1992; Santer 1995; Dlott et al. 1996). These trends contrast strikingly with our experience from the 1970s through the '80s, when "progressive farming" relied heavily on chemical inputs to maximize productivity. IPM can now be viewed as a stepping stone towards more diverse farming systems. Coordinated use of several "soft" pest management techniques disrupt natural processes very little. Organic and middle-path growers (see Häni et al. this volume) have demonstrated that farming is not only possible, but productive and profitable in the absence of broad-spectrum pesticides. In many cases, increased vegetational diversity has played an important role in enabling this. In most cases, the precise role and the overall economics of habitat enhancement in these scenarios remains to be elucidated. Clearly, farmers, consultants, researchers, educators, and facilitators can work together to develop even better biologically integrated farming systems. Habitat management to enhance natural enemies of pests will certainly be a key element in this evolution.

References

Ables, J. R., S. L. Jones & D. W. McCommas, Jr. 1978. Response of selected predator species to different densities of *Aphis gossypii* and *Heliothis virescens* eggs. Environ. Entomol. 7:402-404.

Allen, W. A., E. G. Ragotte, R. F. Kazmeirczak, Jr., M. T. Lambur & G. W. Norton. 1987. The national evaluation of extension's integrated pest management (IPM) programs. Virginia Cooperative Extension Service Publication 4914-010, Blacksburg.

Altieri, M. A. & D. K. Letourneau. 1982. Vegetation diversity and insect pest outbreaks. CRC Critical Rev. Plant Sci. 2:131-169.

Andow, D. A. 1988. Management of weeds for insect manipulation in agro-ecosystems, pp. 265-301. *In* M. A. Altieri & M. Z. Liebman [eds.], Weed management in agro-ecosystems: ecological approaches. CRC Press, Raton, Florida.

Andow, D. A. 1991. Vegetational diversity and arthropod population response. Ann. Rev. Entomol. 36:561-586.

Barbour, J. D., R. R. Farrar & G. G. Kennedy. 1993. Interaction of *Manduca sexta* resistance in tomato with insect predators of *Helicoverpa zea*. Entomologia Experimentalis Et Applicata 68:143-155.

Barbour, J. D., R. R. Farrar & G. G. Kennedy. 1997. Populations of predaceous natural enemies developing on insect-resistant and susceptible tomato in North Carolina. Biol. Control 7. In press.

Barnett, W. W., L. C. Hendricks, W. K. Asai, R. B. Elkins, D. Boquist & C. L. Elmore. 1989. Management of navel orangeworm and ants. California Agric. 43(4):21-22.

Benbrook, C., E. Groth III, J.M. Halloran, M. K. Hansen & S. Marquardt. 1996. Pest management at the crossroads. Consumers Union. Yonkers, N.Y.

Bird, G. W., T. Edens, F. Drummond & E. Groden. 1990. Design of pest management systems for sustainable agriculture, pp. 55-110. *In* C. A. Francis, C. B. Flora, & L. D. King [eds.], Sustainable agriculture in temperate zones. John Wiley & Sons, Inc., N.Y., N.Y.

Braimah, H. & H. F. Van Emden. 1994. The role of the plant in host acceptance by the parasitoid *Aphidius rhopalosiphi* (Hymenoptera: Braconidae). Bull. Entomol. Res. 84:303-306.

Bugg, R. L. 1991. Cover crops and control of arthropod pests of agriculture, pp. 157-163. *In* W. L. Hargrove [ed.], Cover crops for clean water. Proceedings of the Soil and Water Conservation Society, Ankeny, IA. International Conference, April 9-11 1991. West Tennessee Experiment Station, Jackson, Tennessee.

Bugg, R. L. 1992. Using cover crops to manage arthropods on truck farms. HortScience 27:741-745.

Bugg, R. L., J. D. Dutcher, and P. J. McNeill. 1991a. Cool-season cover crops in the pecan orchard understory: effects on Coccinellidae (Coleoptera) and pecan aphids (Homoptera: Aphididae). Biological Control 1:18-15.

Bugg, R. L., and J. D. Dutcher. 1993. *Sesbania exaltata* (Rafinesque-Schmaltz) as a warm-season cover crop in pecan orchards: effects on aphidophagous Coccinellidae and pecan aphids. Biol. Agric. Hortic. 9:215-229.

Bugg, R. L., L. E. Ehler & L. T. Wilson. 1987. Effect of common knotweed (*Polygonum aviculare*) on abundance and efficiency of insect predators of crop pests. Hilgardia 55(7):1-53.

Bugg, R. L., R. T. Ellis & R. W. Carlson. 1989. Ichneumonidae (Hymenoptera) using extrafloral nectar of faba bean (*Vicia faba* L., Fabaceae) in Massachusetts. Biol. Agric. Hortic. 6:107-114.

Bugg, R. L., S. C. Phatak & J. D. Dutcher. 1990. Insects associated with cool-season cover crops in southern Georgia: implications for pest control in the truck-farm and pecan agroecosystems. Biol. Agric. Hortic. 7:17-45.

Bugg, R. L., F. L. Wäckers, K. E. Brunson, J. D. Dutcher & S. C. Phatak. 1991b. Cool-season cover crops relay intercropped with cantaloupe: influence on a generalist predator, *Geocoris punctipes* (Hemiptera: Lygaeidae). J. Econ. Entomol. 84:408-416.

Coll, M. & D. G. Bottrell. 1992. Mortality of European cornborer larvae by natural enemies in different corn microhabitats. Biol. Control 2:95-103.

Coll, M. & D. G. Bottrell. 1994. Effects of nonhost plants on an insect herbivore in diverse habitats. Ecology 75:723-731.

Coll, M. & R. L. Ridgway. 1995. Functional and numerical responses of *Orius insidiosus* (Heteroptera: Anthocoridae) to its prey in different vegetable crops. Ann. Entomol. Soc. Amer. 88:732-738.

Corbett, A. & R. E. Plant. 1993. Role of movement in the response of natural enemies to agroecosystem diversification: a theoretical evaluation. Environ. Entomol. 22:519-531.

Costello, M. J. 1994. Broccoli growth, yield and level of aphid infestation in leguminous living mulches. Biol. Agric. Hortic. 10:207-222.

Costello, M. J. & M. A. Altieri. 1995. Abundance, growth rate and parasitism of *Brevicoryne brassicae* and *Myzus persicae* (Homoptera:Aphididae) on broccoli grown in living mulches. Agric. Ecosystems Environ. 52:187-196.

Costello, M. J., M. A. Mayse, K. M. Daane, W. A. O'Keefe, and C. B. Sisk. 1995. Spiders in San Joaquin Valley grape vineyards. University of California Division of Agriculture and Natural Resources Leaflet No. 21530.

DeBach, P. 1964. Biological control of insect pests and weeds. Reinhold Publishing Corporation, N.Y.

DeBach, P. 1974. Biological control by natural enemies. Cambridge University Press, N.Y., N.Y.

De Charmoy, D. D. 1917. An attempt to introduce scoliid wasps from Madagascar to Mauritius. Bull. Entomol. Res. 13:245-254.

De Lima, J. O. G. 1980. Biology of *Geocoris pallens* Stål on selected cotton genotypes. Ph.D. dissertation, Department of Entomology, University of California, Davis.

Dlott, J., T. Nelson, R. Bugg, M. Spezia, R. Eck, J. Redmond, J. Klein, and L. Lewis. 1996. California, USA: Merced County BIOS Project, pp. 115-126. *In* L.A. Thrupp [ed.], New partnerships for sustainable agriculture. World Resources Institute. Washington, D.C.

Doutt, R. L. & J. Nakata. 1973. The Rubus leafhopper and its egg parasitoid: an endemic biotic system useful in grape pest management. Environ. Entomol. 2:381-386.

Drea, J. J. & R. M. Hendrickson. 1986. Analysis of a successful classical biological control project: the alfalfa blotch leafminer (Diptera: Agromyzidae) in the northeastern United States. Environ. Entomol. 15:448-455.

Dritschilo, W. & D. Wanner. 1980. Ground beetle abundance in organic and conventional corn fields. Environ. Entomol. 9:629-631.

Ellis, P. R. 1990. The role of host plant resistance to pests in organic and low input agriculture, pp. 93-102. *In* Brighton Crop Protection Conference Mono. No. 45. Organic and Low Input Agriculture. Brighton, U.K.

Fleischer, S. J. & M. J. Gaylor. 1987. Seasonal abundance of *Lygus lineolaris* (Heteroptera: Miridae) and selected predators in early season uncultivated hosts: implications for managing movement into cotton. Environ. Entomol. 16:379-389.

Francis, C. A. & M. D. Clegg. 1990. Crop rotations in sustainable production systems, pp. 107-122. *In* Edwards et al. [eds.], Sustainable agricultural systems. Soil and Water Conservation Society, Ankeny, Iowa.

Franklin, J. F. 1988. Structural and functional diversity in temperate forests, pp. 166-175. *In* E. O. Wilson [ed.], Biodiversity. National Academy Press. Washington, D.C.

Frisbie, R. E. & J. W. Smith, Jr. 1989. Biologically intensive integrated pest management: the future, pp. 151-164. *In* J. J. Menn & A. L. Steinhauer [eds.], Progress and perspectives for the 21st century. Entomological Society of America Centennial National Symposium.

González, D., B. R. Patterson, T. F. Leigh & L. T. Wilson. 1982. Mites: A primary food source for two predators in San Joaquin Valley cotton. Calif. Agric. 36:18-20.

Groden, E., F. Drummond, R. A. Casagrande, & D. L. Haynes. 1990. *Coleomegilla maculata* (Coleoptera: Coccinellidae): its predation upon the Colorado potato beetle (Coleoptera: Chrysomelidae) and its incidence in potatoes and surrounding crops. J. Econ. Entomol. 83:1306-1315.

Hagen, K. S. 1974. The significance of predaceous Coccinellidae in biological and integrated control of insects. Entomophaga 7:25-44.

Hauptli, H., D. Katz, B. R. Thomas & R. M. Goodman. 1990. Biotechnology and crop breeding for sustainable agriculture, pp. 141-150. *In* C.A. Edwards, R. Lal, P. Madden, R. H. Miller & G. House [eds.], Sustainable Agricultural Systems. Soil and Water Conservation Society. Ankeny, Iowa.

Hesler, L. S., A. A. Grigarick, M.J. Oraze & A. T. Palrang. 1993. Arthropod fauna of conventional and organic rice fields in California. J. Econ. Entomol. 86:149-158.

Hodek, I., Z. Ruzicka, & M. Hodkova. 1978. Pollinivorie et aphidophagie chez *Coleomegilla maculata* lengi. Ann. Zool. Ecol. Anim. 10:453-459.

Hokkanen, H. M. T. 1991. Trap cropping in pest management. Annu. Rev. Entomol. 36:119-138.

Huang, H. T. & Yang, P. 1987. The ancient cultured citrus ant. Bioscience 37:665-671.

Huffaker, C.B. (ed.). 1974. Biological control. Plenum Publishing Corporation. New York, New York.

Ikerd, J. E. 1990. Agriculture's search for sustainability and profitability. J. Soil Water Conserv. 45:18-30.

Ingels, C. & W. C. Liebhardt. 1991. What is sustainable agriculture? University of California Sustainable Agriculture Research & Education Program, Davis, California.

Janvier, H. 1956. Hymenopterous predators as biological control agents. J. Econ. Entomol. 49:202-205.

Jervis, M. A., N. A. C. Kidd, M.G. Fitton, T. Huddleston & H. A. Dawah. 1993. Flower-visiting by hymenopteran parasitoids. J. Nat. Hist. 27:67-105.

Jones, V. P., N. C. Toscano, M. W. Johnson, S. C. Welter, and R. R. Youngman. 1986. Pesticide effects on plant physiology: integration into a pest management program. Bull. Entomol. Soc. Am. 32:103-109.

Kareiva, P. 1983. Influence of vegetation texture on herbivore populations: resource concentration and herbivore movement, pp. 259-289. In R. F. Denno and M. S. McClure [eds.], Variable plants and herbivores in natural and managed ecosystems. Academic Press, N.Y.

Kareiva, P. 1985. Finding and losing host plants by Phyllotreta: patch size and surrounding habitat. Ecology 66:1809-1816.

Kennedy, G. G. & D. C. Margolies. 1985. Mobile arthropod pests: management in diversified agroecosystems. Bull. Entomol. Soc. Amer. 31:21-27.

Kido, H., D. L. Flaherty, D. F. Bosch & K. A. Valero. 1984. French prune trees as overwintering sites for the grape leafhopper egg parasite. Am. J. Enol. Viticulture 35:156-160.

Kourik, R. 1986. Designing and maintaining your edible landscape naturally. Metamorphic Press, Santa Rosa, California.

Kromp, B. 1990. Carabid beetles (Coleoptera, Carabidae) as bioindicators in biological and conventional farming in Austrian potato fields. Biol. Fertil. Soils 9:182-187.

Kruess, A. & T. Tscharntke. 1994. Habitat fragmentation, species loss, and biological control. Science 264:1581-1584.

Legner, E. F. & G. Gordh. 1992. Lower navel orangeworm (Lepidoptera: Phycitidae) population densities following establishment of Goniozus legneri (Hymenoptera: Bethylidae) in California. J. Econ. Entomol. 85:2153-2160.

Legner, E. F. & R. W. Warkentin. 1988. Parasitizations of Goniozus legneri (Hymenoptera: Bethylidae) at increasing parasite and host, Amyelois transitella (Lepidoptera: Phycitidae), densities. Ann. Entomol. Soc. Am. 81:774-776.

Leon, J. A. & D. B. Tumpson. 1975. Competition between two species for two complementary or substitutable resources. J. Theoretical Biol. 50:185-201.

Litsinger, J. A., V. Hasse, A. T. Barrion & H. Schmutterer. 1991. Response of Ostrinia furnacalis (Guenee) (Lepidoptera: Pyralidae) to intercropping. Environ. Entomol. 20:988-1004.

Lockeretz, W. 1989. Defining a sustainable future: basic issues in agriculture. Northwest Report No. 8.

Luckmann, W. H. & R. L. Metcalf. 1982. The pest-management concept, pp. 1-32. In R. L. Metcalf & W. H. Luckmann [eds.], Introduction to insect pest management. Wiley, N. Y.

Maingay, H., R. L. Bugg, R.W. Carlson & N. A. Davidson. 1991. Predatory and parasitic wasps (Hymenoptera) feeding at flowers of sweet fennel (Foeniculum vulgare Miller var. dulce Battandier & Trabut, Apiaceae) and spearmint (Mentha spicata L., Lamiaceae) in Massachusetts. Biol. Agric. Hortic. 7:363-383.

McCook, H. 1882. Ants as beneficial insecticides, pp. 263-271. In Proceedings of the Academy of Natural Sciences, Philadelphia.

McGrath, D., L. S. Lev, H. Murray & R.D. William. 1992. Farmer/scientist focus sessions: a how-to manual. Working Paper No. 92-104. Graduate Faculty of Economics, Oregon State University, Corvallis, Oregon.

McKee, R. 1910. Orchard green-manure crops in California. U.S. Department of Agriculture, Bureau of Plant Industry, Bulletin No. 190.

Mollison, B. 1988. Permaculture: a designer's manual. Tagari Publications, P.O. Box 1, Tyaltum, New South Wales, Australia 2484. 574 pp.

Murphy, B. C., J.A. Rosenheim, & J. Granett. 1996. Habitat diversification for improving biological control: abundance of *Anagrus epos* (Hymenoptera: Mymaridae) in grape vineyards. Environ. Entomol. 25:495-504.

Norton, A. P., S. C. Welter, J.L. Flexner, C. G. Jackson, J. W. DeBolt & C. Pickel. 1992. Parasitism of *Lygus hesperus* (Miridae) by *Anaphes iole* (Mymaridae) and *Leiophron uniformis* (Braconidae) in California strawberry. Biol. Control 2:131-137.

Noss, R. 1990. Indicators for monitoring biodiversity: a hierarchical approach. Conserv. Biol. 4:355-364.

Obrycki, J. J. & M. J. Tauber. 1985. Seasonal occurrence and relative abundance of aphid predators and parasitoids on pubescent potato plants. Can. Entomol. 117:1231-1237.

O'Connor, B. A. 1950. Premature nutfall of coconuts in the British Solomon Islands Protectorate. Agric. J. [Figi] 21:1-2.

Ouyang, Y., E. E. Grafton-Cardwell & R. L. Bugg. 1992. Effects of various pollens on development, survivorship, and reproduction of *Euseius tularensis* (Acari: Phytoseiidae). Environ. Entomol. 21:1372-1376.

Peacock, L. & G. A. Norton. 1990. A critical analysis of organic vegetable crop protection in the U.K. Agric. Ecosystems Environ. 31:187-197.

Pickett, C. H., S. E. Schoenig & M. P. Hoffmann. 1996. Establishment of the squash bug parasitoid, *Trichopoda pennipes* Fabr. (Diptera: Tachinidae), in northern California. Pan-Pacific Entomol. 72:220-226.

Pickett, C. H., L. T. Wilson, & D. L. Flaherty. 1990. The role of refuges in crop protection, with reference to plantings of French prune trees in a grape agroecosystem, pp. 151-165. *In* N. J. Bostanian, L. T. Wilson, & T. J. Dennehy [eds.], Monitoring and integrated management of arthropod pests of small fruit crops. Intercept Publishing, Great Britain.

Pickett, C. H., L. T. Wilson, & D. González. 1988. Population dynamics and within-plant distribution of western flower thrips, an early season predator of predator of spider of spider mites on cotton. Environ. Entomol. 17:551-559.

Prokopy, R. J. 1991. A small low-input commercial apple orchard in eastern North America: management and economics. Agric. Ecosystems Environ. 33:353-362.

Prokopy, R. J., S. A. Johnson & M. T. O'Brien. 1990. Second-stage integrated management of apple arthropod pests. Entomol. Exp. Appl. 54:9-19.

Rämert, B. 1996. The influence of intercropping and mulches on the occurrence of polyphagous predators in carrot fields in relation to carrot fly (*Psila rosae* [F.]) (Dipt., Psilidae) damage. J. Appl. Entomol. 120:39-46.

Rapport, D. J. 1980. Optimal foraging for complementary resources. Amer. Nat. 116:324-46.

Reed, D. K., W. G. Hart, & S. J. Ingle. 1970. Influence of windbreaks on distribution and abundance of brown soft scale in citrus groves. Ann. Entomol. Soc. Amer. 63:792-794.

Ridgway, R. L. & S. L. Jones. 1968. Field cage releases for *Chrysopa carnea* for suppression of populations of the bollworm and the tobacco budworm on cotton. J. Econ. Entomol. 61:892-898.

Riechert, S. E. & L. Bishop. 1990. Prey control by an assemblage of generalist predators in a garden test system. Ecology 71:1441-1450.

Risch, S. J., D. Andow & M. A. Altieri. 1983. Agroecosystem diversity and pest control: data, tentative conclusions, and new research directions. Environ. Entomol. 12:625-629.

Robinson, R. A. 1996. Return to resistance: breeding crops to reduce pesticide dependence. agAccess. Davis, California.

Rodale, R. 1983. Breaking new ground: the search for sustainable agriculture. Futurist 1:15-20.

Root, R. B. 1973. Organization of a plant-arthropod association in simple and diverse habitats: the fauna of collards (*Brassica oleraceae*). Ecol. Monogr. 43:95-120.

Rose, M., & P. DeBach. 1990. Conservation of natural enemies, pp. 461-472. *In* D. Rosen [ed.], The armored scale insects, their biology, natural enemies and control, Vol. 4B. Elsevier Science Publishers B.V., Amsterdam, The Netherlands.

Rosenheim, J. A. & L. R. Wilhoit. 1993. Why lacewings may fail to suppress aphids: predators that eat other predators disrupt cotton aphid control. Calif. Agric. 47(5):7-9.

Rosenheim, J. A., L. R. Wilhoit & C.A. Armer. 1993. Influence of intraguild predation among generalist insect predators on the suppression of an herbivore population. Oecologia 96:439-449.

Russell, E. P. 1989. Enemies hypothesis: a review of the effect of vegetational diversity on predatory insects and parasitoids. Environ. Entomol. 18:590-599.

Ryan, J., M. F. Ryan & F. McNaeidhe. 1980. The effect of interrow plant cover on populations of the cabbage root fly, *Delia brassicae* (Wiedemann). J. Appl. Ecol. 17:31-40.

Santer, L. (ed.). 1995. BIOS for almonds: a practical guide to biologically integrated orchard systems management. Community Alliance with Family Farmers Foundation, Davis, California/Almond Board of California, Modesto, California.

Sattler, F. & E. vonWistinghausen. 1992. Bio-dynamic farming practice. English translation by A.R. Meuss. Bio-Dynamic Agriculture Association. Client, Stourbridge, West Midlands, England.

Schettini, T. M. 1992. Multilevel habitat management as a paradigm for developing regenerative crop production systems. HortScience 27:736-737.

Scofield, A. M. 1986. Organic farming—the origin of the name. Biol. Agric. Hortic. 4:1-5.

Settle, W.H., L. T. Wilson, D.L. Flaherty, & G.M. English-Loeb. 1986. The variegated leafhopper, an increasing pest of grapes. California Agriculture 40(4):30-32.

Shahjahan, M. 1968. Effect of diet on the longevity and fecundity of the adults of the tachinid parasite *Trichopoda pennipes pilipes*. J. Econ. Entomol. 61:1102-1103.

Sheehan, W. 1986. Response by specialist and generalist natural enemies to agro-ecosystem diversification: a selective review. Environ. Entomol. 15:456-461.

Smith, B. C. 1960. A technique for rearing coccinellid beetles on dry foods, and influence of various pollens on the development of *Coleomegilla maculata lengi* Timb. (Coleoptera: Coccinellidae). Can. J. Zool. 38:1047-1049.

Smith, R. F. & H. T. Reynolds. 1966. Principles, definitions and scope of integrated pest control. vol. 1, pp. 11-17. *In* Proceedings, Food and Agricultural Organization symposium on integrated pest control, October 11-15 1965. Food and Agricultural Organization, Rome.

Solbreck, C. & B. Sillén-Tullberg. 1990. Population dynamics of a seed-feeding bug, *Lygaeus equestris*. 1. Habitat patch structure and spatial dynamics. Oikos 58:199-209.

Stanhill, G. 1990. The comparative productivity of organic agriculture. Agric. Ecosystems Environ. 10:1-26.

Stern, V. M., R.F. Smith, R. Van Den Bosch & K.S. Hagen. 1959. The integration of chemical and biological control of the spotted alfalfa aphid. I. The integrated control concept. Hilgardia 29:81-101.

Stoner, K. A. 1996. Plant resistance to insects: a resource available for sustainable agriculture. Biol. Agric. Hortic. 13:7-38.

Tamaki, G. 1981. Biological control of potato pests, pp. 178-192. *In* J.H. Lashomb and R. Casagrande [eds.], Advances in potato pest management. Hutchison Ross Publishing Company, Stroudsberg, PA. 289 pp.

Tamaki, G. & G. E. Long. 1978. Predator complex of the green peach aphid on sugarbeets: expansion of the predator power and efficacy model. Environ. Entomol. 7:835-842.

Tamaki, G., J. U. McGuire & J.E. Turner. 1974. Predator power and efficacy: a model to evaluate their impact. Environ. Entomol. 3:625-630.

Tamaki, G., M. A. Weiss & G.E. Long. 1981. Evaluation of plant density and temperature in predator-prey interactions in field cages. Environ. Entomol. 10:716-720.

Thomas, M. B., S. D. Wratten, & N. W. Sotherton. 1991. Creation of "island" habitats in farmland to manipulate populations of beneficial arthropods: predators densities and emigration. J. Appl. Ecol. 28:906-917.

Thrupp, L. A. [ed.]. 1996. New partnerships for sustainable sgriculture. World Resources Institute. Washington, D.C.

Tilman, D. 1996. Biodiversity: population versus ecosystem stability. Ecology 77:350-363.

Tilman, D. & J. A. Downing. 1994. Biodiversity and stability in grasslands. Nature 367:363-365.

Tilman, D., D. Wedin, & J. Knops. 1996. Productivity and sustainability influenced by biodiversity in grassland ecosystems. Nature 379:718-720.

United States Department of Agriculture. 1980. Report and recommendations on organic farming. Washington, D.C.

van den Bosch, R. V. 1978. The pesticide conspiracy. University of California Press, Berkeley, California.

Vereijken, P. 1989. From integrated control to integrated farming, an experimental approach. Agric. Ecosystems Environ. 26:37-43.

Völkl, W., H. Zwólfer, M. Romstöck-Völkl, C. Schmelzer. 1993. Habitat management in calcareous grasslands: effects on the insect community developing in heads of *Cynarea*. J. Appl. Ecol. 30:307-315.

Walde, S. J. 1991. Patch dynamics of a phytophagous mite population: effect of number of subpopulations. Ecology 72:1591-1598.

Watson, T. F., L. Moore, & G. W. Ware. 1976. Practical insect pest management. W. H. Freeman and Co., San Francisco, CA.

Way, M. J., & K. C. Khoo. 1992. Role of ants in pest management. Annu. Rev. Entomol. 37:479-503.

Wilson, L. T. and A. P. Gutierrez. 1980. Within-plant distribution of predators on cotton: comments on sampling and predator efficiencies. Hilgardia 48(2):3-12.

Yang, P. 1982. Biology of the yellow citrus ant, *Oecophylla smaragdina* and its utilization against citrus insect pest. Acta Sci. Nat. Univ. Sunyatseni 3:102-105.

Zhao, J. Z. , G. S. Ayers, E. F. Grafius & F.W. Stehr. 1992. Effects of neighboring nectar-producing plants on populations of pest Lepidoptera and their parasitoids in broccoli plantings. Great Lakes Entomol. 25:253-258.

The Importance of Movement in the Response of Natural Enemies to Habitat Manipulation

ANDREW CORBETT—Entomology Department, University of California, Davis, CA 95616

Introduction

Increasing the vegetational diversity of an agroecosystem in order to enhance the abundance of natural enemies is dependent on two important assumptions:

1. Natural enemies that have overwintered in interplanted vegetation will eventually move to the adjacent crop vegetation, subsequently killing herbivorous insects; or,

2. Natural enemies that use resources in interplanted vegetation (regardless of where they have overwintered) will eventually move to, and spend a significant amount of time in, the adjacent crop vegetation, killing herbivorous insects.

These predictions are, in general, quite reasonable: movement is important in the life-history of most insect species. A large number of habitat management approaches to natural enemy augmentation have proven fruitful (Russell 1989), attesting to the basic validity of the predictions stated above. However, insect movement is not an "all-or-nothing" behavior. Movement behavior varies both among species, due to differences in life-histories, and within species, due to environmental heterogeneity. This variability in movement behavior can express itself in a variety of ways:

1. Variation in the tendency to initiate a movement—as a response, for example, to the local abundance of resources (e.g., Kareiva & Odell 1987).

2. Variation in the duration of a movement—due, for example, to the current life-history stage of the insect (e.g., Dingle 1972).

3. Variation in the choice to end a movement at a given location—due, for example, to visual or chemical stimuli perceived from a given vegetation type (e.g., Rauscher 1983).

The expression of these components of movement by a given natural enemy will determine whether interplanted vegetation will act as a source of the natural enemy and will determine the spatial extent of enhancement. Since all of these components of movement vary within and among species, and depend on the vegetation types involved, the potential variation in the success of habitat management systems is tremendous. This is reflected in the variability noted in reviews of studies of natural enemy enhancement (Russell 1989; Andow 1991).

To account for this variability we need to develop a mechanistic framework within which to think about, and perhaps predict, the response of natural enemies to spatial arrangements of vegetation in agricultural systems. The primary objective of this chapter is to present a simple mechanistic context in which to understand how movement behavior will influence a natural enemy's response to vegetational diversity. I shall first review some field results with natural enemy enhancement and discuss the issues raised with respect to natural enemy movement. I shall then propose a mechanistic framework and discuss the predictions of a simple mathematical representation of natural enemy movement in diversified agricultural systems. Finally, I shall discuss ways in which movement considerations should be integrated into habitat management research.

A Review of Field Studies

Kemp and Barrett (1989) conducted one of the more comprehensive studies of the influence of agroecosystem diversity on arthropod abundance. Their objective was to determine if maintaining uncultivated strips in soybeans might serve as "corridors" for natural enemies to enter the crop. They established plots of soybeans under four replicated regimes: two strips of "successional" vegetation within a plot; two strips of grassy vegetation; insecticide treated; and a control. The abundance of natural enemies was not, in general, increased by the diversification of the soybean agroecosystem. In particular, the insidious flower bug (*Orius insidiosus* [Say], Hemiptera: Anthocoridae), an important predator, was more abundant in control plots than in plots with strips of successional vegetation. The temporal changes in abundance of *O. insidiosus* suggest that, although *O. insidiosus* was more abundant in diversified plots early in the season, as the season progressed it moved into the successional vegetation eventually resulting in lower abundance on soybeans. In other words, successional vegetation appeared to be serving as a "sink" for *O. insidiosus* rather than a source, resulting in decreased abundance in the diversified agroecosystem.

The possibility that interplanted vegetation in agroecosystems might actually decrease natural enemy abundance in some situations has been raised by other researchers as well. Bugg et al. (1987) found that, for most of the crops studied, interplanting common knotweed did not result in higher abundance of *Geocoris* spp. (Hemiptera: Lygaeidae) or higher predation rates. This was the case despite the fact that common knotweed harbors large populations of *Geocoris*. The authors concluded that the abundant floral resources and alternate prey on common knotweed may act to hold natural enemies and prevent them from foraging in adjacent crop vegetation. Perrin (1975) voiced similar concerns as a result of his studies with the perennial stinging nettle, suggesting that the abundant aphids on this weed may act as "diversionary prey" for natural enemies and thus reduce their foraging on adjacent crop vegetation. This sink effect observed in some systems is the direct result of the movement behavior of natural enemies and is predictable based on concepts of optimal foraging: natural enemies should exhibit higher residence times in areas that provide a greater abundance of preferred food resources.

In other agroecosystems interplanted vegetation clearly serves as a source of natural enemies. Van Emden (1967) provided perhaps the earliest example: he found that the abundance of coccinellids feeding on aphids on brussels sprouts was highest adjacent to weedy borders. Fye and Carranza (1972) showed that sorghum interplanted in cotton could act as a source of convergent lady beetle (*Hippodamia convergens* [Guérin-Méneville], Coleoptera: Coccinellidae), due to the abundance of green bugs as an alternate food source early in the season. Similarly, Corbett et al. (1991) found that strips of alfalfa interplanted within cotton can result in increased abundance of the predatory mite (*Metaseiulus occidentalis* [Nesbitt], Acari: Phytoseidae) on cotton immediately adjacent to the strips. Strips of grass interplanted into winter wheat serve as an early season source of the beetle *Demetrias atricapillus* (L.) (Coleoptera: carabidae) (see Coll this volume). Numerous other examples can be found in reviews (Russell 1989; Andow 1991).

Thus it appears that increasing the vegetational diversity of agroecosystems can have variable results depending on the natural enemies and vegetation involved (Bugg 1992). In some cases, vegetation management leads to increased natural enemy abundance and improved control of herbivores. In other cases, however, interplanting vegetation decreases the abundance of natural enemies on the crop—in much the same way as "trap cropping" serves as a cultural control for some herbivorous species (e.g.,

Sevacherian & Stern 1974). Understanding how the consequences of increasing vegetational diversity can vary so dramatically is a major challenge in developing vegetation management systems for natural enemy enhancement. Such understanding is needed to predict the characteristics of diversified agroecosystems that will lead to increased, rather than decreased, natural enemy abundance on crop vegetation.

Where interplanted vegetation serves as a source of natural enemies a further question arises: at what spatial scale are natural enemies enhanced? In other words, to what distance from interplanted vegetation is natural enemy abundance greater than it would be in an undiversified agroecosystem? Not all natural enemies will be enhanced at the same spatial scale. For example, Corbett (1991) found that the predatory mite *M. occidentalis* was enhanced on cotton up to 15 meters from alfalfa. This is not surprising given the passive wind dispersal and relatively low mobility exhibited by phytoseiid mites. By contrast, the parasitoid *Anagrus epos* (Hymenoptera: Mymaridae) was enhanced in abundance in grape vineyards up to one kilometer from its overwintering site (Doutt & Nakata 1973). Lady beetles (Coleoptera: Coccinellidae) (van Emden 1965; Fye 1972) and ground beetles (Coleoptera: Carabidae) (Thomas et al. 1991) are probably intermediate, exhibiting enhancement at 30 to 100 m from interplanted vegetation. The spatial scale of enhancement for this small sample of natural enemies is closely related, as expected, to their overall mobility (Figure 1; see "A Mechanistic Framework for Natural Enemy Movement in Diversified Agroecosystems," below).

The spatial scale of enhancement, and therefore the mobility of the "target" natural enemy, have implications for the design of a vegetation management system. The spatial scale of enhancement influences the ratio of crop vegetation to alternate vegetation and their interspersion. While this has been recognized implicitly (see especially Russell 1989), there has been little agreement about the appropriate design of vegetationally diversified agroecosystems. For example, Fye (1972) argued that high mobility should result in enhancement even with large distances between strips of interplanted vegetation. Others have argued that "fine-grained" interspersion of crop and alternate vegetation is necessary to retain highly mobile natural enemies within the system (Pollard 1971; Perrin 1975; Dempster & Coaker 1974; Sheehan 1986).

Clearly the movement behavior of natural enemies has a strong influence over their response to agroecosystem diversification. Movement behavior is probably responsible for much of the variation seen in the results of studies with agroecosystem diversification (Russell 1989; Andow 1991). In the following sections I shall provide a mechanistic context within which to account for some of this variability and will discuss the implications of results from a simple mathematical model.

A Mechanistic Framework for Natural Enemy Movement in Diversified Agroecosystems

Let us imagine a hypothetical, diversified agricultural system and a natural enemy population inhabiting that system. The crop system could be, for example, winter wheat interplanted with strips of perennial grass as an overwintering refuge (see Wratten et al. this volume). Let us focus on one potential natural enemy, the carabid beetle *Demetrias atricapillus,* and imagine its potential behavior in wheat. At any given instant, an individual beetle will be at a location beneath the crop canopy and will have the "choice" of either moving or remaining (hopefully to feed on some ill-fated grub). If it moves, then it will need to "choose" a direction. The distance traversed in that direction before the next "instant" will depend on the speed at which it walks. In the absence of mortality and unintended movements (e.g., in the mouth of a foraging insectivore) the sequence of these behaviors through time—moving or not, direction of movement, and walking speed—will fully determine the position of the beetle at any given instant in the future. This does not mean that we can predict the beetle's location in the future. It would be possible if we knew the choices it would make through time. In practice, however, the time sequence of behaviors is the result of a complex interaction between the beetle and its surroundings. The choice of whether to move will depend on many factors, such as the time elapsed since the beetle last encountered food, the soil temperature, or the drive to mate. Walking speed and direction also vary due to similar factors. When this variation is combined with simple random variation in movement behavior, the future location of the beetle is unknowable.

It is possible, however, to approximate the future location of our beetle by making simplifying assumptions. The simplest assumptions are: (1) there is a fixed probability that the beetle will move from its current location within a specified, short time interval (i.e., a probability that does not change with time or location), (2) walking speed remains constant, and (3) there is an equal probability of moving in any direction away from the current location. We would make the same assumptions in describing, for example, the movement of a dust particle through air—these assumptions underlie descriptions of diffusion (Okubo 1980). Despite the apparent simplicity of

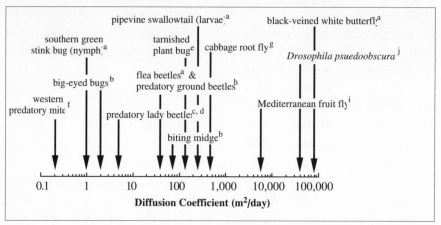

Figure 1. Estimated diffusion rates for a variety of arthropods. Estimates of diffusion rate that are based on immature stages are indicated on graph. Sources: (a) Kareiva 1981; (b) Rudd & Gandour 1985 ("biting midge": Culicoides impunctatus (Goet)); (c) Wetzler & Risch 1984; (d) Kareiva & Odell 1987 (Coccinella septempunctata (L.); approximate middle of range for D shown in Figure 4); (e) Fleischer et al. 1988 (From Table 2, 1984, day 2. D is calculated from hours and B provided in Table 2 using eq. 12 in Rudd & Gandour 1985); (f) Rabbinge & Hoy 1980 (D calculated from data provided on mean displacement using eq. 6.11 in Kareiva 1981); (g) Banks et al. 1988; (h) Gordon & McKinlay 1986 (Pterostichus spp.; D calculated from data provided on mean displacement using eq. 6.11 in Kareiva 1981); (i) Plant & Cunningham 1991; (j) Dobzhansky et al. 1979.

these assumptions, they can be used to approximate the movement of many insects in simple habitats and over relatively short time intervals (Kareiva 1981; Rudd & Gandour 1985). The assumptions lead to a specific mathematical equation (see Appendix A) that predicts the probability that an insect, starting from a specified location, will be in some other location after a fixed amount of time (Kareiva 1981). We cannot predict our beetle's precise location after a given amount of time. However, if we release a large number of beetles in a location, we can predict their distribution after a fixed amount of time if they behave roughly according to our assumptions. Also, if we specify an assumed distribution of individuals within the crop at a given time, we can predict their future distribution. With the passage of time the diffusion equation predicts that individuals will become increasingly evenly distributed through the field.

The diffusion equation includes a variable known as the diffusion coefficient (see Appendix A), which for an insect can be considered a measure of overall mobility (Kareiva 1981). By releasing and recapturing large numbers of marked individuals, the diffusion coefficient has been estimated for several insect species (Figure 1). These estimates represent approximations of the overall mobility for the species and are valid only for the conditions under which they were measured. Variation in resource availability or life-history stage can have large effects on measured mobility (Kareiva 1981).

Returning to our hypothetical diversified field and population of carabids, we would now like to explore how individual carabids might respond to interplanted strips of perennial grass. Following harvest, the strips provide a refuge in which carabids can overwinter (see Wratten et al. this volume). Early in the season these strips provide a more favorable physical environment for carabids than does the exposed soil beneath the recently germinated crop. The strips also provide abundant food compared to that afforded by the newly-established crop. Therefore, our carabid probably prefers the strips, and spends more time there early in the season than in the crop. This inference can be taken into account by making a simple modification to our assumptions. Specifically, we can assume that the probability of our hypothetical beetle "choosing" to move within a specified time period is lower in the strips than it is in the crop. The other assumptions remain unaffected. This modified set of assumptions leads to a specific mathematical equation (see Appendix A) that represents the process of spatially varying diffusion (Okubo 1980; Kareiva 1981). In this new equation the diffusion coefficient is not constant but varies with location: it is lower in the strip vegetation than in the crop. Dobzhansky et al. (1979) used a spatially varying diffusion model to successfully predict the distribution of fruit flies within a heterogeneous habitat.

Spatially varying diffusion can be used as the basis for a model to predict population changes if we make further assumptions concerning the life-history and growth rate of the natural enemy. Let us assume that (1) the natural enemy has two life-stages—a mobile adult stage that follows the assumptions of movement discussed above and an immobile, immature stage; (2) there is colonization of the agroecosystem by adult natural enemies at a rate that is constant through time and for all locations in the field; (3) birth and death rates are constant through time and for all locations within the field—there is no variation in these rates between crop and interplanted vegetation. These assumptions, combined with the assumptions for movement discussed above, lead to a mathematical model that predicts the abundance of natural enemies through time (Corbett & Plant 1993; see Appendix B). This model has further implicit assumptions (and therefore limitations):

Figure 2. Diagram of hypothetical diversified agroecosystem.
Interplanted strips are placed
100m apart within a crop. The
model predicts natural enemy
abundance along a transect
through the field.

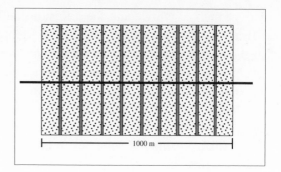

1. The natural enemy has no directional behavior at the interface between the crop and interplantings—strong directional behavior would yield different results from those currently predicted;
2. Both the crop and the interplantings provide essential food resources and an acceptable physical habitat for the natural enemy, thus the population experiences moderate growth throughout the system and there is no tendency for long-range dispersal;
3. Diffusion is constant through time in the crop and interplantings, thus the model does not account for temporal changes in the availability of resources and its influence on movement;
4. Natural enemies enter a "non-migratory" movement phase upon maturation with no long-range pre-ovipositional dispersal.

Despite these simplifying assumptions, the model represents a useful starting point for understanding the response of natural enemies to agroecosystem diversification.

Predictions of the Model
Scenario I

Let us assume that our hypothetical field has 10m wide strips interplanted at 100m intervals (Figure 2). These strips are used solely as an overwintering refuge by three natural enemy species: (1) a predatory mite having very low mobility (specifically, a diffusion coefficient of $1m^2$/day, see Figure 1); (2) a predatory coccinellid beetle having moderate mobility ($10m^2$/day); and, (3) a highly mobile parasitoid ($100m^2$/day). Once the crop has germinated, the strips do not provide resources in any greater abundance than the crop nor do they provide a more favorable physical habitat. The mobility (i.e., the probability of making a move in a given time period) is therefore the same in the strips as it is in the crop. Natural enemies overwinter in the strips at a density of 10 individuals per m^2.

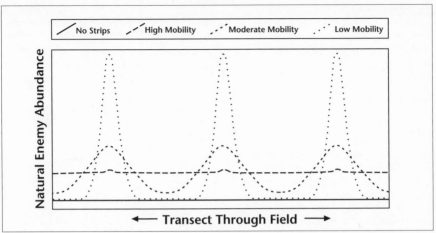

Figure 3. Spatial patterns predicted by model for field with interplanted strips that serve solely as an overwintering refuge. Peak abundance occurs at interplanted strips. Patterns are shown for hypothetical natural enemies of three different mobilities: low mobility (1 m²/day), moderate mobility (10 m²/day), high mobility (100 m²/day).

The spatial patterns in abundance predicted by the model for these three hypothetical natural enemies are illustrated in Figure 3. The model predicts that the natural enemies will spread from the strips, resulting in higher abundance in the crop than would have occurred in a crop monoculture. The distance to which they are enhanced varies substantially, however. For the predatory mite, enhancement is confined to the region immediately adjacent to interplanted strips, producing a steep gradient in density with increasing distance. The highly mobile parasitoid, on the other hand, is enhanced throughout the crop—there is no spatial pattern to suggest that strips influenced abundance. As a result, natural enemies with low mobility exhibit no enhancement beyond 20 m from strips while more mobile natural enemies are enhanced by 4-fold (Figure 4).

In one sense these results simply demonstrate the obvious: highly mobile species disperse farther than less mobile species. However, strikingly disparate spatial patterns (Figure 3) are produced by diffusion rates that differ by only two orders of magnitude, which is much smaller than the range of diffusion rates observed for various insect species (Figure 1). Thus, relatively small differences in mobility result in large differences in the extent of enhancement. Two important implications arise. First, it may not be possible to generalize concerning the spatial patterns that would be exhibited by different taxonomic groups (e.g., parasitoids versus ground

Figure 4. Effect of mobility on the abundance of natural enemies on crop vegetation in a diversified agroecosystem. "Relative Abundance" is the ratio of natural enemy abundance predicted for the diversified system to that predicted for a crop monoculture. Relative abundance is calculated only for crop vegetation more than 20m from interplantings. Effect of diversification is shown for three different situations: where the interplantings act solely as an overwintering refuge; where they provide additional food resources but no overwintering sites; and where interplantings provide both.

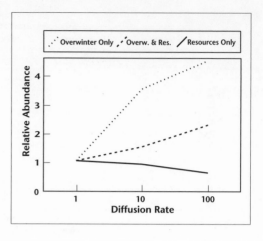

beetles). Second, mobility within species may vary by an order of magnitude or more. Variation in host abundance, environmental conditions, or life-history stage may influence the response of a particular natural enemy species to vegetational diversity.

Scenario II

As a second hypothetical situation, let us assume that we have the same diversified field and the same natural enemies. In this scenario, however, the interplanted vegetational zones are not overwintering refuges: natural enemies must colonize the agroecosystem from external sources. The strips do, however, provide more resources than the crop. Therefore, the probability of making a move in a given time period is lower in the strips than in the crop. The resources in interplantings are assumed to be "substitutable," to some degree, for resources that occur in the crop. They could be either: (1) alternate prey or hosts ("supplementary" resources), or (2) floral resources that are an imperfect substitute for the preferred host but that benefit the natural enemy when available ("complementary" resources) (Rogers 1985; Bugg et al. 1987; see also Bugg 1992).

The spatial patterns in abundance predicted by the model for this scenario are shown in Figure 5. The abundance of our three natural enemies is higher in the interplanted strips than in the crop vegetation. This accumulation of natural enemies in the strips is due to the lower tendency for movement there and results in the strips acting as a sink for the natural enemies (Figure 4), as has been observed in the field studies discussed above. Again, the spatial patterns vary greatly (Figure 5). The parasitoid exhibits a spatially

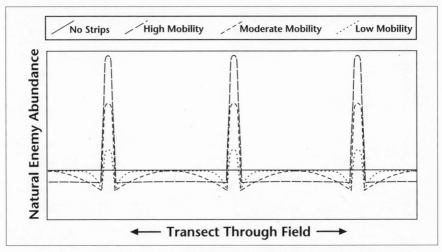

Figure 5. Spatial patterns predicted by model for field with interplanted strips that provide additional food resources but do not serve as an overwintering refuge. Peak abundance occurs at interplanted strips. Patterns are shown for hypothetical natural enemies of three different mobilities: low mobility ($1 m^2$/day); moderate mobility ($10 m^2$/day); high mobility ($100 m^2$/day).

uniform density in the crop vegetation and the greatest accumulation in the strips. This sink effect results in an abundance on the crop that is 60% of what it would be in an undiversified field (Figure 4). The other natural enemies exhibit some spatial patterning within the crop and a milder sink effect.

Scenario III

Strips could serve a dual function: overwintering refuges and sources of abundant supplementary resources. Predicted spatial patterns are shown in Figure 6. As when strips serve only as overwintering refuges, the natural enemies spread from the strips and are enhanced on the crop. However, natural enemies also accumulate within the interplanted strips due to the lower probability of movement there. As a result, the enhancement is not as great as when interplanted strips are solely overwintering refuges (Figure 4). As with the other scenarios, mobility influences spatial pattern, with the parasitoid exhibiting no spatial gradient within the crop vegetation (Figure 6).

Implications

These simulations show that we can reproduce all of the "anomalous" patterns observed in field studies using a single mechanistic framework. A simple mathematical model predicts that interplanted vegetation can act as either a source or a sink of natural enemies that exhibit highly variable

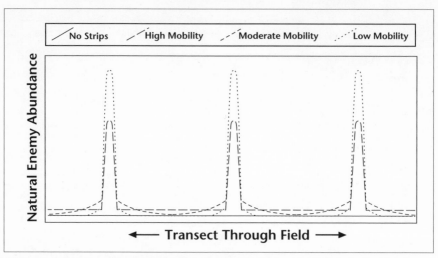

Figure 6. Spatial patterns predicted by model for field with interplanted strips that serve as an overwintering refuge and also provide additional food resources. Peak abundance occurs at interplanted strips. Patterns are shown for hypothetical natural enemies of three different mobilities: low mobility ($1m^2$/day), moderate mobility ($10m^2$/day), high mobility ($100m^2$/day).

spatial patterns dependent on their mobility. The implication is that diversification does not have a single outcome—it is not exclusively a benefit or a detriment—but rather there is a multitude of possible outcomes depending on the design of the system and the natural enemies involved (see also Bugg 1992). Given that real systems are certainly more complicated, the outcomes are likely to be even more variable than those predicted by this simple model. However, the success of this simple model suggests that the outcomes in real diversified systems are understandable and, as importantly, subject to manipulation. With sufficient research it should be possible to identify the characteristics of habitat management systems that are most likely to enhance natural control.

The model suggests that a key characteristic of vegetation management systems is whether natural enemies are present in interplantings at the time of crop germination—i.e., overwintering in the strips or colonizing strips that germinate prior to the crop. When natural enemies simultaneously colonize the crop and interplantings, the latter will act as sinks if they provide more resources (Figure 4). However, when natural enemies are present in the interplantings at crop germination, natural enemies are enhanced, even if strips provide more resources than the crop (Figures 3 & 6). Of the three natural enemies, the parasitoid (with a diffusion rate of $100m^2$/day) had the greatest accumulation in strips when the interplantings and the crop were colonized simultaneously, but the greatest enhancement when it was present in interplantings at crop germination. Thus, a relatively simple change in system design converted a dramatic sink into a major source. In several field studies in which "alternate" vegetation clearly served as a source: (1) the natural enemies overwintered in the alternate vegetation (Doutt & Nakata 1973; Thomas et al. 1991), (2) the natural enemies colonized the inter-plantings prior to crop germination (Bugg et al. 1991; Corbett et al. 1991), or (3) maturation of the interplanted vegetation forced the dispersal of natural enemies into crop vegetation (Fye 1972; Bugg et al. 1991). In contrast, Kemp and Barrett (1989) found that *O. insidiosus* density was lower in diversified soybean plots even though the interplanted strips had been present well before soybeans were sown. It is possible that strips of successional vegetation developed a greater abundance of resources than the soybean crop later in the season, causing *O. insidiosus* to accumulate there. Clearly, temporal changes in resource abundance—in both the crop and interplanted vegetation—will strongly influence the degree to which interplanted vegetation acts as a source or a sink (see also Bugg 1992).

The conclusion that increased vegetational diversity can sometimes decrease natural enemy abundance on a crop has been based solely on considerations of movement. The resources provided by intercrops, however, may also increase population growth (Russell 1989). This raises an important question: Can an increase in population growth due to increased resource abundance compensate for the accumulation of natural enemies in strip vegetation? If strips do not change population growth, the model predicts that abundance of a highly mobile natural enemy (e.g., a parasitoid) would be decreased by about 40% (Figure 4). Let us assume simplistically that resources in interplantings increase fecundity of the parasitoid throughout the field—i.e., periodic feeding in the interplantings results in increased "per capita" fecundity of all natural enemies in the field. This is a reasonable assumption for a highly mobile parasitoid, which should quickly disperse through the crop after leaving interplantings. With this assumption the model predicts that a 5% increase in fecundity would compensate for the sink effect; an increase of 10% or more would result in substantial enhancement (Figure 7). It is not known what increase in population growth might be realized in a commercial scale agroecosystem. While numerous studies have demonstrated increased fecundity and longevity of natural enemies provided with supplementary resources (DeLima 1980; Rogers 1985, 1992), I have found none that evaluate effects on population growth at the field

Figure 7. Effect of increase in fecundity (due to resources in interplantings) on the abundance of natural enemies on crop vegetation in a diversified agroecosystem. Abundance of natural enemies is shown relative to an undiversified system.

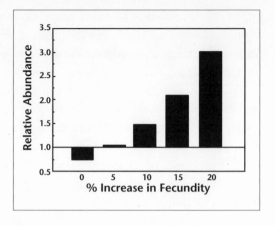

scale. Field-scale effects of resources provided by interplantings will be dependent on the abundance and quality of resources in interplantings, the proportion of time spent in interplantings, the distance traveled from interplantings, and the energy costs of dispersal. The results from this model, however, indicate that it would not require a dramatic increase in natural enemy population growth to offset a sink effect.

Integrating Movement Considerations into Habitat Management Research

Implications for Experimental Design & Interpretation

The mobility of natural enemies, and the resulting spatial scale of enhancement could have important implications for the interpretation of field experiments with agroecosystem diversification (see Altieri & Whitcomb 1980; Russell 1989). Specifically, highly mobile natural enemies may be enhanced throughout an experimental area by vegetational diversity in one of the treatments, resulting in no observed difference among treatments. We can use the model to evaluate how natural enemy mobility might influence experimental results. Imagine that our hypothetical field is now the site of an experiment consisting of two treatments: plots having a 10 m wide strip of alternate vegetation at the center, and homogeneous plots. We assume that natural enemies are present in interplanted vegetation prior to crop germination. The relative abundance of natural enemies in diversified versus homogeneous plots is shown for varying diffusion rate and plot size in Figure 8. The abundance of natural enemies is based on "sampling" being done at six evenly spaced intervals within homogeneous plots and at three evenly spaced intervals on each side of the strip in diversified plots. The

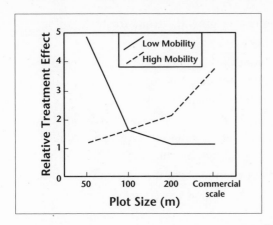

Figure 8. Effect of experimental plot size and natural enemy mobility on the perceived response of natural enemy abundance to agroecosystem diversification. "Relative Treatment Effect" is the ratio of natural enemy abundance observed in diversified plots to that observed in monoculture plots.

enhancement predicted for a commercial-scale field in which strips are spaced 100m apart is also shown.

The model predicts that natural enemy mobility would dramatically affect the observed enhancment due to increased diversity (Figure 8). Natural enemies that are highly mobile would show little enhancement when plots are 50m wide because predators are dispersing among all plots in the experimental field. The observed enhancement increases with plot size since, as strips are farther apart, their effect becomes more detectable. However, even plots 200m in size do not detect the enhancement that would occur in a diversified, commercial-scale field. For less mobile natural enemies enhancement is observed when plots are small but the observed enhancement decreases with increasing plot size. This is because such predators are enhanced only in the area adjoining strips (see Figure 3). Observed enhancement with small plots overestimates the enhancement that would occur in a commercial-scale field even if strips are relatively closely spaced (Corbett 1991).

These simulations show that results of small-scale field experiments may be misleading. More disturbingly, there appears to be no necessary correlation between the enhancement (or lack of enhancement) observed in small-scale experiments and the enhancement that might be obtained in an implemented commercial-scale system. There are a number of conclusions for experimental design:

1. Mobility of the natural enemy should be kept in mind.
2. For natural enemies of high mobility, the only meaningful designs are those in which (a) plot sizes are at the scale of commercial fields and (b) plots are sufficiently separated to minimize colonization of

control plots by individuals originating from treatment plots (Murphy et al. this volume).

3. For natural enemies that are of low to moderate mobility, small-scale experiments should be interpreted as an "assay" rather than as a dependable indication of the enhancement that might occur in commercial-scale fields.

4. An accurate idea of the potential for natural enemy enhancement through vegetation management will be obtained only as a result of large-scale manipulative experiments or through trial and error at the commercial scale.

Quantifying Natural Enemy Movement

Results of field studies and simulations with this simple model make it clear that movement behavior has profound effects on natural enemy response to agroecosystem diversification. Yet, there is a lack of basic information on mobility and variation in mobility for most important natural enemies. Specifically, variation in mobility needs to be quantified with respect to:

1. vegetation type
2. prey/host abundance
3. availability of complementary resources (i.e., nectar, pollen)
4. life-history stage (e.g., pre-ovipositional versus mature adult)
5. seasonal phases (e.g., spring emergence versus mid-season)
6. physical environment (e.g., wind, temperature, etc.)

Obtaining this information for representative species in various taxonomic groups will allow us to answer fundamental questions such as:

1. How variable is mobility within a species and is the variability comparable between species?
2. Is mobility roughly comparable within taxonomic groups and is there a relationship between taxonomic group and mobility?
3. Can mobility be predicted for a natural enemy species if environmental conditions are known?

Such information will be critical for interpreting existing experimental results, designing meaningful experiments, and proposing promising habitat management systems to study.

Quantifying mobility requires the release of large numbers of marked individuals with subsequent recapture at multiple locations around the release point. The recapture data are then compared with the diffusion model

(Appendix A): (1) to determine whether simple diffusion provides an appropriate description of the recapture data, and (2) to find the value of the diffusion coefficient that gives the best fit to the data. This best-fit diffusion coefficient can then be taken as a measure of mobility for the conditions under which the mark-recapture study was conducted (see Figure 1). Kareiva (1981) and Rudd and Gandour (1985) have provided thorough overviews of this methodology. Applying this methodology with various natural enemy species under varying conditions of vegetation type and resource abundance, and with different life-history stages, will begin to fill in the gaps in information described above. Wetzler and Risch (1984) provide an excellent example of the application of mark-recapture techniques to measurement of natural enemy mobility in diversified agroecosystems. The application of diffusion modeling to mark-recapture data must be done with a full appreciation of the assumptions underlying the model. Kareiva (1981) has reviewed the potential complications and design considerations that should be kept in mind when conducting mark-recapture studies.

Parasitic Hymenoptera, because of their typically delicate structure, provide a special challenge; thus, there is a particular lack of information on mobility for this taxonomic group. Releases of new biological control agents provide an ideal opportunity to avoid the need to manually mark individual parasitoids. If recapturing is conducted as part of such releases, it should be done so as to permit estimates of parasitoid mobility (Kareiva 1981). However, it is possible to effectively mark parasitoids. I have successfully marked *Anagrus epos* (Girault) (Hymenoptera: Mymaridae)—which measures less than 0.5mm—by applying fluorescent dust to leaves from which adult parasitoids later emerged (Corbett & Rosenheim 1996). Elemental markers (see below) may also provide a means of performing mark-recapture studies with parasitic hymenoptera (Jackson et al. 1988). Thus, with some ingenuity, it should be possible to obtain mobility estimates for these difficult organisms.

Another critical question is whether certain natural enemies exhibit directional behavior at vegetational boundaries within diversified systems. Directionality may be exhibited, for example, by adult hoverflies (Diptera: Syrphidae) or non-host-feeding parasitoids that are dependent on complementary floral resources. These natural enemies may preferentially move toward interplanted vegetation—perhaps in response to visual or chemical cues—when at or near the boundary with the crop. If present, directional behavior would greatly magnify the sink effect predicted by our mathematical model (Figure 4)—especially if interplantings also provide an abundance of alternate prey or hosts. Directionality can be incorporated into spatially-structured diffusion models (Okubo 1980) and the parameters of these models can be estimated through mark-recapture studies. For example, Banks et al. (1988) utilized "spline-based" parameter estimation to quantify preferential movement of female cabbage root flies, *Delia brassicae* (Bouché) (Diptera: Anthomyiidae), toward *Brassica*. It is also possible to use more accessible—i.e., "user-friendly"—parameter estimation techniques (see Press et al. 1988) to fit mark-recapture data to complex diffusion models (Plant & Cunningham 1991; Corbett & Rosenheim 1996). Using these fitting procedures, it may be possible to use mark-recapture studies within diversified agroecosystems to estimate spatially-varying diffusion coefficients, in addition to quantifying directionality.

Elemental Markers as a Promising Tool

Trace elements that act as chemical analogs of biologically important elements—e.g., rubidium as an analog of potassium—have been used successfully to mark insects (Akey et al. 1991). Application of rubidium to foliage results in increased levels of rubidium in: (1) herbivores feeding on that foliage (Hayes 1991), and (2) natural enemies feeding on those herbivores (Jackson 1991). Captured insects that have rubidium concentrations significantly above naturally occurring levels are considered "labeled" and originating from areas containing rubidium-enriched food sources (Van Steenwyk 1991). In contrast to traditional marking techniques, elemental markers can be applied to naturally occurring populations without disrupting behavior. This methodology has been used, for example, to quantify dispersal of tarnished plant bug (*Lygus lineolaris* [Palisot de Beauvois], Hemiptera: Miridae) into cotton from interplanted mustards (Fleischer et al. 1988).

Elemental marking is a potentially valuable tool in habitat management research in that it provides a means with which to label natural enemies and follow their movements. Application of trace elements to interplantings could be used to answer critical questions such as:

1. Are interplantings the source of natural enemies found on a crop?
2. What proportion of natural enemies in a crop originate from interplantings versus sources external to a system?
3. How quickly do natural enemies spread from interplantings?
4. What is the spatial extent of enhancement of natural enemies around interplantings?

Application to crop foliage could be used to determine what proportion natural enemies found in interplantings were "drawn" out of the crop. Without the ability to label natural enemies, these questions would be answerable only through carefully designed, large-scale manipulative experiments. The use of elemental markers will be most appropriate for systems in which the prey species is either low in vagility or uses exclusively the treated vegetation as a resource: prey that freely disperses between interplantings and the crop will cause labeling of natural enemies throughout the system. Application of rubidium to prune tree refuges has been used to quantify the early-season dispersal of *Anagrus epos* into grape vineyards (Corbett et al. 1996).

The Role of Modeling

Mathematical modeling is a useful tool for increasing our understanding of ecological systems (e.g., Okubo 1980; May 1981). In this chapter I have presented a relatively simple mathematical model for the spatial dynamics of natural enemies in diversified agroecosystems. The model is "heuristic" in nature: it is not intended to represent any particular system, but rather to explore underlying mechanisms and resulting patterns. Simulations with the model have resulted in ideas that remain to be validated (or invalidated!) through research. Yet, these approaches offer a useful mechanistic context within which to evaluate existing data on the response of natural enemies to agroecosystem diversity. Development of more complicated models that include temporal variation, directional movement, and response to resources provided by interplantings will probably yield similar benefits. The results from such efforts should provide a wealth of hypotheses for experimental evaluation and lay the foundation for a theory of natural enemy response to agroecosystem diversification.

An important result from this model is that process interacts strongly with spatial scale in diversified agroecosystems. Thus, the results obtained from small-scale studies do not necessarily reflect the patterns that will occur in larger-scale systems. Detailed experimentation at the scale of interest—that of commercial agricultural fields—requires extensive commitments of land, labor, and years of planning. Such investments are rarely affordable or feasible. Use of mathematical modeling and simulation can provide a bridge between tractable, small-scale studies and the evaluation of habitat management systems at the scale of implementation. Through small-scale studies, the key processes operating in an experimental system can be identified and critical parameters such as population growth and diffusion rates can be estimated. This information can then be used to construct a model of the system at the commercial scale. Simulations with such a model can assist with forming hypotheses concerning the influence of critical aspects of system design such as:

1. The ratio of crop to interplanted vegetation.
2. The spatial arrangement of interplanted vegetation within the crop.
3. The timing of planting of crop versus interplanted vegetation.
4. The proximity of external sources of natural enemies.

Trials at the commercial scale can then be designed with these hypotheses in mind. Investments in large-scale trials can thus be directed toward the most promising designs for the proposed habitat management system. Mathematical modeling cannot replace evaluation at the commercial scale. However, integrating modeling into habitat management research is likely to lead to a clearer understanding of the underlying dynamics and to more efficient progress toward successful systems for natural enemy enhancement.

References

Akey, D. H., J. L. Hayes & S. J. Fleischer [eds]. 1991. Use of elemental markers in the study of arthropod movement and trophic interactions. Southw. Entomol. Supplement #14.

Altieri, M. A. & W. H. Whitcomb. 1980. Weed manipulation for insect pest management in corn. Environ. Manage. 4:483-489.

Ames, W. F. 1977. Numerical methods for partial differential equations. Academic Press, New York, New York.

Andow, D. A. 1991. Vegetational diversity and arthropod population response. Annu. Rev. Entomol. 36:561-586.

Banks, H. T., P. M. Kareiva & L. Zia. 1988. Analyzing field studies of insect dispersal using two-dimensional transport equations. Environ. Entomol. 17:815-820.

Bugg, R. L., L. E. Ehler & L. T. Wilson. 1987. Effect of common knotweed (*Polygonum aviculare*) on abundance and efficiency of insect predators of crop pests. Hilgardia 55:1-51.

Bugg, R. L., F. L. Wäckers, K. E. Brunson, J. D. Dutcher & S. C. Phatak. 1991. Cool-season cover crops relay intercropped with cantaloupe: influence on a generalist predator, *Geocoris punctipes* (Hemiptera: Lygaeidae). J. Econ. Entomol. 84:408-416.

Bugg, R. L. 1992. Using cover crops to manage arthropods on truck farms. HortScience 27 (7):741-745.

Corbett, A. 1991. Spatial dynamics of natural enemies in diversified agroecosystems: experimental studies with an alfalfa/cotton intercropping system and investigations with a theoretical model. Ph.D. thesis. University of California at Davis and San Diego State University, California.

Corbett, A. & R. E. Plant. 1993. The role of movement in the response of natural enemies to agroecosystem diversification: a theoretical evaluation. Environ. Entomol. 22:519-531.

Corbett, A., T. F. Leigh & L. T. Wilson. 1991. Interplanting alfalfa as a source of *Metaseiulus occidentalis* (Acari: Phytoseiidae) for managing spider mites in cotton. Biol. Control 1:188-196.

Corbett, A., B. C. Murphy, J. A. Rosenheim & P. Bruins. 1996. Labeling an egg parasitoid, *Anagrus epos* (Hymenoptera: Mymaridae), with rubidium within an overwintering refuge. Envoiron. Entomol. 25:29-38.

Corbett, A., and J. A. Rosenheim. 1996. Quantifying movement of a minute parasitoid, *Anagrus epos* (Hymenoptera: Mymazidae), using fluorescent dust marking and recapture. Biol. Control 6:35-44.

DeLima, J. O. G. 1980. Biology of *Geocoris pallens* Stål on selected cotton genotypes. Ph.D. dissertation, University of California at Davis.

Dempster, J. P. & T. H. Coaker. 1974. Diversification of crop ecosystems as a means of controlling pests, pp. 106-114. *In* D. P. Jones and M. E. Solomon [eds.]. Biology in pest and disease control. Blackwell Sci. Pub., Oxford, England.

Dingle, H. 1972. Migration strategies of insects. Science 175:1327-1335.

Dobzhansky, T., J. R. Powell, C. E. Taylor & M. Andregg. 1979. Ecological variables affecting the dispersal behavior of *Drosophila pseudoobscura* and its relatives. Am. Nat. 114:325-334.

Doutt, R. L. & J. Nakata. 1973. The *Rubus* leafhopper and its egg parasitoid: an endemic biotic system useful in grape-pest management. Environ. Entomol. 2:381-386.

Fleischer, S. J., M. J. Gaylor & N. V. Hue. 1988. Dispersal of *Lygus lineolaris* (Heteroptera: Miridae) adults through cotton following nursery host destruction. Environ. Entomol. 17:533-541.

Fye, R. E. 1972. The interchange of insect parasites and predators between crops. PANS 18:143-146.

Fye, R. E. & R. L. Carranza. 1972. Movement of insect predators from grain sorghum to cotton. Environ. Entomol. 1:790-791.

Gordon, P. L. & R. G. McKinlay. 1986. Dispersal of ground beetles in a potato crop: a mark-release study. Entomol. Exp. Appl. 40:104-105.

Hayes, J. L. 1991. Elemental marking of arthropod pests in agricultural systems: single and multigenerational marking. Southw. Entomol. Supplement #14:34-37.

Jackson, C. G. 1991. Elemental markers for entomophagous insects. Southw. Entomol. Supplement #14:65-70.

Jackson, C. G., A. C. Cohen & C. L. Verdugo. 1988. Labeling *Anaphes ovijentatus* (Hymenoptera: Mymaridae), an egg parasite of *Lygus* spp. (Hemiptera: Miridae), with rubidium. Annu. Entomol. Soc. Am. 81:919-922.

Kareiva, P. M. 1981. Non-migratory movement and the distribution of herbivorous insects: experiments with plant spacing and the application of diffusion models to mark-recapture data. Ph.D Dissertation. Cornell University, NY.

Kareiva, P. & G. Odell. 1987. Swarms of predators exhibit "preytaxis" if individual predators use area-restricted search. Am. Nat. 130:233- 270.

Kemp, J. C. & G. W. Barrett. 1989. Spatial patterning: impact of uncultivated corridors on arthropod populations within soybean agroecosystems. Ecology 70:114-128.

May, R. M. (Ed.) 1981. Theoretical ecology: principles and applications. Blackwell Sci. Pub., Oxford, England.

Okubo, A. 1980. Diffusion and Ecological Problems: Mathematical Models. Springer-Verlag, Berlin, Germany.

Perrin, R. M. 1975. The role of the perennial stinging nettle, *Urtica dioica*, as a reservoir of beneficial natural enemies. Annu. Appl. Biol. 81:289-297.

Plant, R. E. & R. T. Cunningham. 1991. Analyses of the dispersal of sterile Mediterranean fruit flies (*Ceratitis capitata* Wied.) released from a point source. Environ. Entomol. 20:1493-1503.

Pollard, E. 1971. Hedges VI. Habitat diversity and crop pests: a study of *Brevicoryne brassicae* and its syrphid predators. J. Appl. Ecol. 5:109-123.

Press, W. H., B. P. Flannery, S. A. Teukolsky & W. T. Vetterling. 1988. Numerical recipes in C: the art of scientific computing. Cambridge University Press.

Rabbinge, R. & M. A. Hoy. 1980. A population model for two-spotted spider mite, *Tetranychus urticae*, and its predator *Metaseiulus occidentalis*. Entomol. Exp. Appl. 28:64-81.

Rausher, M. D. 1983. Ecology of host-selection behavior in phytophagous insects, pp. 223-257. *In* R. F. Denno & M. S. McClure [eds.]. Variable plants and herbivores in natural and managed systems. Academic Press.

Rogers, C. E. 1985. Extrafloral nectar: entomological implications. Bull. Entomol. Soc. Amer. 31:15-20.

Rudd, W. G. & R. W. Gandour. 1985. Diffusion model for insect dispersal. J. Econ. Entomol. 78:295-301.

Russell, E. P. 1989. Enemies hypothesis: a review of the effect of vegetational diversity on predatory insects and parasitoids. Environ. Entomol. 18:590-599.

Sevacherian, V. & V. M. Stern. 1974. Host plant preferences of *Lygus* bugs in alfalfa-interplanted cotton fields. Environ. Entomol. 3:761-766.

Sheehan, W. 1986. Response by specialist and generalist natural enemies to agroecosystem diversification: a selective review. Environ. Entomol. 15:456-461.

Thomas, M. B., S. D. Wratten & N. W. Sotherton. 1991. Creation of "island" habitats in farmland to manipulate populations of beneficial insects. J. Appl. Ecol. 28:906-917.

Wetzler, R. E. & S. J. Risch. 1984. Experimental studies of beetle diffusion in simple and complex crop habitats. J. Anim. Ecol. 53:1-9.

van Emden, H. F. 1967. The effect of uncultivated land on the distribution of cabbage aphid (*Brevicoryne brassicae*) on an adjacent crop. J. Appl. Ecol. 2: 171-196.

Van Steenwyck, R. A. 1991. The uses of elemental marking for insect dispersal and mating competitiveness studies: from the laboratory to the field. Southw. Entomol. Supplement #14:15-24.

Appendix A

The equation for diffusion in one dimension is

$$\frac{\partial n}{\partial t} = D\frac{\partial^2 n(x,t)}{\partial x^2}$$

where

n(x,t) = density insects at position x and time t

(Okubo 1980). The equation states that there will be net movement away from regions of high density at a rate that is directly proportional to the rate of change in the spatial density gradient. The proportionality constant, D, represents the coefficient of diffusion. Particles (e.g., insects) obey the equation for simple diffusion if they move independently of one another and if they are not influenced by external forces causing attraction or repulsion.

Assuming that the probability of movement is dependent only on conditions at the individual's current location, the equation for spatially varying diffusion is

$$\frac{\partial n}{\partial t} = \frac{\partial^2}{\partial x^2}\{D(x)\,n(x,t)\}$$

(Kareiva 1981). In this equation D(x) represents the diffusion coefficient as a function of local conditions—e.g., lower rates of diffusion where resources are more abundant.

Appendix B

The assumptions detailed in the text lead to the following equations for the spatial dynamics of a natural enemy population in a diversified agroecosystem:

$$\frac{\partial i}{\partial t} = fn(x,t) - mi(x,t) - di(x,t)$$

$$\frac{\partial n}{\partial t} = di(x,t) - mn(x,t) - sn(x,t) + \frac{\partial^2}{\partial x^2}\{D(x)n(x,t)\} + c$$

where:

i(x,t) = density of immature predators,
n(x,t) = density of adult predators,
D(x) = spatially-varying diffusion coefficient,
   f = oviposition rate,
   d = developmental rate of immatures,
   m = mortality rate due to extrinsic factors,
   s = mortality rate of adults due to "senescence," and
   c = colonization rate.

Parameter values used in the simulations are:

Parameter	Value	Interpretation
f	0.5	One egg every two days
d	0.461	10 days to adult
m	0.077	93% daily survival for all stages
s	0.23	20 day adult life-span
c	0.02	One adult every 50 meters every day

Finite-difference methods were employed to numerically solve this system of equations (Ames 1977).

Weedy Plant Species and Their Beneficial Arthropods: Potential for Manipulation in Field Crops

WOLFGANG NENTWIG—Institute of Zoology, University of Berne, Baltzerstraße 3, CH-3012 Berne, Switzerland

Introduction

It is well known that plants represent important resources for arthropods. This is also true for weeds in cultivated areas as they provide essential sources of food and refuge for many beneficial arthropods (van Emden 1965; Perrin 1975; Curry 1976; Altieri & Whitcomb 1979; Bosch 1987; Frei & Manhart 1992). Such links between plant and insect may take the form of pollen or nectar for syrphid flies (Kugler 1950; Dobrosmyslow 1968; Gilbert 1981; Ruppert & Klingauf 1988; Kevan et al. 1990; Weiss & Stettmer 1991) or parasitic Hymenoptera (Hagvar & Hofsvang 1989), plant hosts for aphids that are additional food for coccinellid beetles (Klausnitzer 1966; Stary & Gonzalez 1991) or plant architecture for spiders (Jennings 1971; Scheidler 1989). The widespread use of herbicides eliminates not only the weeds as competitors for the crop plant, but also the beneficials by robbing them of "their" weeds. Equally, just as insects disappear with the disappearance of their weeds, they can also reappear when the weeds return. Therefore the insect abundance can—to some extent—be augmented and influenced by manipulation of the weed community. In this chapter I present some results—most of which have been obtained by our group in the last 3 years—which concern specific links and interactions between some common weeds and beneficial arthropods.

Most of our work was performed on an 8 ha wheat field near Berne from 1989 to 1991. About 80 plant species (weeds in the broad sense) were sown as mono-cultures in a total of 360 plots (4 replications per plant species, each plot 1.5m x 10m) that were arranged as long weed strips. These strips divided the crop area into strips 24m wide (12, 27 and 36m would accommodate cultivation equipment and were also tested), so that the complete field was subdivided into small weed and wide wheat strips. This regular alternating system introduces a high edge effect into a large agricultural area and represents a special form of habitat management: strip management (Schlinger & Dietrick 1960; El Titi 1987; Nentwig 1988, 1989; Kemp & Barrett 1989; Thomas & Wratten 1990; van Emden 1990; Heitzmann 1995; Heitzmann et al. 1992). Important reviews of the principle are given by Sheehan (1986) and by Russell (1989).

In our weed plots of 15m^2 we studied the floral succession and several arthropod groups. Flower-visiting insects (primarily aphidophagous syrphids [Diptera: Syrphids]) were investigated by observing each of five 1m^2 subsamples per weed plot for 15 minutes during the day every 1–2 weeks throughout the flowering period of the plant (Weiss & Stettmer 1991). The resident arthropods on the green parts of the plants were collected by a D-Vac suction trap (Frei & Manhart 1992). This was done very carefully to avoid flower-visiting insects, so that only arthropods from the vegetative parts of the plants were collected. Overwintering arthropods were studied by sampling plant and litter material during the winter (stored in plastic boxes, extraction at 20° C, LD 18:6 h over several months) and by digging out soil samples (20 cm deep) which were extracted in a MacFadyen apparatus (Bürki & Hausammann 1992). The botanical names follow Binz and Heitz (1986); the common English names are used according to Clapham et al. (1956).

Arthropods and Weeds

General Aspects

The overall impression resulting from our arthropod collections is that the weeds are insect habitats of widely differing quality. Plants such as chervil of France (*Anthriscus cerefolium*), flax (*Linum usitatissimum*), comfrey (*Symphytum officinale*) and gallant soldier (*Galinsoga ciliata*) have extremely low arthropod populations of less than 15 individuals/m^2, whereas most plants have 100–300 arthropods/m^2 according to our sampling method (D-Vac results, i.e. without flower-visiting and ground-layer arthropods). Extremely high values were found on poppy (*Papaver rhoeas*), rape (*Brassica napus*), buckwheat (*Fagopyrum esculentum*) and tansy (*Tanacetum vulgare*), where up to 500 or more arthropods were found per square meter. Considering the trophic structure of the arthropod communities, our results are even more striking. Of all arthropods, phytophagous insects constitute about 65% (most values between 45% and 80%) but the composition of the remaining arthropods varies greatly among predators or parasitoids and phytophagous arthropods or between aphidophagous predators and aphids according to the plant species (Figure 1).

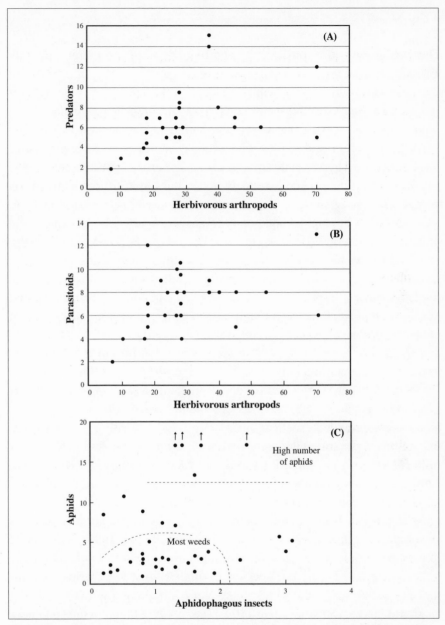

Figure 1. Arthropods at selected weeds in sown strips within a wheat field according to a D-Vac suction trap investigation (average values of numerous collections from 1 m² plots throughout the growing period of the respective plant, vegetative parts only, i.e. without litter layer and flower region). A) predators versus herbivores, B) parasitoids versus herbivores, C) aphids versus aphidophagous insects (the arrows indicate values between 40 and 300) (data modified from Frei & Manhart 1992).

Predators

Our predatory guild consisted of approximately 30% dipterans (mainly Empididae), 25% spiders, 17% ants, 11% beetles (Coleoptera: Coccinellidae, Carabidae, Staphylinidae, Cantharidae), 9% bugs (Hemiptera) and 8% chrysopids green lacewings (Neuroptera: Chrysopidae) (individuals over all weeds, one vegetation period D-Vac suction trap collection). Extremely low densities with average values of 5–20 predators/m^2 were found on chervil of France (*Anthriscus cerefolium*) and corncockle (*Agrostemma githago*) and extremely high values up to 70 predators/m^2 on borage (*Borago offcinalis*), blue knapweed (*Centaurea cyanus*) (many spiders, ants and chrysopids) and poppy (*Papaver rhoeas*) (many Diptera and ants [Hymenoptera: Formicidae]).

Predacious beetles are primarily Carabidae and Cantharidae; for Coccinellidae see below. Carabids are generally considered to be epigeal predators but a few species also climb on plants. In our studies, high carabid densities on plants have been found on borage (*Borago officinalis*), bastard clover (*Trifolium hybridum*), white clover (*T. repens*) and alfalfa (*Medicago sativa*) (2–4/m^2), seven other plant species had densities of 0.5–2/m^2. The carabids were predominantly *Demetrias atricapillus* L. and *Platynus dorsalis*, species that are very effective aphid predators as shown by Griffiths et al. (1985), Hechler (1988) and Janssens and de Clercq (1988). In the winter some weeds provided excellent hibernation sites. According to MacFadyen extractions of the soil under yarrow (*Achillea millefolium*) and wild chamomile (*Matricaria chamomilla*) about 250 carabid beetles/m^2 were found. The corresponding numbers for a dozen other weed species are 100–200/m^2 and for the surrounding wheat field about 60/m^2.

Cantharidae (Coleoptera: Cantharidae), probably important predators of aphids and other insects (Janssen 1964, Wetzel et al. 1991), were found in most cases only in very low densities, but broad bean (*Vicia faba*) had up to 2 cantharids/m^2 (only one species: *Cantharis fusca* L.). During the winter, our weed strips, however, were very attractive for overwintering cantharid larvae. Densities up to 110 larvae/m^2 were found in the soil under several weed mixtures whereas in the wheat field an average of 10 larvae/m^2 occurred.

Among Diptera, dance flies (Diptera: Empididae) were the most important group of predators (compare Stark & Wetzel 1987). Two plants, blue knapweed (*Centaurea cyanus*) and poppy (*Papaver rhoeas*), had very high empidid densities (12–15/m^2), and 8 other plants had medium to high densities. Among these, most Brassicaceae and many clover species (*Trifolium* spp.) were found, so these plants may be attractive to empidid flies.

Only a few predacious heteropterans are found in agroecosystems; however damsel bugs (*Nabis* sp., Nabidae) are important aphid predators (Wetzel et al. 1991) and Lygaeidae may be very numerous as well (e.g. Bugg et al. 1991). Only 5 plants had high nabid densities of 2-5/m^2, i.e. two Brassicaceae, rape (*Brassica napus*) and radish (*Raphanus sativus*), borage (*Borago officinalis*) and the Fabaceae bastard clover (*Trifolium hybridum*) and alfalfa (*Medicago sativa*). On alfalfa, (*Medicago sativa*) nabids constituted 3.2% of all arthropods.

Parasitoids

Most parasitoids were of the Hymenoptera. Our D-Vac analysis showed 34% Ichneumonoidea (mainly Aphidiidae, Braconidae and Ichneumonidae), 30% Proctotrupoidea (mainly Platygasteridae, Ceraphronidae and Proctotrupidae), 25% Chalcidoidea (mainly Pteromalidae, Torymidae and Mymaridae) and 11% Cynipoidea (mainly Eucoilidae). Mean densities lay between 5 and 30 parasitoids per m^2 vegetation. Only a few hymenopterans were found on chervil of France (*Anthriscus cerefolium*), tansy phacelia (*Phacelia tanacetifolia*) and common hemp-nettle (*Galeopsis tetrahit*), whereas alfalfa (*Medicago sativa*), blue knapweed (*Centaurea cyanus*), borage (*Borago officinalis*) and turnip (*Brassica rapa*) contained up to 40 parasitoids/m^2. Papilionaceae, most Asteraceae, and many Brassicaceae seem to have high densities of parasitic wasps, whereas Apiaceae and Caryophyllaceae are not very attractive and harbor only low numbers. This is in some contrast to Berenbaum (1990) and Maingay et al. (1991) who report high numbers of parasitic hymenopterans in Apiaceae. These works refer, however, primarily to flowers, whereas we collected from the vegetative parts of the plants.

At least 17 families of parasitic Hymenoptera were collected from the vegetative parts of the weeds; at least 13 of these families also hibernated in or near the weeds. On alfalfa (*Medicago sativa*), 64% of all hibernating arthropods were Eurytomidae; on scentless mayweed (*Tripleurospermum inodorum*), 46% of the arthropods were Braconidae and 27% were *Pteromalidae*. These two families constituted also more than 12% and 27% of all arthropods on carrot (*Daucus carota*) and yarrow (*Achillea millefolium*). From soil samples, up to 100–160 parasitoids/m^2 were collected under comfrey (*Symphytum officinale*) and under yarrow (*Achillea millefolium*). Our winter investigation clearly demonstrates that plants that are not very attractive for most arthropods in the vegetation period (e.g., *Daucus carota* or *Symphytum officinale*) may be extremely important as hibernation sites for a wide spectrum of arthropod groups.

Phytophagous Insects

Several species of plants harbored abundant aphids (Table 1). For example, buckwheat ((*Fagopyrum esculentum*) hosted several species; poppy (*Papaver rhoeas*), *Aphis fabae*; rape (*Brassica napus*), *Brevicoryne brassicae* L.; tansy (*Tanacetum vulgare*), *Uroleucon tanaceti* L. and others; yarrow (*Achillea millefolium*), *Uroleucon achilleae*; white campion (*Silene alba*), *Brachycaudus lychnidis* L.; and blue knapweed (*Centaurea cyanus*), *Uroleucon jaceae* L. Although aphids of the genera *Aphis* and *Brevicoryne* attack some crop plants, the other aphid species listed above do not. They do, however, represent a good source of food for aphidophagous predators and represent therefore an alternative prey in times of low crop aphid densities (Stary & González 1991).

Besides aphids, the phytophagous beetles (Coleoptera) are the most problematic group of potential pest insects. Brassicaceae always had high densities of *Phyllotreta* sp. (Chrysomelidae) and *Meligethes aeneus* F. (Nitidulidae); evening primrose (*Oenothera biennis*) and poppy (*Papaver rhoeas*) was also very attractive for the nitidulids. During the winter, however, Nitidulidae could be found neither on the dry parts of the weeds nor in the soil under them. The same is true for most curculionid and chrysomelid species that did not hibernate in the weed strips; the only exceptions were the curculionid *Sitona lineatus* L. and halticine chrysomelids. Because these beetles can easily move over long distances, our weed strips cannot be seen as a crucial pest reservoir. Additionally, weed strips bear a high predatory and parasitoid reservoir, so that pest insects are more likely to be controlled by the increased predator pressure as has repeatedly been reported for such habitats (e.g. van Emden 1965, 1990; Pollard 1971; Russell 1989; Stary & González 1991)

The Best Weeds for Aphidophagous Syrphids

Flowering weeds have varying degrees of attractiveness for syrphids having aphidophagous larvae. The comparative investigation of 47 weeds by Weiss and Stettmer (1991) showed remarkable phenological differences (early to late flowering and short to long flowering) and also differing numbers of flower-visiting hoverflies (Table 2). An optimal plant that attracts syrphids for a long period independent of the season does not seem to exist. There

Table 1. Aphid species associated with weedy herbaceous plants near Berne,
Switzerland.

Weed	Aphid (Homoptera: Aphididae)
Blue knapweed (*Centaurea cyanus*)	*Uroleucon jaceae* L.
Buckwheat (*Fagopyrum esculentum*)	Several species
Poppy (*Papaver rhoeas*)	Bean aphid (*Aphis fabae*)
Rape (*Brassica napus*)	Cabbage aphid (*Brevicoryne brassicae* L.)
Tansy (*Tanacetum vulgare*)	*Uroleucon tanaceti* L.
Yarrow (*Achillea millefolium*)	*Uroleucon achilleae*
White campion (*Silene alba*)	*Brachycaudus lychnidis* L.

are, however, a few distinct groups of plants: early flowering naturally
occurring weeds such as field pansy (*Viola arvensis*), red dead-nettle
(*Lamium purpureum*), penny cress (*Thlaspi arvense*) and chickweed
(*Stellaria media*) have only small numbers of flower-visiting insects, but
these plants are probably important for the establishment of the syrphid
population early in the year. A second group is composed of highly attractive Brassicaceae (*Brassica, Raphanus,* or *Sinapis* species) (Figure 2A). A
third group of plants that flower later in the season (summer to autumn)
consists of wild plants such as the weeds common hemp-nettle (*Galeopsis
tetrahit*), white campion (*Silene alba*), treacle mustard (*Erysimum
cheiranthoides*), gallant soldier (*Galinsoga ciliata*) and perennial sowthistle
(*Sonchus arvensis*), and of sown plants such as evening primrose (*Oenothera
biennis*), blue knapweed (*Centaurea cyanus*), chicory (*Cichorium intybus*),
and tansy (*Tanacetum vulgare*) (Figure 2B). In most cases, varying pollen
and nectar supplies cause the differing degrees of attractiveness among
these plants, while the color of the flower is less important. Many but not all
yellow flowers are highly attractive (Kugler 1950), and some white and blue
or violet flowers are as attractive as yellow flowers (e.g. blue knapweed,
[*Centaurea cyanus*]; or chickweed, [*Stellaria media*]). (Weiss & Stettmer
1991). Zygomorphic flowers attract less visitors than flowers with radial
symmetry. In the case of blue knapweed (*Centaurea cyanus*), extrafloral
nectaries on the involucral leaves excrete a liquid with a sugar content of
75% and lead to additional visits by many ichneumonids, coccinellids,
chrysopids (Figure 3), predatory wasps, and ants (Stettmer 1993).

Table 2. Vegetation in artificial weed strips in a wheat field (from Weiss & Stettmer 1991). The flower visits by syrphids are given as average numbers of visiting syrphids per week (observation interval 15 min/m^2).

Symbol:	–	=	≡	■
Range:	< 5	≥ 5 < 10	≥ 10 < 15	≥ 15

Weed Species	Cover (%)	Year	Week (20 · 25 · 30 · 35)
Viola arvensis	30	1990	– – | –
Thlaspi arvensis	70	1990	– – – |– =
Capsella bursa–pastoris	50	1990	– –|= ≡ ≡–≡
Stellaria media	60	1990	– – –|– – ≡ ≡ ≡ –
Symphytum officinalis	80	1990	– –|
Ranunculus arvensis	40	1990	– –| – ≡
Leucanthemum vulgare	70	1990	– –|– – ≡ ■ – – – – = – – – ≡ –
Brassica rapa	50	1989	–= ≡–
Lupinus polyphyllus	20	1990	– – – –
Fagopyrum esculentum	5	1989	≡ ≡ = = –
Sinapis arvensis	80	1989	≡ ■ ■ ■ = –
Sinapis alba	70	1989	≡ ■ ■ ■ = –
Phacelia tanacetifolia	70	1989	= ■ = ≡ ≡ ≡
Lamium purpureum	50	1989	– – = ≡ ≡ = = – –|– – –
Veronica persica	60	1989	– – – ≡ = = ≡ ≡ = = = =
Vicia sativa	5	1989	– –
Vicia faba	5	1989	– –
Linum usitatissimum	50	1989	– – – |
Matricaria chamomilla	30	1989	– – ≡ |–
Calendula arvensis	30	1989	– = ≡ ≡
Raphanus sativus	70	1989	■ ■ ■ ≡ ≡ ≡
Foeniculum vulgare	5	1989	|– – – –
Papaver rhoeas	5	1989	– ≡ ≡ –
Galeopsis tetrahit	70	1990	= |– ≡ – –
Echium vulgare	80	1990	= ≡ = – – – |–
Borago officinalis	60	1989	– ≡ = = = = ≡ –
Silene alba	30	1989	– ≡ = = = = = – –
Erysimum cheiranthoides	50	1989	– |– ≡ ≡ ≡ ■ ■ = = –
Oenothera biennis	100	1990	≡ ■ = = – –|– = – –	–
Galinsoga ciliata	40	1990	– ≡ = ■ ≡ = ≡ – = = – | –
Centaurea cyanus	100	1989	≡ ■ ■ ■ ■ ■ ≡ ■ – = ≡ = –
Galium aparine	70	1990	– –
Linaria vulgaris	60	1990	= – –|– – ≡ – – |– – = =
Reseda lutea	50	1990	– = –|– – ≡ = – –
Trifolium incarnatum	50	1989	– – |–
Daucus carota	5	1989	– –|– – –
Malva sp.	60	1990	≡ |– – – = – ≡ – = = –
Medicago sativa	10	1989	|– – – –
Legousia speculum–veneris	20	1989	≡ ≡ = –
Agrostemma githago	50	1989	–|– ≡ – –
Achillea millefolium	30	1989	–|– = – – –

Table 2. *continued*

Weed Species	Cover (%)	Year	Week			
			20	25	30	35
Cichorium intybus	60	1990		=--	=====	=
Sonchus arvensis	90	1989		▪▪▪	==-==	-=-
Knautia arvensis	40	1990		-		
Solanum nigrum	20	1990		--		
Tanacetum vulgare	90	1990		--|--=--|==	
Rorippa silvestris	30	1990		=|---		

Lacewings—Where to Lay the Eggs

Adult green lacewings are active at night, migrate over long distances, and lay their eggs on vegetation. Some experiments suggest that no specific interactions between lacewing and plant species exist (Duelli 1984a, 1984b). However, like most chrysopids common green lacewing (*Chrysoperla carnea* [Stephens], Neuroptera: Chrysopidae) visits flowers and probably shows preferences for specific plant species. This idea is admittedly based only on field observations and D-Vac collections and has not been tested under experimental field conditions.

The preference of *Chrysoperla carnea* for specific plants during the oviposition period, however, has been investigated by Eichenberger (1991). He found that the attractiveness of 52 plant species, when offered to females in an experimental choice situation, varied strongly (Table 3). Highly attractive plants induced the female lacewings to attach many eggs to the plant surfaces, whereas plants that are not especially attractive received medium to low numbers of eggs. A few plants even had a repellant effect, and the chrysopids preferred to lay their eggs on the cage surface rather than on the plant. To some degree, simple morphological characteristics explain this selective behavior. Hairy leaves as in borage (*Borago officinalis*), evening primrose (*Oenothera biennis*), mullein (*Verbascum densiflorum*), and poppy (*Papaver rhoeas*) or large leaves as in comfrey (*Symphytum officinale*), garden lupine (*Lupinus polyphyllus*), and wild heliotrope (*Phacelia tanacetifolia*) receive more eggs than bare, waxy cuticles or small leaves. An additional chemical stimulus, however, also seems to be very important (see below).

These laboratory results are confirmed by our D-Vac collections where 75% of all chrysopid eggs were found on borage (*Borago officinalis*) and 5-12% on common hemp-nettle (*Galeopsis tetrahit*), poppy (*Papaver rhoeas*) and garden lupine (*Lupinus polyphyllus*). Whereas 97% of all eggs were

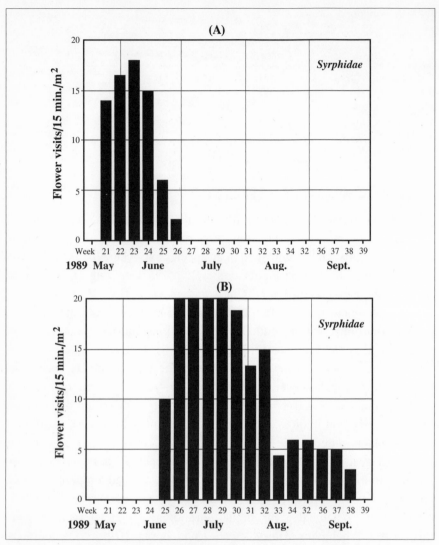

Figure 2. The number of flower visiting syrphids (15 min. observation/m² vegetation) in sown weed strips within a wheat field. (A) Sinapis arvensis, (B) *Centaurea cyanus* (data modified from Weiss & Stettmer 1991).

found on 4 species of plants, larvae occurred on 46 of 52 plants, i.e. they showed a much less pronounced specificity (Frei & Manhart 1992).

The taxonomic specificity of lacewing oviposition seems astonishing: most Boraginaceae and Papaveraceae are highly attractive for lacewings, whereas Lamiaceae, Papilionaceae, and Apiaceae are moderately attractive. Most Brassicaceae, Asteraceae, and Scrophulariaceae are less attractive or not attractive at all. As most of these families are characterized by typical bio-

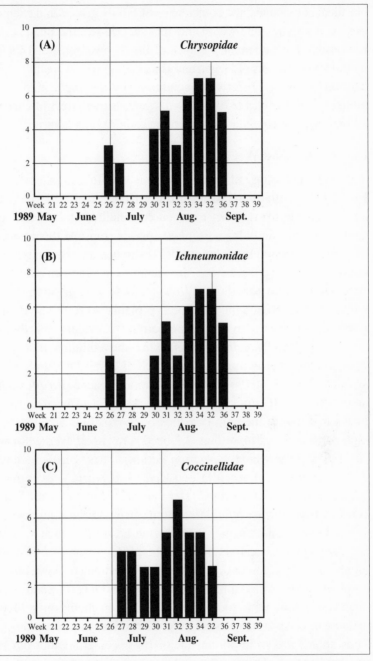

Figure 3. The number of beneficial insects at the extrafloral nectaries of *Centaurea cyanus* in sown strips within a wheat field (counts per 15 min. observation/m² vegetation). (A) Chrysopidae, (B) Ichneumonidae, (C) Coccinellidae (data modified from Weiss & Stettmer 1991).

chemical compounds, the coincidence of family groups and chemotaxonomic aspects is striking, and we assume that species-specific plant compounds are responsible for the female's choice of the oviposition site. This is suggested by the investigations of Hagen and Hale (1974) who were able to induce high lacewing densities under field conditions by the application of some yeast products. Flint et al. (1979) showed the attractive effect of caryophyllene, a secondary plant compound from cotton and Caryophyllaceae.

Coccinellids and Weeds

Although the biology of Coccinellidae is well investigated (for reviews see Hodek [1973, 1986] and Klausnitzer & Klausnitzer [1986]), little is known about possible interactions between coccinellids and plants. An intensive analysis of coccinellids on 76 plant species (most of them common weeds of agroecosystems) by Schmid (1992) showed an extremely uneven distribution of the beetles among the plants. On eighteen plants species no coccinellids were found during an intensive investigation of one season. Among these obviously unattractive plants were 4 out of a total of 7 Fabaceae—mainly clover—2 non-native but widely distributed plants— tansy phacelia (*Phacelia tanacetifolia*) and common lilac (*Syringa vulgaris*)—and other common weeds such as small burnet (*Sanguisorba minor*), chickweed (*Stellaria media*), common orache (*Atriplex patula*), great hedge bedstraw (*Galium mollugo*), or wild thyme (*Thymus serpyllum*). Most plants had low to medium densities of coccinellids and 18 plants showed high densities (i.e., more than 2.1 beetles/m^2 in an average of weekly censuses during the whole vegetation period). These plants included stinging nettle (*Urtica dioica*), alfalfa (*Medicago sativa*), evening primrose (*Oenothera biennis*), carrot (*Daucus carota*), white mustard (*Sinapis alba*), comfrey (*Symphytum officinale*) and mullein (*Verbascum densiflorum*).

Generally, one assumes that coccinellids oviposite at sites where there are many aphids, because adults and larvae feed (some of them exclusively) on aphids. Forty percent of the beetles investigated, however, were found on plants without any aphids (Schmid 1992). Only 18 plants out of 76 had abundant aphids with coccinellids feeding on them; only 11 plants were deemed to be so suitable by the female beetles that they laid eggs on them. Thus aphids are an important factor for coccinellids, but other issues are involved as well. The correlation between aphid and coccinellid abundance is significant, but shows such a high variance that no predictions are possible. Other factors must be almost as important as aphids. These include alternative food sources such as pollen (important in carrot [*Daucus carota*],

Table 3. Different preferences of weeds as oviposition sites for common green
lacewing (*Chrysoperla carnea*). This classification is based on 64 choice tests under
laboratory conditions (each in 12 replications) where one female had the opportunity
to decide between 2 plants as oviposition site. The number of eggs laid within 24 h
and the degree of preference for one plant (chi square test, $p < 0.001$) is the basis for
a ranking where the first 25% of the plants are declared as "very highly attractive,"
the next 25% as "highly attractive" and so on to yield four classes. As some plants
have a similar attractivity, the number of plant species per class differs slightly (data
from Eichenberger 1991)

Very High Preference

Agrostemma githago	Chelidonium majus
Papaver rhoeas	Lupinus polyphyllus
Trifolium arvense	Oenothera biennis
Phacelia tanacetifolia	Borago officinalis
Echium vulgare	Myosotis arvensis
Symphytum officinale	Ajuga reptans
Verbascum densiflorum	Knautia arvensis
Centaurea cyanus	Centaurea jacea

High Preference

Urtica dioica	Alliaria officinalis
Geum urbanum	Sanguisorba minor
Onobrychis viciifolia	Trifolium pratense
Vicia sativa	Galeopsis tetrahit
Salvia pratensis	Achillea millefolium

Medium Preference

Ranunculus acris	Fumaria officinalis
Capsella bursa-pastoris	Raphanus raphanistrum
Sinapis arvensis	Malva moschata
Viola arvensis	Epilobium hirsutum
Anthriscus silvestris	Heracleum sphondylium
Lamium maculatum	Lamium purpureum

Low Preference

Fagopyrum esculentum	Silene vulgaris
Stellaria graminea	Brassica napus
Thlaspi arvense	Reseda lutea
Geranium pyrenaicum	Linaria vulgaris
Veronica chamaedrys	Galium mollugo
Calendula arvensis	Leucanthemum vulgare
Crepis taraxacifolia	Matricaria chamomilla

penny cress [*Thlaspi arvense*], etc.), nectar from extrafloral nectaries as already mentioned for blue knapweed (*Centaurea cyanus*), or good hibernation sites as in white campion (*Silene alba*), royal pigweed (*Amaranthus retroflexus*), common hemp-nettle (*Galeopsis tetrahit*), lesser burdock (*Arctium minus*), or carrot (*Daucus carota*). In the soil under lesser burdock (*Arctium minus*) 180 coccinellids/m^2 were found by MacFadyen extraction whereas under a dozen other weed species the densities were not higher than 40/m^2. Coccinellids did not hibernate in the soil of the wheat field.

It is possible that coccinellids orient themselves by means of secondary plant compounds. This would explain high coccinellid densities on plants with low prey densities or on plants that are otherwise not very suitable. Boldyrev et al. (1969) found an attractive effect for an alcohol extract of the juniper (*Juniperus virginiana*), and Shah (1983) demonstrated that an aquaceous extract of *Berberis* attracts coccinellids independently of aphids. Iperti and Prudent (1986) cite further examples on the repellency of strawberry and raspberry odours and the attractiveness of cypress (*Cypressus* spp.) and fennel (*Foeniculum vulgare*) odour.

Spiders Need Construction Assistance

Nearly all spiders (Araneae) that live in vegetation belong to the web-building guild. Using a D-Vac suction trap we collected about 60% theridiids, 16% linyphiids, 12% araneids, 8% tetragnathids and only 4% other families (lycosids, salticids, thomisids) that do not build webs (Frei & Manhart 1992). Species-specific vegetation requirements of spiders have not—to my knowledge—been investigated yet in a comparative way (but see studies such as Jennings 1971, Morse 1986, Jennings et al. 1989, Scheidler 1989), but a tall and diverse vegetation with sufficient interspaces is likely to enhance the numbers of most web-building spiders. Our results, classified into four density categories, largely confirm this idea (Table 4). Most plants with a tall or dense structure have a medium to high spider density, whereas low and sparse plants support a lower spider population. Exceptions to this general rule are represented in some low plants with a high spider density like clovers (*Trifolium* species) or field pansy (*Viola arvensis*); this, however, can probably be explained by the favourable microclimate that dense stands of these plants produce and that is essential for most small spiders. More interesting are the tall plants with low spider densities. This is the case with, for instance, vetch (*Vicia sativa*) or carrot (*Daucus carota*) with 0.6-0.7 spiders/m^2. Although some of these plants achieved high vegetation densities, they were not attractive for spiders.

Table 4. Density of spiders on different weeds according to D-Vac-collections (modified from Frei & Manhart 1992). Mean value from 8-16 collections during the growing season of the respective plant, vegetative parts only, i.e. without litter layer and flower region. The plants are grouped into 4 classes of spider densities which are considered to reflect a high, medium, low and very low spider density. This relative classification allows a comparison with other collection methods which yield different absolute values.

Spiders/m²	Plant Taxa
> 2.2	*Achillea millefolium, Borago officinalis, Brassica napus, Centaurea cyanus, Echium vulgare, Linum usitatissimum, Malva sp., Matricaria chamomilla, Medicago sativa, Oenothera biennis, Phacelia tanacetifolia, Raphanus sativus, Sinapis alba, Sinapis arvensis, Sonchus arvensis, Symphytum officinale, Tanacetum vulgare, Trifolium hybridum, Trifolium repens, Verbascum densiflorum, Vicia faba*
1.3-2.2	*Anthriscus cerefolium, Arctium minus, Chrysanthemum vulgare, Erysimum cheiranthoides, Fagopyrum esculentum, Foeniculum vulgare, Knautia arvensis, Lupinus polyphyllus, Papaver rhoeas, Pastinaca sativa, Reseda lutea, Silene alba, Solanum nigrum, Thlaspi arvense, Viola arvensis*
0.6-1.2	*Brassica rapa, Calendula arvensis, Capsella bursa-pastoris, Daucus carota, Galeopsis tetrahit, Galinsoga ciliata, Onybrychis viciifolia, Ranunculus arvensis, Stellaria media, Vicia sativa, Veronica persica*
< 0.6	*Agrostemma githago, Lamium purpureum, Trifolium incarnatum, Trifolium pratense*

These results relate to the growing season and neglect the important role that plants can play as hibernation sites during the winter. Comparative investigations in winter revealed that some of the above mentioned unattractive plants such as corncockle (*Agrostemma githago*), carrot (*Daucus carota*) or common hemp-nettle (*Galeopsis tetrahit*) support large populations of overwintering spiders (Bürki & Hausammann 1992). Hollow stems or seed capsules (white campion [*Silene alba*]) and suitable flowerheads (tansy [*Tanacetum vulgare*]) offer ideal hibernation sites: indeed 40–70% of all hibernating arthropods may be spiders as on tansy (*Tanacetum vulgare*), yarrow (*Achillea millefolium*), perennial sowthistle (*Sonchus arvensis*), white mustard (*Sinapis alba*), or chicory (*Cichorium intybus*). According to MacFadyen extractions of the soil under comfrey (*Symphytum officinale*), 240 spiders/m^2 hibernate there. Under a dozen other weed species, the corresponding figures vary between 50 and 170/m^2, whereas fewer than 10 spiders/m^2 were extracted from the soil of a wheat field (Bürki & Hausammann 1992).

Conclusion

A single weed plant can be considered as an island that affords a habitat for a number of arthropods. There are numerous group- or species-specific connections between plants and arthropods. In most cases, we barely understand these links. There may be a clear causal relationship or patterns of association that reflect a correlation due to other yet undiscovered factors. Associations can even be accidental. The latter possibility is certainly only an exception as the results here are reproducible, i.e., subsequent years, different locations, or comparisons of field research and laboratory experiment confirm each other. Possible explanations for the particular attractiveness of a given weed for a predator or parasitoid species are numerous and include the existence of a specific prey species on the weed; alternative food such as nectar, pollen or gland products; secondary plant compounds that are used for orientation; the insect hormone system, etc.; hibernation opportunities; and probably many more.

Thus, it appears feasible to use specific weed species for the manipulation of the arthropod spectrum in the vicinity. Consequently, it should be possible to augment predators by managing weed strips in agroecosystems. Strips with varying compositions could even be arranged in such a way that they are optimized for particular predator groups. However, two problems must be addressed. Firstly, not all plants are compatible with each other. The inclusion of one species may exclude another; tall plants limit lower ground-covering plants and in subsequent years more vigorous weeds may overwhelm other plants. The initial weed strip vegetation therefore will change within a few years. This succession is a natural process but may be channelled or controlled. In the last few years we have investigated many combinations of "beneficial weeds" and are presently testing a weed vegetation that consists of approximately 20 plants (publication in prep.). These weeds are the best candidates for the augmentation of beneficial arthropods in agroecosystems: they form a stable vegetation structure with low, medium and tall plants (about 1m in height), cover a long flowering season and include annual, biennial, and perennial plants. The latter point guarantees an

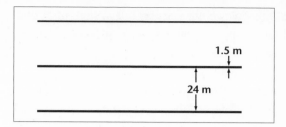

Figure 4. Regular arrangement of weed strips (1.5 m wide) within a field; schematic drawing illustrates strip farming with sown weed strips (compare text).

optimal vegetation from the first year on and stabilizes the plant succession. In addition, a modest tillage, mowing or mulching regime may enable the establishment of stable strip vegetations for 3–5 years or even longer. This point is under investigation at present. Commercial seed distributors are interested in our optimized weed strips and ready-to-use seed mixtures will soon be available.

The second problem concerns the dispersal of arthropods. Investigations of the activity pattern of carabid beetles from our weed strips into cereal fields (Lys & Nentwig 1991, 1992) clearly indicate that the average range of most carabids is limited to a mean daily distance of less than 20m. At locations farther away, the activity density (Heydemann 1956), the number of individuals per m^2, and the number of species decreases to low values. We therefore arranged weed strips at regular distances from each other so that a large cereal field is divided by 1.5m wide weed strips forming 24m wide cereal parts (Figure 4). These weed strips stay for several years; during the first three years of our system we found constantly increasing numbers of beneficials, e.g. carabids (Lys 1992), and we found that the females of the dominant carabid *Poecilus cupreus* L. in such weed strips had a much better nutritional condition than comparable beetles in a wheat area farther away. The beetles had a higher body weight and size, the gut was better filled, and the ovaries produced more eggs. The predatory pressure of these carabids close to weed strips is therefore certainly much higher than farther away (Zangger et al. 1994a, 1994b).

Such strip management requires approximately 5% of the field area. There is some minor time saving in soil tillage and in the application of agrochemicals but an overall financial loss of about 4.5%–5% results (Wingeier 1992). On the other hand our system offers a serious opportunity to combine modern methods of agriculture on large fields with a maximum of nature conservation and species protection. An organic or alternative sustainable production (i.e., on the basis of a low input system) that exceeds the usual scope of an integrated pest management (e.g., integrated production [IP] or integrated farming systems *sensu* Häni et al. 1990, 1991; Häni

1990) should be possible. At present we are starting our system of habitat manipulation on several test farms and are gaining practical experience.

Additional Note

In the last 5 years the idea of sown weed strips has been studied in more detail and is actually propagated in Switzerland as a recommended type of ecological compensation area. The seed mixture consists of approximately 30 species of wild flowers and is available from several commercial seed distributors in Switzerland. This mixture includes a wide range of low, middle sized and tall plants; it contains early to late flowering plants and also annuals, biennials and perennials. This seed mixture and the additional wild flowers from the seed reservoir in the soil form a vegetation that can be maintained for several years at a high floral diversity by mowing at 2-year intervals (Günter 1998). Additionally, cultural practices have been developed that maintain young and old succession weed strips within an agricultural landscape.

For further research we arranged such sown weed strips between fields and within fields and formed small-scaled landscapes. Many arthropod groups increased their species number in these strips and migrated into the adjacent crop area (for spiders: Frank & Nentwig 1995; Jmhasly & Nentwig 1995; for carabid beetles Lys et al. 1994; Zangger et al. 1994a, 1994b; Frank 1997; for syrphids Salveter 1996). Herbivores also increased in species number; however, this concerns only non-pest species that survive due to the increased diversity of weedy plants (Lethmayer 1995, Lethmayer et al. 1997). Predation rate of many predators or parasitization rate of parasitoids on potential pest species was higher near sown weed strips. This resulted in a reduced abundance of herbivores in wheat (Hausammann 1996 a), in rape (Hausammann 1996 b), and in apple orchards (Wyss et al. 1995).

We assume that sown weed strips can keep pest species below their economic threshold and reduce the number of insecticide applications. Integration of sown weed strips into integrated farming systems can make agriculture more sustainable.

Acknowledgments

My thanks go to all my students, assistants and coworkers, to F. Häni, A. Heitzmann, J.A. Lys and E. Wyss for their comments on an earlier draft of this chapter and to the Seva Lotteriefonds, Berne, for financial support.

References

Altieri M. A. & W. H. Whitcomb. 1979. The potential use of weeds in the manipulation of beneficial insects. HortScience 14:12-18.

Berenbaum, M. R. 1990. Evolution of specialization in insect-umbellifer associations. Annu. Rev. Entomol. 35:319-344.

Binz, A., C. Heitz. 1986. Schul- und Exkursionsfora für die Schweiz. Schwabe & Co, Basel.

Boldyrev, M. I., W. H. A. Wilde & B. C. Smith. 1969. Predacious coccinellid oviposition responses to *Juniperus* wood. Can. Entomol. 101:1199-1206.

Bosch, J. 1987. Der Einfuss einiger dominanter Ackerunkräuter auf Nutz- und Schadarthropoden in einem Zuckerrübenfeld. Zeitschift für Pfanzenkrankheiten und Pflanzenschutz 94:398-408.

Bürki, H. M. & A. Hausammann. 1992. Überwinterung von Arthropoden im Boden und an Ackerunkräutern künstlich angelegter Ackerkrautstreifen. Agrarökologie, 7:1-158.

Bugg, R. L., F. L. Wäckers, K. E. Brunson, J. D. Dutcher & S. C. Phatak. 1991. Cool-season cover crops relay intercropped with cantaloupe: influence on a generalist predator, *Geocoris punctipes* (Hemiptera: Lygaeidae). J. Econ. Entomol. 84:408-416.

Clapham, A. R., T. G. Tutin & E. F. Warburg. 1956. Excursion flora of the British Isles. Cambridge University Press.

Curry, J. P. 1976. The arthropod fauna of some common grass and weed species of pasture. Proc. R. Irish Acad. B. 76:1-35.

Dobrosmyslow, P. A. 1968. Nektarbildende Pflanzen und biologischer Schutz des Gartens. Zeitschift für Pflanzenkrankheiten und Pflanzenschutz 77:276.

Duelli, P. 1984a. Dispersal and oviposition strategies in *Chrysoperla carnea*. pp. 133-145. *In* J. Gepp, H. Aspöck, H. Hölzel [eds.]. Progress in world's neuropterology. Graz.

Duelli, P. 1984b. Flight activity patterns in lacewings Planipennia: Chrysopidae, pp. 165-170. *In* J. Gepp, H. Aspöck, H. Hölzel [eds.]. Progress in world's neuropterology. Graz.

Eichenberger, J. 1991. Zur Eiablage von *Chrysoperla carnea* Stephens Planipennia, Chrysopidae an verschiedenen Ackerunkräutern in Wahlversuchen im Labor. Diploma thesis, University of Berne, Switzerland.

El Titi, A. 1987. Environmental manipulation detrimental to pests, pp. 105-121. *In* V. Delucchi [ed.]. Protection intégrée: quo vadis? Parasitis, Zurich, Switzerland.

Flint, H. M., S. S. Salter & S. Walters. 1979. Caryophyllene: an attractant for the green lacewing. Environ. Entomol. 8:1123-1125.

Frank, T. 1997. Species diversity of ground beetles (Carabidae) in sown weed strips and adjacent fields. Biol. Agric. Hortic. (in press).

Frank, T. & W. Nentwig. 1995. Ground dwelling spiders (Araneae) in sown weed strips and adjacent fields. Acta Oecologia 16:179-193.

Frei, G. & C. Manhart. 1992. Nützlinge und Schädlinge an künstlich angelegten Ackerkrautstreifen in Getreidefeldern. Agrarökologie 4:1-140.

Gilbert, F. S. 1981. Foraging ecology of hoverflies: morphology of the mouthparts in relation to feeding on nectar and pollen in some common urban species. Ecol. Entomol. 6:245-262.

Griffths, E., S. D. Wratten & G. P. Vickermann. 1985. Foraging by the carabid *Agonum dorsale* in the field. Ecol. Entomol. 10:181-189.

Günter, M. 1998. A management plan for sown weed strips in arable land with regard to technical and economical restrictions. University of Berne, Ph.D. thesis in preparation.

Häni, F. 1990. Farming systems research at Ipsach, Switzerland—The "third way" project. Schweizer Landwirtschaftliches Forschung 29:257-271.

Häni, F., E. Boller & F. Bigler. 1990. Integrierte Produktion—ein ökologisch ausgerichtetes Bewirtschaftungssystem. Schweizer Landwirtschaftliches Forschung 29:101-115.

1991. Integrated production—A way to achieve ecologically sound agriculture. World Farmers' Times 6:15-18.

Hagen, K. & R. Hale. 1974. Increasing natural enemies through use of supplementary feeding and non-target prey, pp. 170-181. In F. Maxell & F. Harris [eds.]. Proc. Summer Inst. Biol. Control Plant Insect Dis.

Hagvar, E. B. & T. Hofsvang. 1989. Effect of honeydew and hosts on plant colonization by the aphid parasitoid *Ephedrus cerasicola*. Entomophaga 34:495-501.

Hausammann, A. 1996a. The effects of weed strip-management on pests and beneficial arthropods in winter wheat fields. Zeitschift für Pfanzenkrankheiten Pflanzenschutz . 103:70-81.

Hausammann, A. 1996b. Strip-management in rape crop: is winter rape endangered by negative impacts of sown weed strips? J. Appl. Entomol. 120:505-512

Hechler, N. 1988. Untersuchungen zur Nahrungsaufnahme und Biologie von *Platynus dorsalis* Coleoptera, Carabidae. Diploma thesis, Technical University Darmstadt.

Heitzmann, A. 1995. Angesäte Ackerkrautstreifen—Veränderungen des Pflanzenbestandes während der natürlichen Sukzession. Agrarökologie 13:1-152.

Heitzmann, A., J-A. Lys & W. Nentwig. 1992. Nützlingsförderung am Rand—oder: Vom Sinn des Unkrautes. Landwirtschaftliches Schweiz 5:25-36.

Heydemann, B. 1956. Die Biotopstruktur als Raumwiderstand und Raumfülle für die Tierwelt. Verhandlungen der Deutschen Zoologischen Gesellschaft 332-347.

Hodek, I. 1973. Biology of Coccinellidae. Academic Publishing, Prague.

Hodek I. 1986. Ecology of Aphidophaga. Academia, Prague & Dr. W. Junk, Dordrecht.

Iperti, G. & P. Prudent. 1986. Effect of the substrate properties on the choice of oviposition sites by *Adalia bipunctata*, pp. 143-149. In I. Hodek [ed.]. Ecology of Aphidophaga. Academia, Prague & Dr W. Junk, Dordrecht.

Janssen, W. 1964. Untersuchungen zur Morphologie, Biologie und Ökologie von *Cantharis* L. und *Rhagonycha* Eschsch. (Cantharidae, Col.). Zeitschrift für wissenschaftliche Zoologie 169:113-202.

Janssens, J. & R. de Clercq. 1988. Observation on the carabids *Pterostichus melanarius* Illiger and *Platynus dorsalis* (Pontoppidan) (Col., Carabidae) as predators of cereal aphids in winter wheat. Mededelingen Faculteit Lanbouw-wetenschappen Rijksuniversiteit Gent 53/3A:1131-1136.

Jennings, D. T. 1971. Plant associations of *Misumenops coloradensis* Gertsch (Araneae: Thomisidae) in central New Mexico. Southw. Nat. 16:201-207.

Jennings, D. T., F. B. Penfield, R. E. Stevens & F. G. Hawksworth. 1989. Spiders (Araneae) associated with dwarf mistletoes (*Arceuthobium* sp). in Colorado. Southw. Nat. 34:349-355.

Jmhasly, P. & W. Nentwig. 1995. Habitat management in winter wheat and evaluation of subsequent spider predation on insect pests. Acta Oecologia 16:389-403.

Kemp, J. C. & G. W. Barrett. 1989. Spatial patterning: impact of uncultivated corridors on arthropod populations within soybean agroecosystems. Ecology 70:114-128.

Kevan, P. G., E. A. Clark & V. G. Thomas. 1990. Insect pollinators and sustainable agriculture. Am. J. Alternative Agric. 5:13-22.

Klausnitzer, B. 1966. Übersicht über die Nahrung der einheimischen Coccinellidae (Col.) Entomologische Berichte, Berlin, 91-101.

Klausnitzer, B. & H. Klausnitzer. 1986. Marienkäfer. Ziemsen, Wittenberg.

Kugler, H. 1950. Schwebfliegen und Schwebfliegenblumen. Berichte der Deutschen Botanischen Gesellschaft 63:36-37.

Lethmayer, C. 1995. Effects of sown weed strips on pest insects. Ph.D. Thesis, University of Berne, Switzerland.

Lethmayer, C., W. Nentwig & T. Frank. 1997. Effects of weed strips on the occurrence of noxious coleopteran species (Nitidulidae, Chrysomelidae, Curculionidae. J. Plant Dis. Protection 104:75-92.

Lys, J-A & W. Nentwig. 1991. Surface activity of carabid beetles inhabiting cereal fields. Seasonal phenology and the influence of farming operations on five abundant species. Pedobiologia 35:129-138.

1992. Augmention of beneficial arthropods by strip management: 4. Surface activity, movements and density of abundant carabid beetles in a cereal field. Oecologia 92:373-382.

Lys, J.-A., M. Zimmermann & W. Nentwig. 1994. Increase in activity density and species number of carabid beetles in cereals as a result of strip-management. Entomol. Exp. Appl. 73:1-9.

Maingay, H. M., R. L. Bugg, R. W. Carlson & N. A. Davidson. 1991. Predatory and parasitic wasps (Hymenoptera) feeding at flowers of sweet fennel (*Foeniculum vulgare* Miller var Dulce Battandier and Trabut, Apiaceae) and spearmint (*Mentha spicata* L., Lamiaceae) in Massachusetts. Biol. Agric. Hortic. 7:363-383.

Morse, D. H. 1986. Foraging behavior of crab spiders (*Misumena vatia*) hunting on inflorescences of different quality. Am. Midl. Nat. 116:341-347.

Nentwig, W. 1988. Augmentation of beneficial arthropods by strip management 1. Succession of predacious arthropods and long-term change in the ratio of phytophagous and predacious arthropods in a meadow. Oecologia 76:597-606.

Nentwig, W. 1989. Augmentation of beneficial arthropods by strip management 2. Successional strips in a winter wheat field. Zeitschrift für und Pflanzenkrankheiten Pflanzenschutz 96:89-99.

Perrin, R. M. 1975. The role of the perennial stinging nettle *Urtica dioica* as a reservoir of beneficial natural enemies. Ann. Appl. Biol. 81:289-297.

Pollard, E. 1971. Hedges. VI. Habitat diversity and crop pests: a study of *Brevicoryne brassicae* and its syrphid predators. J. Appl. Ecol. 8:751-780.

Ruppert, V. & F. Klingauf. 1988. Attraktivität ausgewählter Pflanzenarten für Nutzinsekten am Beispiel der Syrphinae. Mitteilungen der Deutschen Gesellschaft für Allgemeine und Angewandte Entomologie 6:255-261.

Russell, E. P. 1989. Enemies hypothesis: a review of the effect of vegetational diversity on predatory insects and parasitoids. Environ. Entomol. 18:590-599.

Salveter, R. 1996. Population structure of aphidophagous hoverflies (Diptera: Syrphidae) in an agricultural landscape. Ph.D. Thesis, University of Berne.

Scheidler, M. 1989. Niche partitioning and density distribution in two species of *Theridion* (Theridiidae, Araneae) on thistles. Zoologischer Anzeiger 223:49-56.

Schlinger, E. I. & E. J. Dietrick. 1960. Biological control of insect pests aided by strip-farming alfalfa in experimental program. Calif. Agric. 1:8-15.

Schmid, A. 1992. Untersuchungen zur Attraktivität von Ackerwildkräuter für aphidophage Marienkäfer (Coleoptera, Coccinellidae). Agrarökologie 5:1-122.

Shah, M. A. 1983. A stimulant in *Berberis vulgaris* inducing oviposition in coccinellids. Entomol. Exp. Appl. 33:119-120.

Sheehan, W. 1986. Response by specialist and generalist natural enemies to agroecosystem diversification: a selective review. Environ. Entomol. 15:456-461.

Stark, A. & T. Wetzel. 1987. Fliegen der Gattung *Platypalpus* Diptera, Empididaebisher wenig beachtete Prädatoren im Getreidebestand. Appl. Entomol. 193:1-14.

Stary, P. & D. González. 1991. The chenopodium aphid, *Hayhurstia atriplicis* (L.) (Hom., Aphididae), a parasitoid reservoir and a source of biocontrol agents in pest management. J. Appl. Entomol. 111:243-248.

Stettmer, C. 1993. Blütenbesucher an den extrafloralen Nektarien der Kornblume *Centaurea cyanus* (Asteraceae). Mitteilungen der Schweizerischen Entomologischen Gesellschaft 66:1-8.

Thomas, M. B. & S. D. Wratten. 1990. Ecosystem diversification to encourage natural enemies of cereal aphids, pp. 691-696. *In* Pests and diseases. Brighton Crop Protection Conf.

van Emden, H.F. 1965. The role of uncultivated land in the biology of crop pests and beneficial insects. Sci. Hortic. 17:121-136.

van Emden, H.F. 1990. Plant diversity and natural enemy eficiency in agroecosystems, pp. 63-80. *In* M. Mackauer, L. E. Ehler & J. Roland [eds.], Critical issues in biological control. Intercept Ltd, Andover, U.K.

Weiss, E. & C. Stettmer. 1991. Unkräuter in der Agrarlandschaft locken blütenbesuchende Nützlinge an. Agrarökologie 1:1-104.

Wetzel, T. H., A. Stark, U. Löbner & O. Hartwig. 1991. Zum Auftreten und zur Bedeutung von Weichkäfern Col., Cantharidae und Sichelwanzen Het., Nabidae als aphidophage Prädatoren in Getreidebeständen. Zeitschrift für Pflanzenkrankheiten und Pflanzenschutz 98:364-370.

Wingeier, T. 1992. Agrarökonomische Auswirkungen von in Ackerflächen angesäten Grünstreifen. Agrarökologie 2:1-97.

Wyss, E., U. Niggli & W. Nentwig. 1995. The impact of spiders on aphid populations in a strip-managed apple orchard. J. Appl. Entomol. 119:473-478.

Zangger, A., J.-A. Lys & W. Nentwig. 1994a. Increasing the availability of food and the reproduction of *Poecilus cupreus* in a cereal field by strip-management. Entomol. Exp. Appl. 71:111-120.

Zangger, A., J-A. Lys & W. Nentwig. 1994b. Augmentation of beneficial arthropods by strip management. 5. Increasing the availability of food and the reproduction in *Poecilus cupreus* (Carabidae, Coleoptera). Entomol. Exp. Appl. 71:111-120.

Biological Control of Aphids in Lettuce Using In-Field Insectaries

WILLIAM E. CHANEY—Entomology Farm Advisor, University of California Cooperative Extension, 1432 Abbott St., Salinas, CA 93901

Introduction

There has been very limited successful use of traditional biological control methods in commercial fresh market vegetables. There are many reasons for this, including cost, the availability of effective chemical pesticides, and arguably most importantly, the extremely low tolerance for insect damage or contamination of vegetables destined for retail consumers. The potential to remove the cost issue was examined for a test system of controlling green peach aphid in lettuce in the Salinas Valley enhancing the indigenous population of the natural enemies with insectary plantings in, or directly adjacent to, commercial lettuce fields.

Aphids (Homoptera: Aphididae) are major pests of lettuce in the coastal areas of California. Attempts to control aphids without conventional insecticides have not been successful in most large scale commercial applications. Both novel and time-honored means have been attempted, including mechanical removal of the aphids from the plants by suction and the inundative release of beneficial insects directly into the field (Ferris 1989), but the level of pest pressure due to the climatic conditions in the Salinas Valley has led to repeated failures.

The importance and impact of plant diversity in providing food and shelter to beneficial species are well recognized (van Emden 1964; van den Bosch & Messenger 1973; Price & Waldbauer 1975; Bugg et al. 1987; Patriquin et al. 1988; Bugg & Dutcher 1989; Altieri 1991; Andow 1991) in many systems, but have been little explored in vegetable production. Because many plants are attractive to predaceous insects in one or more of their life stages, (Bugg et al. 1987; Bugg et al. 1989), there is reason to expect them to be beneficial in a short-duration vegetable system. Annuals grow more quickly than perennials and provide floral nectar earlier (Bazzaz 1984). Perennial insectary plants could be expected to be important in perennial cropping systems. Problems arising from tenant farming and the rapidly changing nature of commercial vegetable production make long-term commitment to perennial insectary plants unlikely.

Green peach aphid (*Myzus persicae* [Sulzer]) is the most serious and common of the aphids that attack lettuce (McCalley 1981). This pest is anholocyclic in the coastal areas on a range of secondary hosts (Blackman & Eastop 1984), where it is a contaminant of lettuce and a vector of some 100 plant viruses (Kennedy et al. 1962). Unlike other aphid pests, green peach aphid populations build on lower leaves and then spread throughout the plant (Flint 1985). Control is most important at the initiation of head formation, because once the aphids are inside the head, control is nearly impossible (Flint 1985).

Biointensive control of aphids is a key missing component of a complete insect management system for producing lettuce in the Salinas Valley and other coastal areas without conventional insecticides. It is also critical to reducing the development of resistance in aphids to insecticides (Georghiou 1983, 1986; Tabashnik 1989).

Presented here are the results of a study designed to evaluate the use of in-field insectaries to aid in the biological control of the green peach aphid and other insect pests in lettuce grown in the Salinas Valley. The insectary was an area devoted to a plant species or combination of species that attracts beneficial insects by providing nectar, alternate insect prey, or refuge from any chemical pest controls applied to the field. These plantings were established in strips in or near the field and posed as little inconvenience as possible to the grower while serving their stated purpose.

The immediate objectives were:

1. Identify a plant species or combination of plant species that will act as effective in-field insectaries for beneficial predatory or parasitic organisms.

2. Evaluate each potential insectary species individually in a trial garden to determine positive and negative attributes for suitability in a lettuce production system, e.g., appropriate phenology and growth rates, efficacy as attractant of beneficial insects, the alternate prey provided, its ability to act as a refuge, and its likelihood of serving as a host of pest aphids or lettuce diseases.

3. Seed strips of an optimal insectary plant species into a large commercial lettuce field. Monitor abundance of beneficial and pest insects on each of the insectary species, disease incidence of each plant species, and plant growth and phenology through the lettuce cropping period.

4. Compare the effects of the in-field nursery strips on lettuce production, insect abundance in lettuce, and lettuce disease incidence between adjacent sites grown either with in-field insectary strips or with conventional production methods and pesticide applications.

Materials & Methods

Three trial gardens were planted in areas of no pesticide use at four times in four sites with similar characteristics and soils.

Field Design for Trial Gardens

Plots of each species were planted in beds spaced 1m (40 inch) center to center and separated by one bed of lettuce on each side. Each plot was 1m x 7m (20-foot lengths along a bed). Each plant species was replicated twice per trial; the most promising species were included in all four trials. The standard lettuce management practices (other than pesticides) of the grower were followed in order to determine compatability with conventional lettuce fields.

Trial Garden Field Measurements and Data Collection

Insect populations were monitored weekly using a combination of direct whole-plant counts and suction samplings. Suction samples were taken with a garden leaf blower modified to collect insects through the intake tube. The collection bag was lined with a fine-weave nylon netting. An area 1 x 1 m was sampled in each case. Samples were immediately placed in a cooler and transported to the laboratory for sorting and identification. These vacuum samples were supplemented with visual counts of insects not adequately sampled by this method, such as syrphid fly adults.

Vacuum samples were sorted first into broad categories: Coccinellidae (Coleoptera), *Orius* spp. (Hemiptera: Anthocoridae), *Geocoris* spp. (Hemiptera: Lygaeidae), *Nabis* spp. (Hemiptera: Nabidae), Aphididae (Homoptera), *Lygus* spp. (Hemiptera: Miridae), lacewings (Neuroptera: Chrysopidae, Hemerobiidae), Syrphidae (Diptera), and Hymenoptera of a size and type likely to be parasitic on aphids or other insects of similar size. After counting, the Hymenoptera and Diptera were stored for possible future analysis. Pest species were considered to be the *Lygus* and non-syrphid diptera. Most of the latter were *Liriomyza sativae* Blanchard (Diptera: Agromyzidae), *Lyriomyza huidobrensis* (Blanchard) (Diptera: Agromyzidea), and various root maggots (*Delia* spp., Diptera: Anthomyiidae) species.

Table 1. Plant species examined.

Achillea millefolium—big-leaf yarrow
Anthriscus cerefolium—chervil
Clarkia amoena—farewell-to-spring
Clarkia unquiculata—elegant Clarkia
Coriandrum sativum—cilantro/coriander
Gilia tri-color—bird's eyes
Linanthus grandiflorus—mountain phlox
Linum grandiflorum—scarlet flax
Nemophila menzeisii—baby blue eyes
Phacelia campanularia—wild Canterbury bells
Stylomecon heterophylla—wind poppy
Lobularia maritima (Alyssum maritimum)—sweet alyssum
Ammi majus—Bishop's flower
Clarkia concinna—red ribbons
Collinsia heterophylla—Chinese houses
Fagopyrum esculentum—buckwheat
Layia platyglossa—tidy tips
Linaria maroccana—toadflax
Lupinus albus—white lupine
Papaver rhoeas—Flanders poppy
Phacelia tanacetifolia—tansy leaf
Vicia faba—bell bean

Field Trial Measurements and Data Collection

Sweet alyssum (*Lobularia maritima*) was established in strips on one side of five commercial lettuce fields, four conventionally farmed head lettuce fields, and one organic leaf lettuce field.

The strips were two beds wide and not less than 76m (250 feet) in length. The strip was located on the upwind side of the field whenever possible. The sweet alyssum was seeded in two rows per bed at approximately one seed per inch. Lettuce and sweet alyssum were seeded at the same time and received the same cultural inputs except for a preemergence herbicide which was used in lettuce but omitted from the alysssum beds.

Insect populations were monitored weekly in both the lettuce and the sweet alyssum. Four suction samples were taken from the sweet alyssum to estimate the beneficial insect populations, as previously described. Samples in the lettuce were taken as a function of distance from the insectary plantings. Samples were taken at 1, 3, 6, 11, 20, and 37 beds from the sweet alyssum. At each of these locations, six lettuce plants were examined for aphids and six 1m (3-foot) sections of the bed were vacuumed for beneficials.

Results and Discussion
Trial Gardens

The plants listed in Table 1 were used one or more times in a trial garden and sampled on a weekly basis. Candidate species were carefully researched to eliminate those that were unsuitable for one or more reasons. Figures 1–4 show which species were examined in each of the four trial gardens and reports the number of beneficial and pest insects found in the samples from each species averaged over the number of times it was sampled.

From these data, we have concluded that the potential for using in-field insectaries does exist with several of the plant species we have tested. The plant that showed the greatest potential was sweet alyssum. It had nearly all of the characteristics we were seeking. It did not, however, provide any wind protection, which would be desirable. Sweet alyssum will reseed itself, but as the plant is not tall or very aggressive, it is not expected to become a weed problem. The very fast bloom by this plant from seeding was a primary consideration in the decision to concentrate on this plant further. No other plant tested bloomed this quickly or attracted as many desirable species.

The presence of alternate prey in some of the plants tested is of both interest and concern. As the alternate prey of greatest interest are various aphid species, the issue of these aphids acting as vectors of plant viruses must be considered. Although a given species may not normally colonize lettuce or other vegetable crops, spread of nonpersistent viruses is still possible. Unfortunately, many of the aphid species found on the plants tested have not been tested for their ability to vector important crop diseases such as lettuce mosaic.

The regular disease sampling of the test plants did not indicate any serious problem with sweet alyssum, although some powdery mildew was noted. Sweet alyssum tested negative as a host of lettuce mosaic.

Field Trials

The field trials demonstrated that the density of beneficial insects could be increased near the insectary planting and that the aphid population could be reduced. The distance over which this effect was seen appears to be in the order of 11m or so, one direction from the insectary area. If this could be expected also to occur in the other direction, it would be appropriate to assume that insectary strips every 33m or so should be effective. This would correspond to every 20th bed in a 1m (40 inch) bed system. This is within

Figure 1. Mean number of pests and beneficials found per sample by species from trial 1.

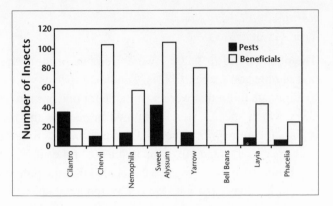

Figure 2. Mean number of pests and beneficials found per sample by species from trial 2.

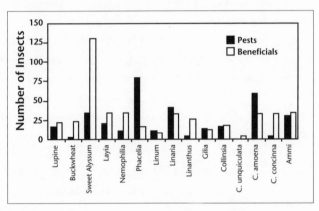

Figure 3. Mean number of pests and beneficials found per sample by species from trial 3.

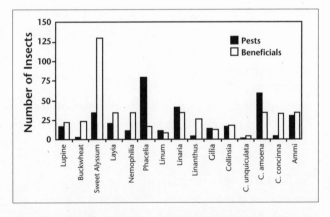

the guidelines suggested by the growers as possibly being an acceptable proportion of the ground to devote to such a system.

Figure 5 presents both aphid and beneficial insect densities as a function of distance from the sweet alyssum strip. Samples taken amid sweet alyssum are considered to have been at distance zero. In three of the fields,

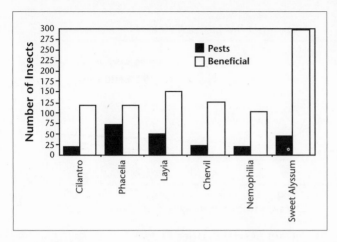

Figure 4. Mean number of pests and beneficials found per sample by species from trial 4.

aphid and beneficial insect densities were also measured from the opposite side of the field so that any edge effect might be separated. In Figures 6 and 7, the left side of the graph represents the edge of the field where sweet alyssum was planted, and the right side the opposite edge of the field. The data presented to theses obtained a mirror image from the center of the field.

Figure 8 presents the typical composition of the beneficial insect complex found in the sweet alyssum by date. The high numbers of parasitic Hymenoptera, *Geocoris*, and *Orius* were especially consistent. It should be assumed that many of the predators (*Geocoris* and *Orius*) were feeding on the flower thrips present in the flowers, especially in the immature stages. The Hymenoptera can be assumed to have been feeding on the nectar and pollen, as there were few if any suitable prey for them to be seeking in the sweet alyssum. They were also frequently found in the lettuce near the sweet alyssum, presumably seeking hosts. Parasitic Hymenoptera, adult *Orius,* adult *Geocoris,* and spiders were the most commonly found beneficials in the samples vacuumed from the lettuce.

Discussion

Changes in the pest complex in lettuce led to the pea leafminer, *Liriomyza huidobrensis* becoming a significantly more important pest in lettuce than the green peach aphid by the conclusion of this work (Chaney 1994; Dlott & Chaney 1995). For that reason, this technique has not been adopted by growers. The use of a nectar-producing plant was shown to increase leafminer populations (Chaney & Heinz unpublished data). Although this work showed promise for the technique, application will be practical only in an area where the leafminer pressure is not as severe. In such a situation

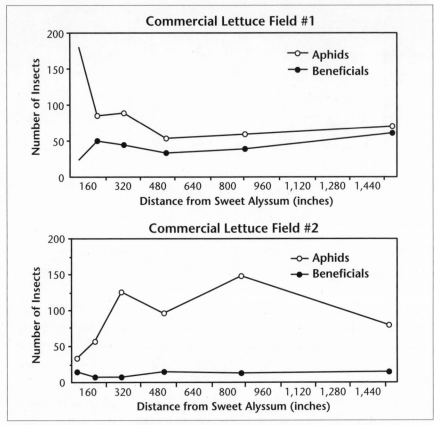

Figure 5 . The mean number of aphids and beneficial insects per sample as a function of distance from strips of sweet alyssum at one edge of commercial lettuce fields.

many improvements on this concept are suggested by this work and observations made while conducting these investigations.

The possibility of combining sweet alyssum with a taller plant, perhaps a cereal grain, should be explored. The wind protection of the cereal and the potential alternate prey (e.g., various pests of cereals) it might provide could prove useful. The possibility of the alternate hosts acting as a detrimental "trap crop" for the beneficials would have to be examined, however (see Corbett this volume). Also worth exploring would be using sweet alyssum and possibly a cereal seeded into bed at pre-irrigation, to give the insectary plants a head start on the crop. This would be compatible with some growers' practices. The possibility of taking advantage of small areas of winter cover crops left undisturbed through the crop production period might also be explored. The data suggest that given the limited distance effect of the

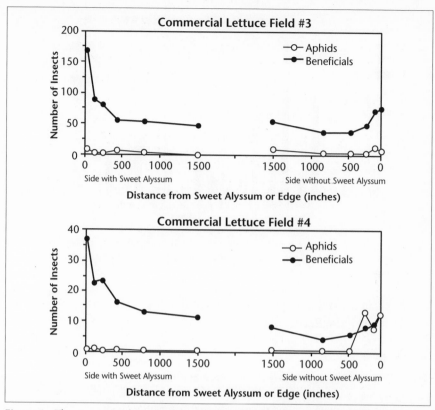

Figure 6. The mean number of aphids and beneficial insects per sample as a function of distance from strips of sweet alyssum at one edge of commercial lettuce fields and from the opposite edge without sweet alyssum of the same fields.

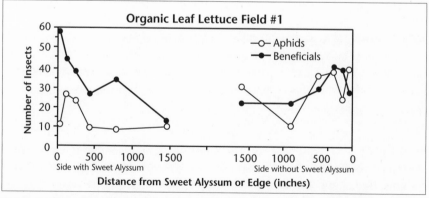

Figure 7. The mean number of aphids and beneficial insects per sample as a function of distance from a strip of sweet alyssum at one edge of an organic leaf lettuce field and from the opposite edge without sweet alyssum of the same field.

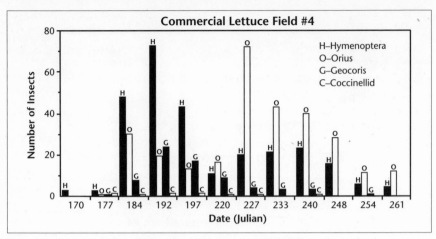

Figure 8. Mean number of beneficial insects per sample belonging to the four most common groups vacuumed from sweet alyssum strips in a commercial lettuce field by date.

insectary areas, single large areas devoted to insectary plantings will not be as effective as smaller scattered ones.

References

Altieri, M. A. 1991. Increasing biodiversity to improve insect pest management in agro-ecosystems, pp.165-182. *In* D. L. Hawksworth [ed.], The biodiversity of microorganisms and invertebrates: its role in sustainable agriculture. CAB International.

Andow, D.A. 1991. Vegetational diversity and arthropod population response. Annu. Rev. Entomol. 36:561-586.

Bazzaz, F. A. 1984. Demographic consequences of plant physiological traits: some case studies, pp. 324-346. *In* R. Dirzo & J. Sarukhan [eds.], Perspectives on plant population ecology. Sinauer Assoc, Inc., Sunderland, Massachusetts.

Blackman R. L. & V. F. Eastop. 1984. Aphids on the world's crops. Wiley & Sons, New York, New York.

Bugg, R. L. & J. D. Dutcher. 1989. Warm-season cover crops for pecan orchards: horticultural and entomological implications. Biol. Agric. Hortic. 6:123-148.

Bugg, R. L., L. E. Ehler & L. T. Wilson. 1987. Effect of common knotweed (*Polygonum aviculare*) on abundance and efficiency of insect predators of crop pests. Hilgardia 55(7):1-52.

Bugg, R. L., R. T. Ellis & R. W. Carlson. 1989. Ichneumonidae (Hymenoptera) using extrafloral nectar of faba bean (*Vicia faba* L., Fabaceae) in Massachusetts. Biol. Agric. Hortic. 6:107-114.

Chaney, W.E. 1994. Factors affecting leafminer populations. Monterey County Crop Notes, pp 1-2. Oct. 1994.

Dlott, J. W. & W.E. Chaney. 1995. Identifying management techniques and research needs for pea leafminer: cultural, chemical, and biological controls in celery, pp. 79-91. *In* 1994-95 Annual Report. California Celery Research Advisory Board.

Ferris S. H. 1989. Organic produce a growing trend in the Salinas Valley, p. 4A. Salinas Californian. Saturday May 20

Flint, M. L. 1985. Green peach aphid, pp. 39-42. *In* M. L. Flint [ed.], Integrated pest management for cole crops and lettuce. Publ. 3307.

Georghiou, G. P. 1983. Management of resistance in arthropods, pp. 769-792. *In* G. P. Georghiou & T. Saito [eds.], Pest resistance to pesticides. Plenum, N.Y.

Georghiou, G. P. 1986. The magnitude of the resistance problem, pp. 14-43. *In* Pesticide resistance: strategies and tactics for management, National Research Council. National Academy of Sciences, Washington D.C.

Kemp, J. C. & G. Barrett. 1989. Spatial patterning: impact of uncultivated corridors on arthropod populations within soybean agroecosystems. Ecology 70: 114-128.

Kennedy, J. S., M. F. Day & V. F. Eastop. 1962. A conspectus of aphids as vectors of plant viruses. Commonwealth Inst. of Entomology, London.

McCalley, N. F. 1981. Identification of aphids commonly associated with crops in the Salinas Valley. Univ. Calif. Agric. Ext. Service, Salinas, California.

Patriquin, D. G., D Baines, J. Lewis & A. MacDoughall. 1988. Aphid infestation of faba beans on an organic farm in relation to weeds, intercrops and added nitrogen. Agric. Ecosyst. Environ. 20:279-288.

Price, P. W. & G. P. Waldbauer. 1975. Ecological aspects of pest management, pp. 37-73. *In* R.L. Metcalf & W. Luckman [eds.], Introduction to insect pest management. John Wiley & Sons, New York.

Tabashnik, B. E. 1989. Managing resistance with multiple pesticide tactics: theory, evidence, and recommendations. J. Econ. Entomol. 82:1263-1269.

Theunissen, J. & H. den Ouden. 1980. Effects of intercropping with *Spergula arvensis* on pests of brussels sprouts. Entomol. Exp. Appl. 27:260-268.

van den Bosch, R. & P. S. Messenger [eds.]. 1973. Biological control. Intext Educational Publishers, New York.

van Emden, H. F. 1964. The role of uncultivated land in the biology of crop pests and beneficial insects. Sci. Hortic. 42:121-136.

Parasitoid Activity and Plant Species Composition in Intercropped Systems

Moshe Coll—Department of Entomology,
University of Maryland, College Park MD 20742, USA[1]

Introduction

Many ecologists have assumed that populations of herbivorous arthropods fluctuate less around a mean equilibrium level in diverse than in simple plant assemblages (Elton 1927, 1958; Odum 1953; Pimentel 1961a; Margalef 1968). Elton (1958), Pimentel (1961a), Litsinger and Moody (1976), and Perrin (1977) extended the diversity-stability hypothesis, suggesting that the reduction in plant diversity in managed habitats (e.g., monocultures) could result in pest outbreaks. Watt (1964), Southwood and May (1970), Huffaker (1974), Murdoch (1975), and others have challenged this thesis as simplistic. Nonetheless, insect herbivores often reach higher population densities in mono- than multi-specific stands of their host plants (reviewed by Root 1973; Dempster & Coaker 1974; van Emden & Williams 1974; Litsinger & Moody 1976; Perrin 1977; Cromartie 1981; Altieri & Letourneau 1982; Kareiva 1983; Risch et al. 1983; Stanton 1983; Andow 1991).

Questions raised by van Emden and Williams (1974), in particular, stimulated studies on the mechanisms and processes that underlie population interactions in diverse habitats. Root (1973) incorporated the possible factors that underlie the response of herbivores to habitat diversification into two hypotheses, the resource concentration hypothesis and the "enemies hypothesis." The resource concentration hypothesis proposes that herbivores are more likely to find and remain on their host plants in dense or pure than in diverse stands. The enemies hypothesis proposes that natural enemies are more abundant and/or efficient in taxonomically diverse plant habitats and thus reduce herbivore densities in such habitats (Root 1973).

This chapter evaluates the impact of plant species composition on parasitoid activity. I discuss the need for a better understanding of the mechanisms that underlie the response of parasitoids to plant diversification. A mechanistic understanding of vegetation-parasitoid interactions is essential

---

[1] Current address: Dept. Entomology, Hebrew University of Jerusalem, P.O. Box 12, Rehovot 76100, Israel.

for the development of ecological theorem and for a better enhancement of parasitoids through habitat management. Specifically, the chapter is focused on the following questions: Do changes in plant community texture, that is, plant density, structural complexity, and species richness, have a consistent impact on the rate with which herbivores are attacked by parasitoids? If not, what can be learned from this apparent variability in parasitoid response to plant diversification? What mechanisms can cause parasitoids to be favored by a plant habitat? How do pest control practices and nonhost pests affect parasitoids in intercropped habitats? And finally, can an understanding of the mechanisms that underlie parasitoid response to their host habitat be used to design agricultural systems that are more likely to enhance parasitoid activity?

Agricultural plant diversity can be increased by growing two or more crop species simultaneously in the same field (intercropping). Intercropping systems are particularly suitable for investigating effects of plant diversity on natural enemies because plant diversity, dispersion (spatial arrangement), and phenology can be controlled. Although intercropping is commonly practiced in many parts of the world (Kass 1978; Francis 1986; Vandermeer 1989), the terminology is confused (Table 1). For example, Vandermeer's (1989) definition of intercropping may include sequential planting of two crops, where the crop planted first affects the second crop. Perrin (1977) refers to intercropping as mixed cropping, and Ruthenberg (1971) and Harwood (1975) defined mixed cropping just as Andrews and Kassam (1976) and Perrin (1977) defined mixed intercropping and mixed planting, respectively. Other terms used include polycropping, interculture, mixed farming, and simultaneous polycropping (Ruthenberg 1971; Glass & Thurston 1978). In this chapter, I follow the terminology used by Andrews and Kassam (1976). For a more complete treatment of intercropping see Kass (1978), Francis (1986), and Vandermeer (1989).

My discussion is focused on mixed and row intercrops of annual field and vegetable crops. A few studies that compared parasitoid activity in row and strip intercrops indicate that parasitism rates tend to be higher when the two crops are closely associated spatially. For example, parasitism rate for *Ostrinia furnacalis* (Guenee) (Lepidoptera: Pyralidae) eggs in maize (*Zea mays*) tended to be higher when one row of sweet potato (*Ipomoea batatas*) was alternated with three rows of maize than with seven rows of maize (Nafus & Schreiner 1986). Thus, a maximal enhancement level of parasitoid activity may be expected in row and mixed intercrops.

The response of natural enemies to plant diversity in forest, orchard, and weedy cultures was reviewed elsewhere (Pimentel 1961a; Altieri et al. 1977; Way 1977; Zandstra & Motooka 1978; Cromartie 1981; Altieri & Letourneau 1982; Altieri & Liebman 1986). Occasionally, however, examples from other systems (e.g., weedy culture, intercrops with beneficial noncrops such as living mulches) are used to illustrate mechanisms that may underlie the response of parasitoids to plant species diversity. Finally, the literature on host-parasitoid and host plant–parasitoid interactions is vast, and the choice of examples is illustrative, not exhaustive.

Host-Parasitoid Interactions in Diverse Plant Habitats

Intercropping could affect parasitoid-host interactions in several ways. The increase in habitat complexity in intercrops could provide refuge for hosts escaping certain parasitoids (Pimentel et al. 1963) and thus could potentially stabilize both host and parasitoid populations. However, a higher degree of herbivore population regulation in diverse than simple habitats has rarely been measured directly. Instead, researchers have inferred pest regulation when lower herbivore density and/or higher activity of natural enemies are found in taxonomically diverse vegetation than in monospecific plant stands.

Host-parasitoid interactions on one plant species can be influenced not only by the presence of associated plants, but also by the presence of other hosts on the associated plants (Price et al. 1980). Letourneau (1987) found that *Trichogramma pretiosum* Riley (Hymenoptera: Trichogrammatidae) exhibited a density-dependant response to *Diaphania hyalinata* L. (Lepidoptera: Pyralidae) eggs in a squash (*Cucurbita pepo*) monoculture but recorded no such relationship in a squash-maize-cowpea (*Vigna unguiculata* ssp. *unguiculata*) triculture. She proposed that these differences were due to the presence of alternative hosts in maize. Use of alternative hosts by parasitoids may depend not only on their host range but also on their tendency to switch between host species (e.g., Gardner & Dixon 1985). Both specialist and generalist parasitoids often show host preference that may reflect learning as well as genetic make-up (van Alphen & Vet 1986; Lewis & Tumlinson 1988; Turlings et al. 1989; Lewis et al. 1990; Vet & Dicke 1992).

In addition to the effects of biotic factors, host-parasitoid interactions in intercrops may also be influenced by abiotic factors. In Uganda, for example, *Heliothis armigera* (Hübner) (Lepidoptera: Noctuidae) reproduces throughout the year on a wide range of cultivated and wild plants. In such favorable climate, the continuity of host-parasitoid interaction is thought to

Table 1. Selected definitions of intercropping and related cropping practices.

Term	Definition	Reference
Intercropping	Two or more crops grown simultaneously in alternate rows in the same area	Ruthenberg 1971
	Growing two or more crops simultaneously on the same field	Andrews & Kassam 1976; Batra 1982
	The cultivation of two or more species of crops in such a way that they interact agronomically (biologically)	Vandermeer 1989
	Two or more crops are grown in close proximity, either in discrete patches or strips, or intermixed	Speight 1983
	The planting of one crop into another crop, either between the rows or into the stubble of a previous crop	National Research Council 1989
	The cultivation of a secondary short-duration crop that is sown after the main crop and harvested before it	Taylor 1976
	The growing of two or more crops such that most if not all of their growth overlaps in time and space (same as mixed cropping)	Perrin 1977
Mixed intercropping	Growing two or more crops simultaneously with no distinct row arrangement (a type of intercropping)	Andrews & Kassam 1976
Row intercropping	Growing two or more crops simultaneously where one or more crops are planted in rows (a type of intercropping)	Andrews & Kassam 1976
Strip intercropping	Growing two or more crops simultaneously in different strips wide enough to permit independent cultivation but narrow enough for the crops to interact agronomically (a type of intercropping)	Andrews & Kassam 1976
Relay intercropping	Growing two or more crops simultaneously during part of the life cycle of each (a type of intercropping)	Andrews & Kassam 1976
Alternate row cropping	The cultivation of two crops in alternate rows thus giving a mixed culture (the rows may be one or more but not large enough to constitute strip cropping)	Taylor 1976
Interplanting	Long-term annual or biennial crops interplanted with short-term annual crops during early stages of plant development	Ruthenberg 1971
	The cultivation of a secondary crop that is sown after the main crop but harvested after it	Taylor 1976

Table 1. *continued*

Term	Definition	Reference
Interplanting	The confinement of each crop to a uniform row (single, double or multiple rows) (a type of intercropping)	Perrin 1977
Mixed interplanting	The confinement of one crop in rows and the random arrangement of others (a type of intercropping)	Perrin 1977
Mixed cropping	Same as intercropping	Steiner 1982
	The sowing of a piece of land with two or more plant types so that there is temporal and spatial overlap of the growth and development of some or all of the plant types	Taylor 1976
	Two or more crops grown simultaneously and intermingled with no row arrangement	Ruthenberg 1971; Harwood 1975
	Same as intercropping	Batra 1982
Mixed planting	The random arrangement of species in a given proportion, to achieve a totally haphazard mix or a mosaic of individual blocks (a type of intercropping)	Perrin 1977
Relay cropping	The intercropping of one crop in another before the first is harvested	Perrin 1977
Multiple cropping	Growing two or more crops on the same field in a year	Andrews & Kassam 1976
	The growing of more than one crop on the same piece of ground in one year	Harwood 1973
Companion cropping	The simultaneous cultivation of crops in a mixed culture, sowing is usually done at the same time (crops are used mainly as forage)	Taylor 1976
Polyculture	A system in which two or more crops are simultaneously planted within sufficient spacial proximity to result in interspecific competition and complementation	Hart 1974 in Altieri et al. 1978
	Two or more crops grown simultaneously in the same area (includes mixed cropping, intercropping, and interplanting)	Kass 1978
	Same as intercropping	Batra 1982

lead to stability and non-pestiferous status. In southern Tanzania, however, dry season diapause of *H. armigera* hampers the action of biocontrol agents (Reed 1965). Under these conditions, planting maize with cotton (*Gossypium hirsutum*) further aggravates *H. armigera* attack because the pest multiplies on maize and migrates to cotton without a check by natural enemies (Bhatnagar & Davies 1981).

Effect of Plant Texture on Parasitoid Species Diversity

Different plant species may differ structurally and thus may provide a variety of microhabitats for parasitoids to colonize. Therefore, more parasitoid species are expected to inhabit interplanted crops than crops grown in monocultures. Additionally, in diverse habitats, a host may escape parasitism by occupying microhabitats inaccessible to one of its parasitoids. Thus, the increase in habitat complexity may reduce the ability of a given parasitoid species to dominate the parasitoid guild on that host, permitting more parasitoid species to exist in intercropped systems. However, Nafus and Schreiner (1986) reported that parasitoid species richness in a diculture was the sum of the number of species in the component monocultures. Sixty-two species of parasitic Hymenoptera were found in a maize-sweet potato diculture system and 36 and 26 were found in monocultures of maize and sweet potato, respectively. Nonetheless, about 25% more individual parasitoids were found in the intercropped than monocultured systems.

Parasitoid guild structure may differ even between two closely related host species in the same habitats. For example, in safflower (*Carthamus tinctorius*) monocultures, Hymenopteran parasitoids caused 91% of total *Heliothis armigera* larval parasitism but only 67% of total parasitism in *Heliothis peltigera* (Schiff) larvae (Pawar et al. 1985). A similar trend was found in intercropped fields although the proportional abundance of the two herbivore species differed greatly in mono- and dicultures (*H. armigera* consisted 27% of all *Heliothis* spp. found in safflower monocultures but was about 50% of the ones found in dicultures; Pawar et al. 1985). It therefore appears that variations in herbivore abundance in simple and diverse habitats may obscure effects of plant diversity on parasitoid species composition.

Factors That Affect Parasitoids in Diverse Plant Habitats

Nonhost plants may affect parasitoids, both directly and indirectly. Direct effects may involve, for example, interference with parasitoid movement. Indirect effects by nonhost plants may impact the reproduction, development, and survival of parasitoids through other trophic levels (e.g., host plants, herbivorous hosts) and through alteration of the habitat's microclimate.

Factors That Affect Parasitoid Movement and Host Finding

Obscured Host-Finding Cues and Altered Host to Leaf Area Ratio

Nonhost plants may impede parasitoids' host and host habitat location by obscuring visual and olfactory cues. Monteith (1960), for example, found that plants in the forest understory hindered the ability of *Bessa harveyi* (Townsend) (Diptera: Tachinidae) to find its host (sawfly larvae) by masking host and host plant odors. Nonhost plants may also affect host finding by altering the leaf surface area searched by parasitoids (Need & Burbutis 1979; Parkman & Shepard 1982; Andow & Prokrym 1990) and by changing host plant apparency (Dempster 1969; Speight & Lawton 1976; Brust et al. 1986).

Physical Barriers and Entrapment of Adult Parasitoids

Nonhost plants may alter the rate and pattern of parasitoid movement, especially when plants of different heights are in close association (i.e., structurally diverse plant habitats). Like herbivores, parasitoids are expected to colonize pure stands of host plants with a greater ease than multispecies stands (cf. Risch 1981; Bach 1984; Sheehan 1986; Russell 1989). Coll and Bottrell (1996) compared immigration and emigration rates of *Pediobius foveolatus* (Riley) (Hymenoptera: Eulophidae) in bean (*Phaseolus vulgaris*) monocultures and in bean-maize dicultures. When released outside the plots, half as many female wasps were recaptured in a tall maize-bean intercrop than in a short maize-bean intercrop and two bean monocultures suggesting that the tall maize plants lowered the parasitoid immigration rate. Further, the effect of the tall maize appeared to be independent of host plant density and plant species composition because similar numbers of wasps were recaptured in low and high density bean monocultures and in a short maize-bean intercrop. Therefore, generalizing statements, such as lower colonization occurs in diverse habitats, may not be useful in predicting the response of parasitoids to habitat diversity. Instead, it appears that various factors affect the movement of parasitoids in different habitats.

Parasitoids that disperse passively may also be affected by the presence of nonhost plants. For example, Risch (1979) suggested that maize stalks could act as a windscreen, causing drifting insects, such as parasitic Hymenopteran, to passively accumulate in plots with maize. However, nonhost plants are not necessarily physical barriers to parasitoids in intercrops. Coll and Bottrell (1996) found in a release-recapture experiment that tall maize plants did not hamper the across-row movement rate of a parasitoid, *P. foveolatus*, in a maize-bean intercrop.

Much like herbivorous insects, female *P. foveolatus* had a shorter tenure time in dicultural than monocultural habitats. Nonetheless, 25 days after its release, the parasitoid was more abundant in the tall maize-bean habitat apparently because of a more favorable microclimate in plots shaded by tall maize plants (Coll 1991; Coll & Bottrell 1996). Because countervailing effects may be at play, it seems that the immediate movement patterns of parasitoids (and other arthropods) in response to plant diversification may not determine their ultimate (even within generation) densities.

Certain associated plants can have an adverse effect on parasitoids by acting as traps for searching adults. For example, *Lysiphlebus testaceipes* (Cresson) (Hymenoptera: Aphidiidae), a parasitoid of cotton aphid (*Aphis gossypii*) (Glover) (Homoptera: Aphidae) was entrapped by glandular exudates of petunia, and thereby prevented from protecting nearby okra (*Abelmoschus esculentus*) plants from aphid attacks (Marcovitch 1935); *Trichogramma minutum* (Riley) (Hymenoptera: Trichogrammatidae) adults could not parasitize hornworms *Manduca* sp. (*Lepidoptera: Sphingidae* on tomato (*Lysopersicon esculentum*) plants because of entrapment by the sticky glandular trichomes on the leaves of the associated tobacco (*Nicotiana tabacum*) plants (Marcovitch 1935); and although high parasitism of *Heliothis armigera* eggs by *Trichogramma* sp. occurred in sorghum (*Sorghum halepense* ssp. *bicolor*), a low parasitism rate was recorded in sorghum associated with chickpea (*Cicer arietinum*) because adult wasps were trapped in its sticky trichomes (ICRISAT 1980).

On the other hand, associated plants may attract parasitoids to a habitat. The addition of sesame (*Sesamum* spp.) to cotton fields attracted *Campoletis sonorensis* (Cameron) that attack *Heliothis virescens* (F.) (Lepidoptera: Noctuidae) and *Helicoverpa zea* (Boddie) (Lepidoptera: Noctuidae) (Pair et al. 1982). However, it is not clear whether such attractiveness will increase (act as a source) or decrease (act as a sink) parasitoid activity on the target host plants (see Corbett this volume). For example, population buildup of *Trichogramma confusum* (Viggiani) and *Diadegma* sp. (Hymenoptera: Ichneumonidae) occurred on *Heliothis armigera* eggs and larvae, respectively, in sorghum. However, this buildup did not benefit the intercropped pigeonpea (*Cajanus cajan*) because the parasitoids on sorghum did not attack *H. armigera* on pigeonpea (Bhatnagar & Davies 1980). Dipteran parasitoids, however, moved readily between cotton and pigeonpea. Thus, cotton served as a source of parasitic flies but as a sink for the hymenopteran parasitoids.

Factors That Affect Parasitoid Reproduction and Survival

Provision of Food Sources

Flowers offer important food sources for adult parasitoids. Many female parasitoids must feed on nectar or pollen before they can deposit viable eggs (Leius 1960; van Emden 1963), and the addition of pollen and nectar to the diet has increased the longevity and fecundity of several parasitoid species (Leius 1961; Syme 1975, 1977; Foster & Ruesink 1984, 1986). Therefore, introducing flowering plants to the field could increase parasitism rates (Leius 1967) by minimizing the time parasitoids devote for food searching and by increasing their fecundity and longevity.

Most monocultural annual crops do not provide an adequate supply of food sources for adult parasitoids. As a result, parasitoids must travel long distances between food and host habitats. Intercropping can sometimes solve this problem. Risch (1979) observed chalcid wasps feeding on maize pollen and recorded three and four times as many parasitoid individuals in a maize monoculture and maize-sweet potato diculture, respectively, than in a sweet potato monoculture.

Another important food source for adult parasitoids is extrafloral nectaries (Bugg et al. 1989). Cotton, for example, was recognized as a valuable plant for parasitoids because of the abundant production of nectar in two types of extrafloral nectaries (Marcovitch 1935). Intercropping, however, may lower flower production. For example, lower flower density was found when cowpea was intercropped with maize than in a cowpea monoculture (Perfect et al. 1979). Although nectar and pollen in many legume flowers are probably inaccessible to parasitoids, similar effect (i.e., lower flower production) of intercropping on other crops (e.g., Apiaceae, Asteraceae) could reduce parasitoids attraction to these systems.

Finally, herbivores may also provide adult parasitoids with food sources, such as honeydew (secreted by aphids in cotton (Marcovitch 1935) and body fluids when host feeding (e.g., Aphidiidae and Aphelinidae, Stary 1988a,b). Because as much as 30% of the world's parasitoid fauna is composed of host-feeding species (Kidd & Jervis 1989) and many species are attracted to honeydew, parasitoids may be favored simultaneously by the high herbivore densities often found in monocultures and by the large variety of honeydew-secreting insects in polycultures. Clearly, there is a lack of critical information about the effect of parasitoid food sources on parasitism rate and herbivore density in the field.

Provision of Alternate and Alternative Hosts

Associated crops may provide parasitoids with alternative hosts and support parasitoids when the target crop lacks the primary host. For example, the presence of an alternative host, *Aphis helianthi* (Monell) (Homoptera: Aphididae) on sunflower (*Helianthus annuus*) leads to an increase in parasitism rate in *Schizaphis graminum* (Rondani) (Homoptera: Aphidae) on nearby sorghum (Eikenbary & Rogers 1974). Similarly, parasitoids, particularly *Lysiphlebus testaceipes*, were able to maintain high population densities on *A. helianthi* during periods of *S. graminum* population scarcity. Further, some parasitoids require alternate (obligatory) hosts to persist in a habitat. For example, alternate host species are required for the tachinid *Lydella thomsonii* (Herting) (Diptera: Tachinidae) because its life cycle is not synchronized with that of its host, European corn borer (*Ostrinia nubilalis* [Hübner]), (Lepidoptera: Pyralidae) larva (Hsiao & Holdaway 1966).

Provision of other hosts in a habitat, however, may not necessarily result in increased parasitism rate in the target pest. For example, Nafus and Schreiner (1986) intercropped sweet potato with maize to allow the population of *Trichogramma chilonis* (Ishii) to increase on *Agrius convolvuli* (L.) eggs in sweet potato before attacking *Ostrinia furnacalis* eggs in maize. No significant enhancement of parasitism rate on *O. furnacalis* was found. Likewise, parasitoids may not transfer to other crop species even when the same host species is present on the two associated crops. For example, hymenopteran parasitoids readily attack *Heliothis armigera* larvae on sorghum but not on the intercropped pigeonpea (Bhatnagar & Davies 1980; 1981). Similarly in the same system, up to 80% of *H. armigera* eggs in sorghum were parasitized by *Trichogramma confusum* whereas in pigeonpea the rate was always less than one percent (Bhatnagar & Davies 1980).

Altered Microclimate

An increase in habitat structural complexity often accompanies the increase in plant taxonomical diversity. In turn, parasitoids in structurally complex habitats may experience a lower temperature, higher relative humidity, and greater shade than those in structurally simple habitats (Altieri et al. 1978; Litsinger et al. 1991). Air turbulence may also differ greatly in structurally simple and complex habitats. Bottenberg and Irwin (1991), for example, found that maize reduced wind speed at bean canopy height, causing aphids to depart more readily from bean plants in maize-bean intercrop than in bean monoculture. Although air movement pattern may differ in monocultures and polycultures, its effect on parasitoid activity is not clear. Many parasitoids land once the wind speed has exceeded a certain threshold. This threshold may be quite low and is often exceeded in the field (e.g., Casas 1989). However, the influence of microclimatic conditions on parasitoids may differ greatly among taxa. Ichneumonids are most active at moderate temperature and high relative humidity whereas braconids are most active at high temperature and low humidity (Juillet 1964).

Differences in temperature, humidity, and air flow in a habitat may alter host-finding behavior by parasitoids. For example, the rate and the pattern at which chemical cues are being emitted and spread in the habitat may differ under different physical conditions. Thus, perception of these cues by searching parasitoids may differ in structurally simple and diverse habitats.

Light intensity may influence parasitoid activity levels (e.g., Barbosa & Frongillo 1977) and their ability to locate and parasitize their hosts in a habitat. In field experiments, where host density was maintained at the same level in different habitats, shade influenced parasitism rate and offspring clutch size (number of parasitoid larvae per host) and sex ratio (Coll 1991). Whereas only about 33% of Mexican bean beetle larvae were parasitized by *Pediobius foveolatus* in bean monocultures, 67% of hosts were parasitized in shaded monocultures and in a tall maize-bean intercrop. (Artificial shading of monocultures simulated shading by tall maize plants in dicultural plots.) Further, more parasitoids and proportionally more female wasps emerged per host in tall maize-bean and shaded monocultural plots (Coll 1991).

Altered Host and Host Plant Quality

Associated nonhost plants often influence the physiology of herbivores' food plants. For example, flowering is delayed and reduced in cowpea intercropped with maize (Perfect et al. 1979). Further, the architecture of plants grown in multispecific stands often differs from that of plants grown in monospecific stands. Like other shaded plants, beans intercropped with maize grew taller and had a larger leaf area than bean plants in monocultures (Coll & Bottrell 1994). Soil nutrient profile may also differ greatly in intercrops and monocultures. Nitrogen yield and nitrogen content of maize-legume intercrops are often higher than those of maize monocultures (Kass 1978; Lipman 1912). These differences in plant physiology, architecture, and nutritional value may affect resource availability (e.g., nectar), host searching behavior (e.g., through changes in leaf area), and development (e.g., through changes in host quality [Loader & Damman 1991]) of parasitoids. However, it is hard to generalize whether such effects enhance or hamper parasitoid activity.

Effect of Pest Control Practices on Parasitoids in Intercrops

Host Plant Resistance

Crop breeding can be used to enhance the effectiveness of parasitoids in intercrops. The presence of nonhost plants in intercrops may affect soil nutrient profile and habitat shade, which in turn could alter the architecture, allelochemical content, attractancy, nutritional value, and phenology of host plants. Such effects on host plants may greatly alter searching efficiency, development and survival of parasitoids (Pimentel 1961b; Schuster & Starks 1975; McGovern & Cross 1976; Campbell & Duffey 1979; Greenblatt & Barbosa 1981; Orr & Boethel 1985; Barbosa & Saunders 1985; Barbosa et al. 1986, 1991). If host plant resistance and biological control are to be compatible in intercrops, the effect of all associated crops on parasitoid activity must be considered in plant breeding programs.

Pesticide Use

In intercropped subsistence farming systems, where insecticidal control is often impractical, the conservation and encouragement of parasitoids could provide sufficient pest control. However, in many areas of the world, use of pesticides has replaced intercropping as a pest suppression practice (Matteson et al. 1984).

The few studies on the effect of pesticides on parasitoids in intercrops suggest that parasitoids in these habitats may be more susceptible to insecticides than those in monocultures (e.g., Bhatnager & Davies 1980). There are several possible reasons why pesticide applications in intercrops may have a greater adverse effect on parasitoids than those in monocultures. Differences in crop height, architecture, and maturity often lead to inefficient and thus multiple insecticide applications in intercrops. Pesticide application machinery, designed for use in monocultures, is often unsuitable for use in intercrops (Banta 1973). Thus, cheap hand-held ultra-low-volume spray equipment was adapted for treating different plants in intercrops (Perrin 1977). Van Emden (1977) cautioned, however, that the use of such equipment often leads to an increase in pesticide applications in tropical intercropped farming systems. The escalation in insecticide use may in turn be detrimental to parasitoids. Also, it is difficult to prevent pesticide drift from reaching nontarget crops because of the proximity of the intercropped plants. Thus, associated crops can no longer serve as reservoirs of parasitoids.

In some cases, pesticide application is more effective in diverse then simple habitats. Chlorfenvinphos provided better control of cabbage maggot (*Delia brassicae* [L.], Diptera: Anthomyiidae) on brassicaceous plants undersown with clover than in monocultures, probably because the microclimate in the more diverse plots enhanced the activity of this insecticide (e.g., Dempster & Coaker 1974). Finally, fewer insecticide applications may be required in intercrops than in monocultures because pests tend to reach lower densities in diverse than simple habitats. However, insecticide application could still reduce parasitoid densities by direct toxicity to adults and by eliminating host populations.

Pesticide application may also affect parasitoids indirectly through phytotoxic effects. For example, pesticide-treated pigeonpea intercropped with sorghum flowered earlier and was harvested about one hundred days before its unsprayed counterpart (Bhatnagar & Davies 1980). Similarly, cowpea sprayed before blooming peaked in flower abundance ten days later than plants sprayed after blooming (Perfect et al. 1979). Although legume pollen and floral nectar may be inaccessible to parasitoids, similar phytotoxic effects in other crops could disrupt the synchrony between parasitoid populations and their resources (e.g., nectar and pollen) and may reduce parasitoids' density and efficiency.

Therefore, the use of pesticides in intercropped systems needs to be reevaluated with respect to its adverse effects on parasitoids. Problems with drift, phytotoxicity, and residual effects on nontarget hosts in associated crops need to be assessed.

Interactions Between Parasitoids and Nonhost Pests in Intercrops

Plant Pathogens

Because plant pathogens are often less widely spread in intercropped than in monocultural systems (Browning 1975; Altieri & Liebman 1986), the intensity of plant pathogen-parasitoid interactions may differ in simple and diverse habitats. Plant pathogens often induce changes in defensive products (e.g., toxic allelochemicals and physical defenses such as lignification, cell wall thickening, and deposition of tannin-like compounds [Akai & Fukutomi 1980]) and the quality and quantity of nutritional compounds (e.g., total nitrogen, amino acids, soluble proteins, carbohydrates) of infected plants. Such changes in host plants may affect the behavior, physiology, and survival of herbivores and their parasitoids. These effects on parasitoids may be direct when, for example, physiological changes in the host plant alter the chemical cues used by parasitoids to locate their hosts. The effects may also be indirect when altered host characteristics, such as size, nutritional value, developmental rate, allelochemical content, and feeding site, make hosts unsuitable for parasitoid attack and development.

Weeds

Effect of weed management on parasitoids has been reviewed extensively (Altieri et al. 1977; Zandstra & Motooka 1978; Altieri & Whitcomb 1979; Altieri & Letourneau 1982; Altieri & Liebman 1986). Because weeds may provide parasitoids with food (nectar and pollen), alternate hosts, and shelter and overwintering sites, weeds could alter the response of parasitoids to intercropping. For example, parasitism rate in *Diaphania hyalinata* eggs was lower in weedy than weed-free squash monocultures but was higher in weedy than weed-free squash-maize-cowpea tricultures (Letourneau 1987).

Moody and Shetty (1981) stated in their review that the growth of weeds in an intercrop may be severely depressed or hardly affected, relative to monocultures. Compared to the component monocultures, intercropping may suppress weeds because of a greater light interception (i.e., greater shade), allelopathy, and, maybe, belowground resource competition. Therefore, factors such as crop species and cultivars, relative proportion of component crops, spatial arrangement of crops, crop density, and fertilization rate may all influence weed occurrence and, in turn, parasitoid activity in intercropped systems.

Nematodes

Little work has been conducted on nematode suppression in intercrops (but see, for example, Egunjobi [1984]). Like plant pathogens, nematode attack may cause physiological and chemical changes in plants. Additionally, nematodes may vector soil-borne viruses, and plants injured by feeding nematodes are often more susceptible to infection by root rots and wilts. In turn, these changes in the host plant may affect arthropod pests and their parasitoids.

The Effect of Intercropping on Parasitoid Activity: A Quest for Patterns

This section evaluates the ability of various habitat (e.g., host and nonhost plant taxa), host (e.g., density, stage), and parasitoid (e.g., taxonomy, host range) factors to predict differences in parasitism rate and parasitoid abundance in simple and diverse habitats.

The study of parasitoid populations in intercropped systems has been hampered by parasitoid species diversity. Perfect et al. (1979), for example, were able to trap about 150 hymenopteran parasitoid species in a three-year study of a maize-cowpea intercrop. However, many species occur sporadically and in low numbers and little is known about their host associations. As a result, researchers tend to treat all hymenopteran parasitoids as a group. Different taxa, however, probably respond differently to changes in plant diversity. For example, whereas parasitoids in the Bethyloidea, Proctotrupoidea, and Chalcidoidea were more abundant in a maize-cowpea intercrop than in maize monoculture, density of ichnemonoid parasitoids was similar in the two systems (Perfect et al. 1979).

Another limiting factor in the study of parasitoids in simple and diverse habitats is the interaction between sampling method and habitat type. Parasitoid activity is usually estimated by trapping or by recording host parasitism rates. These methods are appropriate only when habitat type does not affect the precision of the sampling. However, sampling may not be equally effective in different habitats because differences in plant height and architecture may alter sampling effort and/ or effectiveness. For example, Perfect et al. (1979) used traps to compare parasitoid density when cowpea and maize were grown alone or intercropped. They found that when traps were placed 0.5m above ground, almost twice as many chalcidoid wasps were trapped in the intercrop than in monocultures. However, when the traps were situated 2m above the ground, similar numbers of chalcidoids were trapped in the two monocultures and in the intercrop. Similarly, maize height affected tachinid fly catch in Malaise traps (Hansen 1983). Thus, in both cases the traps are inappropriate for comparing parasitoid abundance in monocultural and intercropped habitats. Colored traps may be more apparent in one habitat, resulting in a greater attractiveness and capture of parasitoids. Estimating parasitism rate may also depend on habitat type if, for example, host spatial distribution (i.e., clumpedness) differs among habitats but the same sampling protocol is used in different habitats.

Parasitoids in intercrops may also be affected by vegetation in neighboring fields. For example, in cotton-growing areas, parasitism rate in *Heliothis* spp. by dipterans was higher in a sorghum-pigeonpea intercrop than in a sorghum monoculture (ICRISAT 1978; Bhatnagar & Davis 1980, 1981). The reverse, however, was recorded in non-cotton-growing areas (i.e., a higher parasitism rate in monocultured than intercropped pigeonpea). Although the vegetation in areas that bordered experimental fields may play an important role in supplying parasitoids with food, hosts, and shelter, this vegetation is rarely described or controlled.

Data Collection

To find patterns in the response of parasitoids to plant species diversity, I searched the literature for field studies that compared parasitoid activity in monocultures to that in row or mixed intercrops of annual field and vegetable crops (Table 2). Thus, complicating effects of plant dispersion, phenology, persistence, and structural complexity were minimized (i.e., no perennials or trees). In other intercrop types, such as strip intercropping, the spatial scale of plant arrangement could range from several rows to relatively large blocks of the alternated crops. Thus, intercropped blocks may not differ greatly from a situation where several crops are being cultivated next to each other in monocultures (i.e., fine-grained vs. coarse-grained interspersion of the components).

Studies in which the highest reported parasitism rate in any of the treatment plots was less than five percent were not included in the review because these parasitoids appear to be insignificant pest mortality factors. Also, low levels of parasitoid activity do not allow a meaningful evaluation of parasitoid response to plant diversity. Finally, the low number of available studies did not allow much stricter choice of studies to be included in the analysis. Thus, studies vary greatly in use of proper experimental designs (e.g., proper replication), unbiased sampling methods, and appropriateness of statistical analyses.

Results

Overall, parasitoid density or parasitism rates were compared in 42 different monoculture-intercrop systems (Table 2). In two thirds of the comparisons, the parasitoids were more abundant or attacked more hosts in the intercropped than monocultured habitats (Table 3). However, in about a third of the comparisons no consistent differences were recorded in parasitoid density or parasitism rate among habitats. A lower rate of parasitoid enhancement was found when data were analyzed by parasitoid species or group of species. Only 54% of the 31 studied species had a higher parasitism rate or density in intercrops compared to monocultures (39% showed similar or variable activity levels in simple and diverse habitats). These data suggest that the response of some species to intercropping differs with different crop combinations, geographical location, and experimental procedures.

Experimental Methods

When parasitoid activity was estimated by capturing adults (using traps or sweep nets), 72% of the studies reported higher parasitoid density in inter-crops than monocultures (none reported lower density in diverse compared to simple habitats). Lower rate of parasitoid enhancement (54%) was reported when parasitoid activity was inferred from host parasitism rate. It seems that although parasitoids may visit intercrops in quest of food sources, vegetation diversity may hinder the ability of some species to attack their hosts. Thus, parasitoid impact on pest populations may not necessarily correlate with parasitoid density in simple and diverse habitats.

Level of parasitoid enhancement was similar when studies used small (<400 m^2) and large plots (59% and 67%, respectively; Table 3). However, parasitoids in small plots were twice as likely to have an inconsistent or no response to plant diversification than parasitoids in large plots (35% and 17% of comparisons in small and large plots, respectively, found same levels of parasitoid activity in sole and intercropped plots; Table 3). It appears that although the spatial scale of experiments did not fundamentally alter parasitoid enhancement by intercropping, increased plot size led to decreased variability in the results.

An increase in the variability in parasitoid response to plant diversification in small adjacent plots may result, for example, from the ability of the wasps to move between plots in quest for resources that become available at different times (e.g., nectar, alternate hosts). Therefore, reported changes in parasitoid density or parasitism rate in small plots arranged in close proximity may reflect nothing more than a temporary choice of one particular habitat over the other. In large plots, parasitoids are less likely to colonize habitats lacking essential resources. As a result, in 84% of the studies that used large plots, parasitoid activity was either enhanced or diminished by plant diversification. Kareiva (1983) cautioned that small scale experiments may not be suitable for predictions about the response to herbivores to agroecosystem diversification. The same appears to be true for parasitoids. It is therefore important to identify the mechanisms that underlie the response of parasitoids to vegetation texture. At a minimum, the mobility of the studied parasitoid species (or any organism) must be considered when the size of experimental plots and the distance between different treatments (e.g., mono- and polycultures) are being decided in manipulative studies.

Table 2. Studies that compared parasitism rates or parasitoid abundance in monocultural and intercropped systems.

Parasitoid	Host (stage)	Cropping system	Host range[a]	Scale[b]	Method[c]	Host density[d]	Parasitoid activity[d]	Suggested factors involved	Ref[e]
HYMENOPTERA, ICHNEUMONOIDEA, BRACONIDAE									
Cotesia plutellae	Plutella xylostella (larva)	cabbage-tomato	G	SM	% paras	lower*	higher*		1
Opius melanagromyzae	3 Agromyzidae (Diptera) species (larva)	bean-maize	S	SM	% paras	lower*	higher	favorable micro-climate	9
Rogas laphygmae	Spodoptera frugiperda (larva)	maize-bean	S	LG	% paras	lower*	lower*		10
Meteorus sp.	Spodoptera frugiperda (larva)	maize-bean	G	SM	% paras	lower*	same*		23
ICHNEUMONIDAE									
Diadegma sp.	Heliothis armigera (larva)	pigeonpea-sorghum	G	LG	% paras	higher	same*	host migration & escape	6
Campoletis chlorideae	Heliothis armigera (larva)	sorghum-pigeonpea	G	LG	% paras	N/A	higher		24
		safflower-sorghum	G	LG	% paras	higher	lower		21
		safflower-chickpea	G	LG	% paras	higher	lower		21
		safflower-linseed (Linum usitatissimum)	G	LG	% paras	higher	lower		21
		safflower-chillies (Capsicum annuum)	G	LG	% paras	higher	lower		21
		safflower-sorghum-chickpea	G	LG	% paras	higher	lower		21
Eriborus argenteopilosus	Heliothis armigera (larva)	safflower-sorghum	?	LG	% paras	higher	lower		21
		safflower-chickpea	?	LG	% paras	higher	lower		21
		safflower-flax	?	LG	% paras	higher	lower		21
		safflower-chillies	?	LG	% paras	higher	higher		21
		safflower-sorghum-chickpea	?	LG	% paras	higher	same		21

Campoletis chlorideae	*Heliothis peltigera* (larva)	safflower-sorghum	G	LG	% paras	higher	lower		21
		safflower-chickpea	G	LG	% paras	higher	lower		21
		safflower-linseed	G	LG	% paras	higher	lower		21
		safflower-chillies	G	LG	% paras	higher	lower		21
		safflower-sorghum-chickpea	G	LG	% paras	higher	lower		21
Eriborus argenteopilosus	*Heliothis peltigera* (larva)	safflower-sorghum	?	LG	% paras	higher	higher		21
		safflower-chickpea	?	LG	% paras	higher	higher		
		safflower-linseed	?	LG	% paras	higher	same		
		safflower-chillies	?	LG	% paras	higher	same		
		safflower-sorghum-chickpea	?	LG	% paras	higher	higher		
Tersilochinae	coleopterans (larva)	oat-faba bean (*Vicia faba*)	?	SM, LG	trap	N/A	higher*	bean flowers	2
APHIDIIDAE									
Diaeretiella rapae, Aphidius spp. & *Lysiphlebus* spp.	*Brevicoryne brassicae* (adult)	brussels sprouts-fava bean	S	SM	% paras	lower*	same* (but lower in high density monoculture)		11
Lysiphlebus sp.	*Aphis gossypii*	melon (*Lulumis melo*) -turnip (*Brussica rapa*)	S	SM	% paras	lower	higher		14
CHALCIDOIDEA TRICHOGRAMMATIDAE									
Trichogramma minutum	*Ostrinia nubilalis* (egg)	maize-bean-squash	G	SM	% paras	N/A	lower*	increased leaf area	4
Trichogramma pretiosum	*Diaphnia hyalinata* (egg)	bean-maize-squash	G	SM	% paras	N/A	higher		4
		squash-maize-cowpea	G	SM	% paras	lower*	higher*	alternative hosts in maize	22
Trichogramma pretosium & *T.* sp.	*Helicoverpa* (*Heliothis*) *zea* (egg)	soybean (*Glycine max*) -maize	G	SM, LG	% paras	same	higher*	chemical attraction/ retention	5
Trichogramma contusum	*Heliothis armigera* (egg)	pigeonpea-sorghum	G	LG	% paras	higher	same*	host migration & escape	6
		sorghum-pigeonpea	G	LG	% paras	N/A	higher		24

Continued on next page

Table 2. continued

Parasitoid	Host (stage)	Cropping system	Host range[a]	Scale[b]	Method[c]	Host density[d]	Parasitoid activity[d]	Suggested factors involved	Ref[e]
Trichogramma sp.	Helicoverpa (Heliothis) zea (egg)	tomato-maize	G	SM	% paras	N/A	same	parasitoid preference: tomato>>bean>maize	12
		tomato-bean	G	SM	% paras	N/A	same		12
		tomato-bean-maize	G	SM	% paras	N/A	same		12
		maize-tomato	G	SM	% paras	N/A	same		12
		maize-bean	G	SM	% paras	N/A	same		12
		maize-bean-tomato	G	SM	% paras	N/A	higher		12
		bean-maize	G	SM	% paras	N/A	same		12
		bean-tomato	G	SM	% paras	N/A	same		12
		bean-tomato-maize	G	SM	% paras	N/A	same		12
Trichogramma chilonis	Ostrinia furnacalis (egg)	maize-sweet potato	G	SM	% paras	lower	same*		13
EULOPHIDAE									
Pediobius foveolatus	Epilachna varivestis (larva)	bean-maize	S	SM	D-vac	same*	higher*	favorable micro-climate	20
Ceranisus menes	Megalurothrips sjostedti	bean-maize	S	SM	% paras	same*	higher*		20
		maize-cowpea	S	SM	trap	lower	higher		7
MYMARIDAE									
Anagrus sp.	Empoasca kraemeri (egg)	bean-maize	S	SM	% paras	lower	higher		23
Anagrus sp.	Empoasca sp. (egg)	squash-maize-cowpea	S	SM	visual count	lower*	same*		19
		squash-maize-cowpea	S	SM	% paras	same*	same*		19

Continued on next page

VARIOUS

Natural enemy	Host (stage)	Cropping system		Size	Method			Notes	Ref
2 ichneumonids, 4 braconids, &1 perilampid	Diaphania hyalinata (larva)	squash-maize-cowpea	?	SM	trap	lower*	higher*		22
		maize-squash-cowpea	?	SM	trap	lower*	same*		22
aphidiids & aphelinids	Rhopalosiphum padi (nymph)	oat-fava bean	S	SM, LG	% paras	same*	lower*		2
Amitus aleurodinus & Eretmocerus aleyrodiphaga	Aleurotrachelus socialis (nymph)	cassava(Manihot esculenta)-cowpea	S	SM	% paras	lower*	same		3
Trichogramma minutum &Prospaltella sp. (egg)	Argyrotaenia sphaleropa & Platynota sp. (egg)	cassava-maize	S	SM	% paras	lower	same		3
		cotton-maize	G	LG	% paras	higher	higher	preference to maize	8
unidentified	?	maize-bean	?	SM	trap	same	higher*		15
		maize-squash	?	MS	trap	same	same*		15
		squash-maize	?	SM	trap	same	higher*		15
		maize-bean-squash	?	SM	trap	same	higher*		15
		squash-maize-bean	?	SM	trap	same	higher*		15
unidentified	?	maize-sweet potato	?	SM	sweep-net	N/A	higher		16
		sweet potato-maize	?	SM	sweep-net	N/A	higher		16
unidentified	?	maize-soybean	?	SM	sweep-net	N/A	same*		18
unidentified	?	squash-maize-cowpea	?	SM	trap	N/A	higher*		22
		maize-squash-cowpea	?	SM	trap	N/A	same*		22
unidentified	?	maize-cowpea	?	SM	trap	N/A	higher		7

DIPTERA TACHINIDAE

Natural enemy	Host (stage)	Cropping system		Size	Method			Notes	Ref
Archytas marmoratus	Spodoptera frugiperda (larvae)	maize-bean	S	LG	% paras	lower*	higher*		10
Celatoria sp.	6 chrysomelid species (adult)	maize-bean-squash	S	LG	% paras	lower*	same		17

Table 2. *continued*

Parasitoid	Host (stage)	Cropping system	Host range[a]	Scale[b]	Method[c]	Host density[d]	Parasitoid activity[d]	Suggested factors involved	Ref[e]
Carcelia illota	*Heliothis peltigera* (larva)	safflower-sorghum	G	LG	% paras	higher	lower		21
		safflower-chickpea	G	LG	% paras	higher	same		21
		safflower-linseed	G	LG	% paras	higher	higher		21
		safflower-chillies	G	LG	% paras	higher	higher		21
		safflower-sorghum-chickpea	G	LG	% paras	higher	lower		21
Carcelia illota	*Heliothis armigera* (larva)	safflower-sorghum	G	LG	% paras	higher	same		21
		safflower-chickpea	G	LG	% paras	higher	same		21
		safflower-linseed	G	LG	% paras	higher	same		21
		safflower-chillies	G	LG	% paras	higher	higher		21
		safflower-sorghum-chickpea	G	LG	% paras	higher	same		21
Lespesia archippivora	*Spodoptera frugiperda* (larva)	maize-bean	G	LG	% paras	lower*	higher*		10
unidentified	*Heliothis armigera* (larva) pigeonpea-sorghum	?	LG	% paras	higher	higher	host migration & escape	6

a G—generalist (parasitizes hosts in more than one family); S—specialist (parasitizes hosts in one family).

b SM—small (plots smaller than 400 m²); LG—large (plots/fields larger than or equal to 400 m²).

c Method used to estimate parasitoid activity.

d In intercrop compared to monoculture of the first crop listed. * indicates that a statistical test was performed.

e 1. Bach & Tabashnik 1990; 2. Helenius 1990; 3. Gold & Altieri 1989; 4. Andow & Risch 1987; 5. Altieri & Todd 1981, Altieri et al. 1981; 6. Bhatnagar & Davis 1980, 1981; 7. Perfect et al. 1979; 8. Beingolea 1957 and references therein; 9. Karel 1991; 10. van Huis 1981; 11. Altieri 1984; 12. Nordlund et al. 1984; 13. Nafus & Schreiner 1986; 14. Marcovitch 1935; 15. Hansen 1983; 16. Risch 1979; 17. Risch 1981; 18. Wrubel 1984; 19. Letourneau 1990; 20. Coll 1991; 21. Pawar et al. 1985; 22. Letourneau 1987; 23. Altieri et al. 1978; 24. ICRISAT 1978.

Table 3. Percent of studies that show higher, lower, or the same (or variable) parasitoid density or parasitism rate in intercrops compared to monocultures. Based on the studies listed in Table 2[a].

Parameter		n	Higher	Lower	Same
Total comparisons		42	61	7	32
Experimental methods					
Parasitoid activity determined as:	% parasitism	26	54	11	33
	Trap/net	18	72	0	28
Spatial scale[b]	small plots	34	59	8	35
	large plots	12	67	17	17
Parasitoid characteristics					
Parasitoid species[c]		31	54	7	39
Parasitoid taxa	Ichneumonoidea	9	56	11	33
	Chalcidoidea	12	67	8	25
	Tachinidae	4	75	0	25
Parasitoid host range[d]	Specialists	12	58	9	33
	Generalists	13	61	8	31
Host characteristics					
Host order	Homoptera	8	25	13	62
	Coleoptera	4	75	0	25
	Lepidoptera	18	61	11	28
Host stage attacked	Egg	10	60	10	30
	Larva	20	60	10	29
Host density[e]	Lower	18	50	6	44
	Higher	4	0	50	50
	Same	10	70	10	20
Plant characteristics					
Plant species diversity	2 crops	29	65	4	31
	3 crops	13	54	15	31
Host crop family	Poaceae	20	50	15	35
	Fabaceae	8	75	0	25
	Other	14	69	0	31
Associated crop family[f]	Poaceae	11	83	0	17
	Fabaceae	15	53	14	33
	Other	3	59	6	35

[a] To remove bias by large number of comparisons per study, references 12 and 21 were excluded from the analysis

[b] Small plots are smaller than 400m^2; Large plots are larger than or equal to 400m^2.

[c] Or a group of species.

[d] Specialists parasitize hosts in one family; generalists parasitize hosts in more than one family. Determined from the primary literature or from Arnaud 1978 and Krombein et al. 1979.

[e] In intercrop compared to monoculture.

[f] In diculture

Parasitoid Characteristics

The three major parasitoid groups (i.e., Ichneumonoidea, Chalcidoidea, and Tachinidae) responded similarly to increased plant species diversity (Table 3). Although only four tachind species were studied, their response to intercropping shows a similar trend to that of hymenopteran parasitoids (Table 3).

Extending the "resource concentration hypothesis" to the third trophic level, Hansen (1983) and Sheehan (1986) proposed that an increase in habitat diversity should favor generalist (i.e., attack a wide range of host species) natural enemies more than specialist ones (i.e., attack one or few related host species). Much like monophagous herbivores, natural enemies with narrow host ranges should be able to locate their prey/host more easily in simple than in diverse habitats. Further, specialist parasitoids, which often use host cues, are expected to search monocultures more intensively because herbivores tend to be more abundant in simple than diverse habitats (Andow 1991).

Some generalist parasitoids may be more effective in simple than diverse habitats. Many generalists are plant or niche specialists, attacking a large variety of hosts in a particular type of microhabitat (see Townes 1960). Such parasitoids may be more efficient in mono- than multi-specific stands of their hosts' food plant because host plant cues, often used by plant specialists, may be more concentrated.

In the past, Hansen and Sheehan's proposal was tested by comparing the response of predators and parasitoids (viewed to be generalists and specialists, respectively; Clausen 1940; Elton 1966; Askew 1971; Huffaker 1971; Root 1973; Price 1980; DeBach & Rosen 1991) to increased habitat diversity (Hansen 1983; Andow 1991). However, because life histories of parasitoids and predators are so different, other factors, such as nutritional requirements of adults and immatures, may be confounded with diet breadth/ host range. It was therefore of interest to test Hansen and Sheehan's proposal by comparing the effect of habitat diversity on specialist and generalist parasitoids.

Data do not support the proposal that generalist natural enemies should be favored more than specialists by an increase in habitat plant diversity (Table 3). Generalist (i.e., those attacking hosts in more than one family) and specialist (i.e., those attacking hosts in one family) parasitoids were equally enhanced by intercropping. It is possible, however, that different mechanisms underlie the response of generalist and specialist parasitoids to plant habitat diversity. Generalists may be more abundant in intercrops than monocultures because, for example, they use alternative host species, whereas both specialists and generalists may be attracted to a greater availability of food sources in diverse habitats.

Host Characteristics

It appears that host taxonomy has an important impact on parasitoid enhancement by intercropping. Whereas parasitoids of lepidopteran and coleopteran hosts were more abundant or caused higher parasitism in intercrops than monocultures, the reverse was true for parasitoids of Homoptera (i.e., a higher parasitoid activity in monocultures than intercrops; Table 3). This effect is probably not the result of differences in experimental methods (parasitoid activity was estimated more often by direct counts in parasitoids of lepidopteran and coleopteran than in those of homopteran hosts). Instead, differences in life histories of herbivores (e.g., population rate of increase) may influence the response of parasitoids to intercropping. Clearly, more research is needed to determine if and how life history parameters of herbivores affect the activity of their natural enemies.

The stage of the attacked host does not appear to affect parasitoids' response to plant diversification (Table 3). Data do not support the idea that differences in host mobility, defense behavior, or dispersion are important in influencing the effect of intercropping on parasitoids.

Host density is an important factor that affects parasitoid-host interactions. Nonetheless, host density is often confounded in the response of parasitoids to plant diversification. In 10 of the reviewed studies, host density was similar (intentionally or by chance) in monocultures and intercrops, thus providing the opportunity to assess the effect of intercropping on parasitoids independently of host density. It appears that parasitoid enhancement by intercropping is not affected greatly by lower herbivore density in intercrops than in monocultures (Table 3). Instead, it appears that low host density, often reported in intercrops (Andow 1991), may greatly increase parasitism rate and/or parasitoid density (although in only four of the reviewed studies were hosts less abundant in monoculture than intercrops; Table 3). Such an inverse density-dependent response of parasitoids to their hosts suggests that intercropping is a useful pest control practice in two ways: it decreases herbivore densities (for example, by means of lower colonization) and simultaneously increases the impact of parasitoids (increases percent parasitism) on pest populations.

Plant Characteristics

The level of plant species diversity appears to have some effect on parasitoids. In tricultures, more parasitoids were negatively affected by intercropping than in dicultures (15 and 4%, respectively, Table 3). However, this effect was small and more studies, specifically designed to test the effect of plant species diversity level on parasitoids, are needed before any conclusion can be drawn.

Cereal crops (primarily maize, sorghum, and rice [*Oryza sativa*]) are commonly intercropped with legumes (primarily soybean [*Glycine max*], groundnut [*Arachis hypogaea*], common bean [*Phaseolus vulgaris*], mung bean [*Vigna radiata*], pigeonpea [*Cajanus indicus*], and cowpea [*Vigna sinensis*]) in tropical as well as temperate regions (Rao 1986). Therefore, it was of interest to determine whether these two components of intercrops (i.e., cereals and legumes) equally enhance parasitoid activity in intercropped habitats. When grasses were the host crop, intercropping with other plant families enhanced parasitoid activity in only 50% of the cases (Table 3). However, when legumes were the host crop, parasitoid activity was enhanced in 75% and 69% of the cases, respectively. Therefore it appears that the addition of grasses to the habitat is more important in enhancing parasitoid activity than the addition of legumes or other crops. Thus, increasing plant species diversity *per se* was not as important as the addition of particular resources, provided by cereal crops, to the habitat. Marcovitch (1935), Risch (1979), and others noted that maize and sorghum are attractive to parasitoids because of the abundant supply of pollen, whereas pollen and nectar in legume flowers are probably inaccessible to parasitoids.

Concluding Remarks and Directions for Future Research

So far, there is only a limited research effort to integrate intercropping into pest management programs in general and biological control in particular. It seems that broad generalizations, such as the statement that diversity is stabilizing, are of little value for the design of agroecosystems to enhance biological control. Furthermore, agroecosystem diversification may not necessarily lead to pest population stability (van Emden & Williams 1974), and could promote pest outbreaks because of interference with searching natural enemies (e.g., Kareiva 1987).

Two basic strategies are distinguished in cultural control: manipulation of the environment to make it less suitable for the pest or manipulation of the environment to make it more favorable for natural enemies. If intercropping is to be used for biological pest control, the two strategies must be compatible. Thus, pests could be controlled in intercrops, for example, by reducing their colonization and increasing their mortality by natural enemies. We therefore need to study the response of both pests and their natural enemies to increased habitat diversity.

It appears that although intercropping tends to enhance parasitoid activity, this effect is highly variable. Several factors may promote the recorded inconsistency in parasitoid response to intercropping. Variation in experimental conditions (e.g., plot size, neighboring fields, additive vs. substitutive designs) may influence parasitoid response to the habitat. Factors in the habitat may counteract or reduce each other's impact on parasitoid population density. Much like other natural enemies, hyperparasitoids (parasitoids that attack other, primary, parasitoids) could also be favored by increased plant species diversity. Similarly, parasitoids (e.g., immatures within hosts) may be more vulnerable to predator attack in diverse habitats than in monocultures. Thus, it is conceivable that enhancement of primary parasitoids in diverse habitats could be counteracted by the increase in density and/ or efficiency of natural enemies at the fourth trophic level.

The short-term (within growing seasons) effect of habitat diversification (particularly in annual cropping systems) on parasitoids (and other natural enemies) will be determined primarily by their behavioral responses to resources and plant texture (e.g., plant density, presence of nonhost plants, relative height of associated plants). It is therefore important that, rather than seeking to increase crop diversity *per se,* agroecosystems be designed to provide natural enemies with specific resources that will augment their efficiency. Factors such as flower production by companion crops and the spatial and temporal arrangement of crops in the habitat could be manipulated.

Although information on the effect of habitat plant diversity on natural enemies (and herbivores) has increased greatly in the last 20 years, most studies do not attempt to explain why population densities differ in monocultures and polycultures. Thus, these studies do not contribute to our understanding of the mechanisms involved. Without a better mechanistic understanding of the influence of plant diversity on natural enemies, there is little hope for making useful predictions about three trophic level interactions in diverse habitats. Such predictions are essential if agroecosystems are to be designed to enhance biological control by natural enemies.

There is a need for detailed behavioral and demographical studies of natural enemies in simple and diverse habitats. We know little about mortality factors that shape population densities of natural enemies in different habitats; how plant texture affects movement patterns of natural enemies in simple and diverse habitats; and the extent to which enemies' behavioral plasticity and learning ability affect their efficiency in monocultures and polycultures. Such information may allow us to manipulate agroecosystems to increase the effectiveness of natural enemies and to understand why the response of natural enemies to plant diversity varies greatly among species, intercropped systems, and geographical locations.

Acknowledgments

I thank Robert Bugg, Paul Gross, Charles Pickett, and William Sheehan for helpful comments on a draft of this manuscript and David Andow and Patricia Matteson for their help in obtaining some of the obscure literature.

References

Akai, S. & M. Fukutomi 1980. Preformed internal physical defenses, pp. 139-159. *In* J. G. Horsfall & E. B. Cowling, [eds.], Plant Disease, Vol. V. Academic Press, New York, N.Y.

Altieri, M. A. 1984. Patterns of insect diversity in monocultures and polycultures of brussels sprouts. Protection Ecol. 6:227-232.

Altieri, M. A. & M. Liebman. 1986. Insect, weed, and plant disease management multiple cropping systems, pp. 183-218. *In* C. A. Francis [ed.], Multiple Cropping Systems. Macmillen Pub., London.

Altieri, M. A. & D. K. Letourneau. 1982. Vegetation management and biological control in agroecosystems. Crop Protection 1:404-430.

Altieri, M. A. & J. W. Todd. 1981. Some influences of vegetational diversity on insect communities of Georgia soybean fields. Protection Ecol. 3:333-338.

Altieri, M. A. & W. H. Whitcomb. 1979. The potential use of weeds in the manipulation of beneficial insects. HortScience 14:12-18.

Altieri, M. A., A. van Schoonhoven & J. D. Doll. 1977. The ecological role of weeds in insect pest management systems: a review illustrated by bean, *Phaseolus vulgaris*, cropping systems. PANS. 23:195-205.

Altieri, M. A., C. A. Francis, A. van Schoonhoven & J. D. Doll. 1978. A review of insect prevalence in maize (*Zea mays* L.) and bean (*Phaseolus vulgaris* L.) polycultural systems. Field Crop Res. 1:33-49.

Altieri, M. A., W. J. Lewis, D. A. Nordlund, R. C. Guelner & J. W. Todd. 1981. Chemical interactions between plants and *Trichogramma* wasps in Georgia soybean fields. Protection Ecol. 3:259-263.

Andow, D. A. 1991. Vegetational diversity and arthropod population response. Annu. Rev. Entomol. 36:561-586.

Andow, D. A. & D. R. Prokrym. 1990. Plant structural complexity and host-finding by a parasitoid. Oecologia 82:162-165.

Andow, D. A. & S. J. Risch. 1987. Parasitism in diversified agroecosystems: phenology of *Trichogramma minutum* (Hymenoptera: Trichogrammatidae). Entomophaga 32:255-260.

Andrews, D. J. & A. H. Kassam. 1976. The importance of multiple cropping in increasing world food supplies, pp. 1-10. *In* R. I. Papendick, P. A. Sanchez & G. B. Triplet [eds.], Multiple cropping systems. Am. Soc. Agron., Special Pub. No. 27.

Arnaud, P. H., Jr. 1978. A host-parasite catalog of North American Tachinidae (Diptera). U.S. Dept. Agric., Misc. Pub. 1319.

Askew, R. R. 1971. Parasitic insects. Elsevier, New York.

Bach, C. E. 1984. Plant spatial pattern and herbivore population dynamics: plant factors affecting the movement patterns of a tropical cucurbit specialist, *Acalymma innubum.* Ecology 65:175-190.

Bach, C. E. & B. E. Tabashnik. 1990. Effects of nonhost plant neighbors on population densities and parasitism rates of the diamondback moth (Lepidoptera: Plutellidae). Environ. Entomol. 19:987-994.

Banta, G. R. 1973. Mechanization, labor, and time in multiple cropping. Agric. Mech. Asia 4:27-30.

Barbosa, P. & E. A. Frongillo Jr. 1977. Influence of light intensity and temperature on the locomotor and flight activity of *Brachymeria intermedia* (Hym.: Chalcididae) a pupal parasitoid of the gypsy moth. Entomophaga 22:405-411.

Barbosa, P., P. Gross & J. Kemper. 1991. Influence of plant allelochemicals on the tobacco hornworm and its parasitoid, *Cotesia congregata.* Ecology 72:1567-1575.

Barbosa, P. & J. A. Saunders. 1985. Plants allelochemicals: linkages between herbivores and their natural enemies, pp. 107-137. *In* G. A. Cooper-Driver, T. Swain & E. E. Conn [eds.], Chemically mediated interactions between plants and other organisms. Plenum Press, New York.

Barbosa, P., J. A. Saunders, J. Kemper, R. Trumbule, J. Olechno & P. Martinat. 1986. Plant allelochemicals and insect parasitoids: the effects of nicotine on *Cotesia congregata* (Say) (Hymenoptera: Braconidae) and *Hyposoter annulipes* (Cresson) (Hymenoptera: Ichneumonidae). J. Chem. Ecol. 12:1319-1328.

Batra, S. W. T. 1982. Biological control in agroecosystems. Science 215:134-139.

Beingolea, G. O. 1957. El sembrio del maiz y la fauna beneficia del algodonero. Informe 104. Estacion Experimental Agricola de la Molina, Lima, Peru.

Bhatnagar, V. S. & J. C. Davies. 1980. Entomological studies in intercropped pigeonpea systems at ICRISAT center, future developments and collaborative research needs. International workshop on pigeonpea, 15-19 December 1980. ICRISAT/ICAR Center, Patancheru, India.

1981. Pest management in intercrop subsistence farming, p. 249-257. *In* Proc., International Workshop on Intercropping, 10-13 January, 1979. Hyderabad, India.

Bottenberg, H. & M. E. Irwin. 1991. Influence of wind speed on residence time of *Uroleucon ambrosiae* alatae (Homoptera: Aphididae) on bean plants in bean monocultures and bean-maize mixtures. Environ. Entomol. 20:1375-1380.

Browning, J. A. 1975. Relevance of knowledge about natural ecosystems to development of pest management programs for agroecosystems. Proc. Am. Phytopathol. Soc. 1:191-194.

Brust, G. E., B. R. Stinner & D. A. McCartney. 1986. Predation by soil-inhabiting arthropods in intercropped and monoculture agroecosystems. Agric. Ecosyst. Environ. 18:145-154.

Bugg, R. L., R. T. Ellis & R. W. Carlson. 1989. Ichneumonidae (Hymenoptera) using extrafloral nectar of faba bean (*Vicia faba* L., Fabaceae) in Massachusetts. Biol. Agric. Hortic. 6:107-114.

Campbell, B. C. & S. S. Duffey. 1979. Tomatine and parasitic wasps: potential incompatibility with biological control. Science 205:700-702.

Casas, J. 1989. Foraging behaviour of a leafminer parasitoid in the field. Ecol. Entomol. 14:257-265.

Clausen, C. P. 1940. Entomophagous Insects. McGraw-Hill, New York.

Coll, M. 1991. Effects of vegetation texture on the Mexican bean beetle and its parasitoid, *Pediobius foveolatus*. Ph.D. dissertation, University of Maryland.

Coll, M. & D.G. Botrell. 1994. Effect of nonhost plants on an insect herbivore in diverse habitats. Ecology 75:723-731.

Coll, M. & D.G. Bottrell. 1996. Movement of an insect parasitoid in simple and diverse plant assemblages. Ecol. Entomol. 21:141-149.

Cromartie, W. J., Jr. 1981. The environmental control of insects using crop diversity, pp. 223-25. *In* D. Pimentel [ed.], CRC handbook of pest management in agriculture, Vol. I. CRC Press, Boca Raton.

DeBach P. & D. Rosen. 1991. Biological control by natural enemies, 2nd ed. Cambridge Univ. Press., N.Y.

Dempster, J. P. 1969. Some effects of weed control on the numbers of the small cabbage white (*Pieris rapae* L.) on brussels sprouts. J. Appl. Ecol. 6:339-346.

Dempster, J. P. & T. H. Coaker. 1974. Diversification of crop ecosystems as a mean of controlling pests, pp. 106-114. *In* D. Price-Jones & M. E. Solomon [eds.], Biology in pest and disease control. John Wiley & Sons, New York.

Egunjobi, O. A. 1984. Effects of intercropping maize with grain legumes and fertilizer treatment on populations of *Pratylenchus brachyurus* (Nematoda) and on the yield of maize *(Zea mays)*. Protection Ecol. 6:153-167.

Eikenbary, R. D. & C. E. Rogers. 1974. Importance of alternate hosts in establishment of introduced parasites, pp. 119-133. *In* Proceedings, Tall Timbers Conference on Ecological Animal Control by Habitat Management, Talahassee, Florida.

Elton, C. S. 1927. Animal ecology. Sidgwick & Jackson, London.

1958. The ecology of invasions by animals and plants. Methuen, London.

1966. The pattern of animal communities. Methuen, London.

Foster, M. A. & W. G. Ruesink. 1984. Influence of flowering weeds associated with reduced tillage in corn on a black cutworm (Lepidoptera: Noctuidae) parasitoid, *Meteorus rubens* (Nees von Esenbeck). Environ. Entomol. 13:664-668.

1986. Impact of common chickweed, *Stellaria media,* upon parasitism of *Agrotis ipsilon* (Lepidoptera: Noctuidae) by *Meteorus rubens* (Nees von Esenbeck). J. Kans. Entomol. Soc. 59:343-349.

Francis C. A. 1986. Multiple cropping systems. Macmillan Pub. New York, N.Y.

Gardner, S. M. & A. F. G. Dixon. 1985. Plant structure and the foraging success of *Aphidius rhopalosiphi* (Hymenoptera: Aphidiidae). Ecol. Entomol. 10:171-179.

Glass, E. H. & H. D. Thurston. 1978. Traditional and modern crop protection in perspective. Biol. Sci. 28:109-114

Gold, C. S. & M. A. Altieri. 1989. The effects of intercropping and mixed varieties on predators and parasitoids of cassava whiteflies (Hemiptera: Aleyrodidae) in Colombia. Bull. Entomol Res. 79:115-121.

Greenblatt, J. A. & P. Barbosa. 1981. Effects of host's diet on two pupal parasitoids of the gypsy moth: *Brachymeria intermedia* (Nees) and *Coccygomimus turionellae* (L.). J. Appl. Ecol. 18:1-10.

Hansen, M. K. 1983. Interactions among natural enemies, herbivores, and yield in monocultures and polycultures of corn, bean, and squash. Ph.D. dissertation, University of Michigan, Ann Arbor.

Harwood, R. R. 1973. The concepts of multiple cropping: an introduction to the principles of cropping systems design. International Rice Research Institute, Los Banos, Philippines.

    1975. Farmer-oriented research aimed at crop intensification, pp. 12-31. *In* Proceedings, Cropping Systems Workshop. International Rice Research Institute, Los Banos, Philippines.

Helenius, J. 1990. Incidence of specialist natural enemies of *Rhopalosiphum padi* (L.) (Hom., Aphididae) on oats in monocrops and mixed intercrops with faba bean. J. Appl. Entomol. 109:136-143.

Hsiao, T. H. & F. G. Holdaway. 1966. Seasonal history and host synchronization of *Lydella grisescens* (Diptera: Tachinidae) in Minnesota. Ann. Entomol. Soc. Am. 59:125-133.

Huffaker, C. B. [ed.] 1971. Biological control. Plenum, New York

    1974. Some implications of plant-arthropod and higher level arthropod-arthropod food links. Environ. Entomol. 3:1-9.

International Crops Research Institute for the Semi-Arid Tropics. 1978. Cropping entomology, pp. 205-210. *In* Annu. Report 1977-1978. Hyperabad, India.

International Crops Research Institute for the Semi-Arid Tropics. 1980. Cropping Entomology, Annu. Report 1979-1980. Andhra Pradesh, India.

Juillet, J. A. 1964. Influence of weather on flight activity of parasitic Hymenoptera. Can. J. Zool. 42:1133-1141.

Kareiva, P. 1983. The influence of vegetation texture on herbivore populations: resource concentration and herbivore movement, pp. 259-289. *In* R. F. Denno & M. S. McClure [eds.], Variable plants and herbivores in natural and managed systems. Academic Press, New York.

Kareiva, P. 1987. Habitat fragmentation and the stability of predator-prey interactions. Nature 326:388-340.

Karel, A. K. 1991. Effects of plant populations and intercropping on the population patterns of bean flies on common beans. Environ. Entomol. 20:354-357.

Kass, D. C. L. 1978. Polyculture cropping systems: review and analysis. Cornell Int. Agric. Bull. 32.

Kidd, N. A. C. & M. A. Jervis. 1989. The effects of host-feeding behaviour on the dynamics of parasitoid-host interactions, and the implications for biological control. Res. Pop. Ecol. 31:235-274.

Krombein, K. V., P. D. Hurd Jr., D. R. Smith & B. D. Burks [eds.] 1979. Catalog of Hymenoptera in America north of Mexico. vol. 1. Symphyta and Apocrita (Parasitica). Smithsonian Institution Press, Washington, D.C.

Leius, K. 1960. Attractiveness of different foods and flowers to adults of some hymenopterous parasites. Can. Entomol. 92:369-376.

1961. Influence of food on fecundity and longevity of adults of *Itoplectis conquisitor* (Say) (Hymenoptera: Ichneumonidae). Can. Entomol. 93:771-780.

1967. Influence of wild flowers on parasitism of tent caterpillar and codling moth. Can. Entomol. 99:444-446.

Letourneau, D. K. 1987. The enemies hypothesis: tritrophic interactions and vegetational diversity in agroecosystems. Ecology 68:1616-1622.

1990. Abundance patterns of leafhopper enemies in pure and mixed stands. Environ. Entomol. 19:505-509.

Lewis, W. J. & J. H. Tumlinson. 1988. Host detection by chemically mediated associative learning in a parasitic wasp. Nature 331:257-259.

Lewis, W. J., L. E. M. Vet, J. H. Tumlinson, J. C. van Lentern & D. R. Papaj. 1990. Variation in parasitoid foraging behavior: essential element of a sound biological control theory. Environ. Entomol. 19:1183-1193.

Lipman, J. G. 1912. The associative growth of legumes and non-legumes. New Jersey Agric. Exp. Stat. Bull. No. 253.

Litsinger, J. A. & K. Moody. 1976. Integrated pest management in multiple cropping systems, pp. 293-316. *In* R. I. Papendick, P. A. Sanchez & G. B. Triplet [eds.], Multiple cropping systems. Am. Soc. Agron., Special Pub. No. 27.

Litsinger, J. A., V. Hasse, A. T. Barrion & H. Schmutterer. 1991. Response of *Ostrinia furnacalis* (Guenee) (Lepidoptera: Pyralidae) to intercropping. Environ. Entomol. 20:988-1004.

Loader, C. & H. Damman. 1991. Nitrogen content of food plants and vulnerability of *Pieris rapae* to natural enemies. Ecology 72:1586-1590.

Marcovitch, S. 1935. Experimental evidence on the value of strip farming as a method for the natural control of injurious insects with special reference to plant lice. J. Econ. Entomol. 28:62-70.

Margalef, R. 1968. Perspectives in ecological theory. University of Chicago Press, Chicago.

Matteson, P. C., M. A. Altieri & W. C. Gagne. 1984. Modification of small farmer practices for better pest management. Annu. Rev. Entomol. 29:383-402.

McGovern, W. L. & W. H. Cross. 1976. Affects of two cotton varieties on levels of boll weevil parasitism. (Col: Curculionidae). Entomophaga 2:123-125.

Monteith, L. G. 1960. Influence of plants other than the food plants of their host on host-finding by tachinid parasite. Can. Entomol. 92:641-652.

Moody, K. & S. V. R. Shetty. 1981. Weed management in intercropping systems, pp. 229-237. *In* Proc., International Workshop on Intercropping, 1975. International Crops Research Institute for the Semi-Arid Tropics, Patancheru, India.

Murdoch, W. W. 1975. Diversity, complexity, and pest control. J. Appl. Ecol. 12:795-807.

Nafus, D. & I. Schreiner. 1986. Intercropping maize and sweet potatoes: effects on parasitization of *Ostrinia furnacalis* eggs by *Trichogramma chilonis*. Agric. Ecosyst. Environ. 15:189-200.

National Res. Council. 1989. Alternative agriculture. National Academy Press, Washington, D.C.

Need, J. T. & P. P. Burbutis. 1979. Searching efficiency of *Trichogramma nubilale*. Environ. Entomol. 8:224-227.

Nordlund, D. A., R. B. Chalfant & W. J. Lewis. 1984. Response of *Trichogramma pretiosum* females to extracts of two plants attacked by *Heliothis zea*. Agric. Ecosys. Environ. 12:127-133.

Odum, E. P. 1953. Fundamentals of ecology. Saunders, Philadelphia.

Orr, D. B. & D. J. Boethel. 1985. Comparative development of *Copidosoma truncatellum* (Hymenoptera: Encyrtidae) and its host, *Pseudoplusia includens* (Lepidoptera: Noctuidae), on resistant and susceptible soybean genotypes. Environ. Entomol. 14:612-616.

Pair, S. D., M. L. Laster & D. F. Martin. 1982. Parasitoids of *Heliothis* spp. (Lepidoptera: Noctuidae) larvae in Mississippi associated with sesame interplantings in cotton, 1971-1974:implications of host-habitat interactions. Environ. Entomol. 11:509-512.

Parkman, P. & M. Shepard. 1982. Searching ability and host selection by *Euplectus plathypenae* Howard (Hymenoptera: Eulophidae). J. Georgia Entomol. Soc. 17:150-156.

Pawar, C. S., V. S. Bhatnagar & D. R. Jadhav. 1985. *Heliothis* species and their larval parasitoids on sole and intercrop safflower in India. Insect Sci. Appl. 6:701-704.

Perfect, T. J., A. G. Cook & B. R. Critchley. 1979. Effects of intercropping maize and cowpea on pest incidence. *In* Internal report 1976-1978. International Institute of Tropical Agriculture, Ibadan, Nigeria.

Perrin, R. M. 1977. Pest management in multiple cropping systems. Agro-Ecosystems 3:93-118.

Pimentel, D. 1961a. Species diversity and insect population outbreaks. Ann. Entomol. Soc. Am. 54:76-86.

1961b. An evaluation of insect resistance in broccoli, brussels sprouts, cabbage, and kale. J. Econ. Entomol. 54:156-158.

Pimentel, D., W. P. Nagel & J. J. Madden. 1963. Space-time structure of the environment and the survival of parasite-host systems. Am. Nat. 94:141-167.

Price, P. W. 1980. Evolutionary biology of parasites. Princeton University, Princeton, N.J.

Price, P. W., C. E. Bouton, P. Gross, B. A. McPherson, Z. N. Thompson & A. E. Weide. 1980. Interactions among three trophic levels: influence of plants on interactions between insect herbivores and natural enemies. Annu. Rev. Ecol. Syst. 11:41-65.

Rao, M. R. 1986. Cereals in multiple cropping, pp. 96-132. *In* C. A. Francis [ed.], Multiple cropping systems. Macmillen Pub., London.

Reed, W. 1965. *Heliothis armigera* (Hb.) (Nuctuidae) in Western Tanganyika. II. Ecology and natural and chemical control. Bull. Entomol. Res. 56:127-140.

Risch, S. J. 1979. A comparison, by sweep sampling, of the insect fauna from corn and sweet potato monocultures and dicultures in Costa Rica. Oecologia 42:195-211.

1981. Insect herbivore abundance in tropical monocultures and polycultures: an experimental test of two hypotheses. Ecology 62:1325-1340.

Risch, S. J., D. A. Andow & M. A. Altieri. 1983. Agroecosystem diversity and pest control: data, tentative conclusions, and new research directions. Environ. Entomol. 12:625-629.

Root, R. B. 1973. Organization of a plant-arthropod association in simple and diverse habitats: the fauna of collards *(Brassica oleracea)*. Ecol. Monogr. 43:95-124.

Russell, E. P. 1989. Enemies hypothesis: a review of the effect of vegetational diversity on predatory insects and parasitoids. Environ. Entomol. 18:590-599.

Ruthenberg, H. 1971. Farming systems in the tropics. Oxford, London.

Schuster, D. J. & K. J. Starks. 1975. Preference of *Lysiphlebus testaceipes* for greenbug resistant and susceptible small grain species. Environ. Entomol. 4:887-888.

Sheehan, W. 1986. Response by specialist and generalist natural enemies to agroecosystem diversification: a selective review. Environ. Entomol. 15:456-461.

Southwood, T. R. E. & M. J. May. 1970. Ecological background to pest management, pp. 6-28. *In* R. L. Rabb & F. E. Guthrie [eds.], Concepts of pest management. North Carolina State University Press, Raleigh.

Speight, M. R. 1983. The potential of ecosystem management for pest control. Agric. Ecosyst. Environ. 10:183-199.

Speight, M. R. & J. H. Lawton. 1976. The influence of weed cover on the mortality imposed on artificial prey by predatory ground beetles in cereal fields. Oecologia 23:211-223.

Stanton, M. L. 1983. Spatial patterns in the plant community and their effects upon insect search, pp. 125-157. *In* S. Ahmad [ed.], Herbivorous insects. Academic Press, New York.

Stary, P. 1988a. Aphidiidae, pp. 171-184. *In* A. K. Minks & P. Harrewijn [eds.], Aphids, their biology, natural enemies and control. Elsevier, Amsterdam.

1988b. Aphelinidae, pp. 185-188. *In* A. K. Minks & P. Harrewijn [eds.], Aphids, their biology, natural enemies and control. Elsevier, Amsterdam.

Steiner, K. G. 1982. Intercropping in tropical smallholder agriculture with special reference to West Africa. German Agency for Tech. Cooperation (GTZ), No. 137.

Syme, P. D. 1975. The effects of flowers on the longevity and fecundity of two native parasites of the European pine shoot moth in Ontario. Environ. Entomol. 4:337-346.

1977. Observations on the longevity and fecundity of *Orgilus obscurator* (Hymenoptera: Braconidae) and the effects of certain foods on longevity. Can. Entomol. 109:995-1000.

Taylor, T. A. 1976. Mixed cropping as an input in the management of crop pests in tropical Africa. *In* 15th International Congress of Entomology, 19-27 August, 1976, Washington, D.C.

Townes, H. 1960. Host selection patterns in some nearctic ichneumonids (Hymenoptera), pp. 738-741. *In* Proc., XI Int. Cong. Entomol., Vienna, Austria.

Turlings, T. C. J., J. H. Tumlinson, W. J. Lewis & L. E. M. Vet. 1989. Beneficial arthropod behavior mediated by airborne semiochemicals. VII. Learning of host-related odors induced by a brief contact experience with host by-products in *Cotesia marginiventris* (Cresson), a generalist larval parasitoid. J. Insect Behav. 2:217-225.

van Alphen, J. J. M. & L. E. M. Vet. 1986. An evolutionary approach to host finding and selection, pp. 23-62. *In* J. Waage & D. Greathead [eds.], Insect Parasitoids. 13th Symposium, Royal Entomol. Soc. London. Academic Press, New York.

Vandermeer, J. 1989. The ecology of intercropping. Cambridge Univ. Press, Cambridge.

van Emden, H. F. 1963. Observations on the effect of flowers on the activity of parasitic Hymenoptera. Entomol. Mon. Mag. 98:265-270.

1977. Insect-pest management in multiple cropping systems: a strategy, pp. 325-343. *In* Proc., Symposium on Cropping Systems Research and Development for the Asian Rice Farmer, 21-24 Sept. 1976. The International Rice Research Institute, Los Banos, Philippines.

van Emden, H. F. & G. F. Williams. 1974. Insect stability and diversity in agro-ecosystems. Annu. Rev. Entomol. 19:455-475.

van Huis, A. 1981. Integrated pest management in the small farmer's maize crop in Nicaragua. H. Veenman & Zonen, Wageningen, Netherlands.

Vet, L. E. M. & M. Dicke. 1992. Ecology of infochemical use by natural enemies in a tritrophic context. Annu. Rev. Entomol. 37:141-172.

Watt, K. F. E. 1964. Comments on fluctations of animal populations and measure of community stability. Can. Entomol. 96:1434-1442

Way, M. J. 1977. Pest and disease status in mixed stands vs. monocultures; the relevance of ecosystem stability, pp. 127-138. *In* J. M. Cherrett & G. R. Sagar [eds.], Origins of pest, parasite, disease, and weed problems. The 18th Symp. British Ecol. Soc., 12-14 April, 1976, Bangor. Blackwell Sci. Pub.Oxford, England

Wrubel, R. P. 1984. The effect of intercropping on the population dynamics of the arthropod community associated with soybean *(Glycine max)*. M.S. thesis, University of Virginia, Charlottesville, VA.

Zandstra, B. H. & P. S. Motooka. 1978. Beneficial effects of weeds in pest management—a review. PANS 24:333-338.

Enhancement of Predation Through Within-Field Diversification

JUHA HELENIUS—Professor, Agroecology, Department of Plant Production, P.O Box 27, University of Helsinki FIN-00014 Helsinki, Finland.

Introduction

This chapter addresses management of crop habitats to enhance naturally occurring predation within fields of annual arable crop plants. The development of biologically based pest management systems is especially important for annual crops because they take up most of the agricultural land in the world (Anonymous 1991). The discussion is restricted to the action of indigenous rather than imported natural enemies.

The term predation is used herein to describe arthropods that consume several prey as they develop from egg to adult. Parasitoids are not true parasites and are usually treated as predators *sensu lato,* e.g., in theories of predator-prey dynamics. However, for habitat management the essence of the predator/parasitoid dichotomy arises from the fact that predators are mostly generalists, whereas parasitoids tend to be specialists (Price 1984). Parasitoids are not dealt with in this chapter.

Agroecosystem diversity can be viewed at different spatial and temporal scales. The large-scale, between-habitat diversity affects the dynamics of metapopulations (which were defined as systems of local populations by Levins 1970) of pests and natural enemies. These aspects are dealt with by A. Corbett and S. Wratten (this volume).

An annual arable field is a piece of land that is being managed as a unit by the farmer, and in which the turnover rate of crop biomass is once per year. I first examine annual field crops as habitats for arthropods, then predators as biological control agents. Next, I discuss vegetational diversity from the point of view of enhancing predation. In the final section I review some issues of management and decision-making. A case study illustrates how intercropping can be used to favor aphid predators in small grain cereals.

The Annual–Field Crop Habitat

Annually, the vegetation cycle typically includes one or more events of destruction of most if not all of the above-ground biomass at harvest and soil preparation, followed by periods of relatively undisturbed growth and development of the crop and associated weeds. Even the below-ground environment undergoes cyclic perturbations (ploughing, harrowing) within the cultivation layer. Thus, the crop habitat is temporally discontinuous and variable, its management preventing ecological succession of the vegetation cover. The ecotone from the field community to any neighbouring natural community is always sharp, and the fields are surrounded by margins that form very characteristic, semi-managed habitats with high agroecological significance (e.g., Mader 1990; see also A. Corbett, and S. Wratten this volume).

Crop rotations and patterning of the fields affect spatial availability and predictability of the habitats. In this respect, a vast and continuous monoculture of wheat in Canadian prairie forms a stable environment (Turnbull & Chant 1961), while wheat in mixed rotations in relatively small and scattered fields in Scandinavia form an unstable environment.

Many arthropods that use annual crop habitats for foraging, breeding, or cover migrate seasonally into and out of the field. Many others also disperse into fields, but most fail to establish, and even successful colonists suffer high rates of local extinction (Price 1976). This may result in crop fields acting as a sink for individuals dispersing from a core population located in a more suitable habitat (for discussion of the source/sink effects on metapopulations see Harrison 1991). However, there are major pest species whose life cycle strategies rely on crop fields for net recruitment. For example, cultivated Brassicaceae rather than wild brassicaceous weeds are no doubt responsible for maintaining cabbage root fly (*Delia radicum* [L.], Diptera: Anthomyiidae) populations in England (Finch 1988).

Some species survive in the relatively stable soil environment during unfavorable above-ground conditions. Here, migration may not be necessary, but the species still must overcome the unpredictability of the cropping pattern and crop rotation. For example, a larval population of orange wheat blossom midge (*Sitodiplosis mosellana* [Géhin], Diptera: Cecidomyiidae) cocoons into soil in autumn and an ever-decreasing proportion of the population pupates and emerges each year during the following approximately ten years (Barnes 1956). The midge is a relatively poor disperser and is adversely affected by crop rotation. However, the variable duration of diapause increases the chances of at least some specimens finding a host crop without dispersing.

After successful colonization, herbivores can enjoy temporarily ideal conditions comprising a uniform stand of nutritionally suitable even-aged host plants in a monoculture or near-monoculture. This kind of habitat structure especially favors specialists, because within-field effort in searching for food is minimized (Root 1973; Risch et al. 1983; Andow 1991a), but polyphagous pests may similarly find it favorable (Andow 1991a).

The situation is less suitable for colonizing natural enemies of the herbivores. If the natural enemy is coupled with a particular prey species, as many parasitoids are, it cannot colonize before its prey and there may be a delay of days or weeks in the colonization. However, if habitat cues rather than prey cues are used in prey finding, colonization need not depend on presence of prey species. At the early stages of development of the crop and of its populations of herbivores, food resources are discontinuous (Price 1976). For searching enemies, visiting several plants is necessary, whereas, for sit-and-wait predators, periods of food shortage are likely. At later crop developmental stages, prey may be more abundant but not necessarily more apparent, as the canopy has developed in complexity and the leaf-area is greater. However, searching several consecutive plants may become easier via leaf-bridges.

The Arthropod Predators

Predator Attributes

In annual field crops, what kind of natural enemies, if any, would be successful in biocontrol of pests? From various studies in natural and agricultural environments it is relatively safe to conclude that natural enemies are able to regulate populations of herbivores, although it has been surprisingly difficult to prove (Price 1987; Wratten 1987).

In ecological jargon, regulation is usually discussed in the context of long-term dynamics of predator-prey interactions, and the central issue has been the ability of natural enemies to maintain the populations of herbivores at low, stable equilibrium. Much of the research tradition relies on analytical models of discrete-time single-predator, single-prey systems. For reviews of this theory, see Murdoch and Oaten (1975), Hassell (1978), and May and Hassell (1988). Recent developments include the introduction of continuous-time, stage-structured analytical models (see, Murdoch 1990). Several refinements of the equilibrium-model exist, but properties of a successful (i.e., capable of regulating at low stable equilibrium) natural enemy remain as summarized by Beddington et al. (1978): the enemy is host-specific; it has synchronous emergence with the host (to suit the life-history stages), high attack rate, and high rate of reproduction; and if the enemy is very efficient, distribution of the host must be clumped for stability to emerge. These attributes fit better to parasitoids than to predators, and furthermore, this traditional approach gives poor prospects for biological control in annual crops.

In arable croplands, regulation of pest populations at stable equilibrium is meaningless. Local populations are typically too short-lived in these transient habitats for stability to emerge. Because individual fields are maintained in nonequilibrium, there are no grounds for exploring equilibrium solutions for pest-enemy interactions (Murdoch 1975; Risch et al. 1983; Pimm 1984; Kareiva 1986a; Letourneau 1990). However, an alternative modeling approach presented by Murdoch et al. (1985) does not require stable equilibrium and allows for local extinctions and environmental stochasticity. The model gives predictions on desirable predator attributes in contrast to those listed above: advantageous properties of a natural enemy include polyphagy, asynchrony in the life cycles with the prey, and consumption of several prey individuals in lifetime (Murdoch et al. 1985). In addition, successful species must have high powers of dispersal: this facilitates their rapid colonization (Southwood 1962) and spreads the risk of extinction over local populations by increasing their exchange of individuals (Den Boer 1981, 1990). Non-synchronous population fluctuations should favor stability of the composite population (Den Boer 1981; Murdoch et al. 1985).

The present theories do not give definite answers on the desirable attributes. Confrontation between proponents of specialist enemies and proponents of generalist predators is not needed: the strategy of enhancing "background mortality" (*sensu* Beddington et al. 1978) of the pests through assemblages of generalists should be integrated into crop management regimes, and control of the remaining outbreak species should be complemented by specific, tactical solutions as in classical biological control. Carroll and Risch (1990), when discussing the role of ants in tropical annual systems, suggested that if pest diversity is high, relatively more may be gained by enhancing generalist predators than by introducing a specific enemy against just one pest species. They supported the view that at the level of individual crop stands, stochastic rather than regulatory dynamics may govern the fate of local populations of pests and their specific enemies.

To successfully use polyphagous predators as biological control agents in annual crops (see, e.g., Ehler & van den Bosch 1974; Ehler 1977, and references therein) one must create high, stable *total* densities of these species (i.e., stable guild density rather than stable population density) in individual fields of crops. Pest mortality imposed by natural enemies should be maximized, and even extinction of the local pest population is desirable.

In this approach, even if the requirement of stable equilibrium is rejected and replaced by possibility of local extinction, the concept of pest stability is applicable at a spatial scale larger than an individual field, i.e., at the agroecosystem level (see Kareiva 1990a).

Although the focus of theoretical population ecology is on long-term dynamics, several more general (i.e., not postulated from stability in predator-prey interaction) results concerning identification of an efficient predator are applicable to annual crop situations.

A predator/parasitoid may become more efficient in catching and handling prey when prey density increases, i.e., the individual's so-called functional response improves. The predator's rate of reproduction may also increase. There is a lag in the reproductive numerical response, which may prevent it from playing a role in suppression of prey populations in an annual system. An immediate effect may be achieved through the aggregative numerical response, where as a result of nonrandom movement (searching), the predators/parasitoids aggregate into patches of high prey density. In a modeling exercise, Crawley (1975) concluded that aggregative numerical responses through immigration and emigration can be very significant in bringing about large, rapid changes in predator numbers. This conclusion has since been corroborated (Kareiva 1990a). There is naturally a saturation point in the numerical response, beyond which no additional predators move into the patch, and the population of the predators is limited, i.e., there is a maximum total number available. Aggregation of generalist predators may be enhanced by "switching" if the predator becomes trained to attack abundant prey species at a disproportionately higher rate than less abundant species (Murdoch 1969).

A synopsis of how predation may interact with habitat stability in the population dynamics of a pest species was given by Southwood and Comins (1976). In their model, the increase in the maximum number of available predators effectively widened the "natural enemy ravine," i.e., the range of population growth rate of the prey within which biocontrol is effective (and where enhancement should be targeted). Another prediction relevant in this context is that for pest species with "boom and bust" dynamics (i.e., "r-strategists" with high reproductive rate) in ephemeral habitats, the reproductive numerical response of natural enemies is limited. Thus, within the natural enemy ravine of population growth rate, the significant predators are often polyphagous and large relative to their prey, and occur at relatively low densities (Southwood & Comins 1976). In other words, the key species would be large voracious generalists rather than delicate specialists.

An important dimension to the theory of predator-prey dynamics is the effect of environmental patchiness on a local scale (e.g., within field). In traditional analytical models the spatial patterns are assumed, rather than incorporated as processes. Kareiva (1984) developed and argued for spatial models that explicitly incorporate dispersal and movement. These are especially suitable for annual field-situations in which detecting a transient density-dependent process necessitates following through time individual predators and cohorts of prey (Kareiva 1986b). Again, in annual fields, aggregation of prey is important in terms of its impact on rate of predation, rather than the stability of the predator-prey interaction. In these systems, predators should be able to aggregate rapidly and control local outbreaks early in the season when pest patches are relatively scarce (Kareiva 1990b).

The real life multi-predator, multi-prey systems are beyond the predictive potential of present theories, and it is important to maintain some pragmatism and even empiricism in identifying biocontrol options. Effects of the whole predator/parasitoid complex, interactions between predators, variability of alternative prey, and interplay of generalist predators and specialist parasitoids require an experimental approach.

Predator Taxa

The most common generalist arthropod predators in farmland are ground beetles (Coleoptera: Carabidae), rove beetles (Coleoptera: Staphylinidae), predatory bugs (Heteroptera), ants (Hymenoptera: Formicidae) and spiders (Araneae). Some examples are given in Table 1. Characteristically, they are true generalists feeding on wide ranges of prey taxa and inhabiting many kinds of crops. In most systems, there are additional generalist predators of other taxa that may significantly contribute to predation but are much fewer in numbers.

Adult ground beetles and rove beetles are mainly soil surface–dwelling species as are the spiders. Spiders may also colonize the foliage, and ants move freely on both soil surface and in the foliage. Many of the predatory bugs disperse aerially into crop foliage, but epigeal species may also be important.

Ground Beetles and Rove Beetles

Among the true generalists in farmland, ground beetles have been best studied in the temperate regions. Typically in Eurasian and North American fields 40–60 different species of carabids can be found, and for many, annual fields are their breeding habitat. Most species have usually been

Table 1. Examples of common generalist predators of major pests in the five most important annual crops. Note that usually several species within a taxa are reported to contribute to the overall predation pressure.

Crop[1]	Predators	Prey (major pests)	Area	References[2]
1. Wheat, barley, oats	Ground beetles	Aphids	Europe	Sunderland (1975), Edwards et al. (1979), Ekbom & Wiktelius (1985), Scheller (1984), Chiverton (1987), Gravesen & Toft (1987), Basedow (1989), Helenius (1990a), Winder (1990), Holopainen & Helenius (1991)
	Rove beetles	Aphids	Europe	Sunderland (1975), Edwards et al. (1979)
	Spiders	Aphids	Europe	Sunderland et al. (1986, 1987), Nyffeler & Benz (1981, 1988), Chiverton (1987)
2. Rice	Predatory bugs	Planthopper	India	Peter (1988)
		Several taxa	S-E Asia	van Vreden & Ahmadzabidi (1986)
	Ground beetles	Planthopper	India	Samal & Misra (1978)
		Several taxa	S-E Asia	van Vreden & Ahmadzabidi (1986)
	Spiders	Leafhoppers and Planthoppers	East Asia	Kiritani et al. (1972), Sasaba et al. (1973), Kiritani & Kakiya (1975), van Vreden & Ahmadzabidi (1986)
		Planthoppers	India	Samal & Misra (1975), Thomas et al. (1979), Peter (1988)
3. Corn	Predatory bugs	Pyralid corn borer	USA	Knutson & Gilstrap (1989)
		Pyralid corn borer	India	Singh et al. (1975), Sharma & Sarup (1980)
	Spiders	Pyralid corn borer	USA	Knutson & Gilstrap (1989)

continued

Table 1. continued

Crop[1]	Predators	Prey (major pests)	Area	References[2]
4. Soya bean	Predatory bugs	Noctuid pod borer	USA	Godfret et al. (1989)
		Epilachna beetles	USA	O'Neill (1989)
	Ground beetles	Noctuid pod borer	USA	Godfret et al. (1989)
	Ants	Noctuid pod borer	USA	Godfret et al. (1989)
	Spiders	Noctuid pod borer	USA	Gregory et al. (1989)
5. Cotton	Predatory bugs	Bollworms	USA	Whitcomb & Bell (1964), van den Bosch & Hagen (1966)
			China	Wu et al. (1981)
	Ants	Mirid bugs	USA	Breene et al. (1989)
		Boll weevil	USA	Whitcomb & Bell (1964), Sturm & Sterling (1990)
	Spiders	Cotton aphid	USA	Kagan (1943), Nyffeler et al. (1988)
		Mirid bugs	USA	Breene et al. (1989)
		Bollworms	USA	Kagan (1943), Whitcomb et al. (1963)
			China	Wu et al. (1981)
		Boll weevil	USA	Whitcomb et al. (1963)

[1] Ranking of the crops is based on the area occupied (Anon., 1991)
[2] The reference list is not exhaustive.

found in cereal fields and least often in root crops (Scherney 1960; Luff 1990). Most of the temperate field species are nocturnal. Ecology and behavior of ground beetles were recently reviewed by Lövei and Sunderland (1996).

Carabids are typically univoltine, precluding the reproductive numerical response as a mechanism contributing to regulation in annual fields. Aggregative numerical response has been demonstrated for some species, however (Bryan & Wratten 1984). The larvae develop in the cultivated layer of soil. At any one time a small number of species dominate, the hierarchy changing with time of season and crop developmental stage. Spring breeders are abundant early in the growing season and autumn breeders dominate later.

Many species are capable of sustained flight, but even among these, movement and dispersal are primarily by running on the soil surface. With some exceptions, carabids are poor climbers and catch their prey on the ground and by searching stem bases of plants. Farmland species typically migrate annually between the overwintering sites on field borders and breeding sites in the open field (Sotherton 1984, 1985; Coombes & Sotherton 1986; Wallin 1987). In spring, colonization of the field starts from the borders, resulting in a wave-like spatial pattern in the density from edge to center of the field. Unless the field is very large, the distribution gradually becomes uniform over the whole field (see, A. Corbett, and S. Wratten this volume). Some large species like *Pterostichus* spp. disperse very quickly, and can cover distances of hundreds of meters in one day (Wallin & Ekbom 1988). Many of the arable land species have high power of dispersal, differently fluctuating local populations, and thus, relatively stable and persistent composite populations (Den Boer 1990).

The role of carabids among other epigeal generalists as natural enemies of cereal aphids became an area of extensive research in Europe, following the pioneering work of Vickerman and Sunderland (1975), Sunderland (1975), and Edwards, and Sunderland and George (1979) in the United Kingdom. High rates of predation by epigeal predators on other cereal pests were earlier reported by Basedow (1973). Predation on eggs of cabbage root fly has also been repeatedly demonstrated since the publications by Coaker and Williams (1963) and by Mitchell (1963). Scherney (1959) showed that at high densities, *Carabus* spp. can significantly improve potato yield by preying on potato beetle (*Leptinotarsa decemlineata* Say, Coleoptera: Chrysomelidae).

The most dominant and geographically widespread genera of carabids in the temperate region are *Pterostichus*, *Harpalus*, *Agonum*, *Trechus* and *Bembidion*, and the fauna can be fairly consistent over large regions (Thiele 1977; Luff 1987). Peak population densities of the dominant species can be relatively high. For example, Scheller (1984) estimated a mean density of 61 (±10 S.E.) per m² of *Bembidion lampros* (Herbst) adults during their breeding period in early summer in a spring barley field in Denmark. Densities up to 10/m² of dominant species are often found, and these sum up to fairly high total densities of carabids in a field.

Carabids feed on both vegetation and arthropods. Damage to seedlings of crop plants has sometimes been noted but is usually insignificant. Large seed predators like the common *Amara* spp. remove seeds of weed plants, but to what extent is not known. Many species tolerate starvation if water is available. The adults are relatively long-lived as compared to most pest species. Rate of daily food intake may far exceed the body weight, large species being most voracious.

Predation behavior of *Pterostichus versicolor* (Sturm) was modeled by Mols (1987) in a detailed simulation. Walking velocity and direction, duration of area-restricted search, and success ratio in prey capture were governed by hunger, measured by size of gut. Predation rate was highest when simulations allowed for aggregated prey and adaptation to varying degrees of prey clustering.

Reviews on carabids in agricultural fields are given by Thiele (1977), Allen (1979, with emphasis on North America) and by Luff (1987). Many topics of agricultural interest in carabid ecology are covered in an issue edited by Stork (1990).

In many respects rove beetles or staphylinids resemble ground beetles in their ecology. Equally high numbers of specimens and species can be found in farmland, and predation on pest insects like aphids (e.g., Sunderland 1975) or root flies (e.g., Mowat & Martin 1980) is common. Adults are epigeal, and search for prey by running on the soil surface or sometimes by climbing plants. Larvae develop in the soil. Staphylinids are fluid feeders and predation in most cases cannot be detected by dissection of the gut for prey remains (Sunderland 1988). Many species are small and predominantly fungus feeders. Dispersal patterns of rove beetles in farmland are less known than of ground beetles.

Predatory Bugs

Many heteropteran species especially in the families Nabidae, Anthocoridae and Reduviidae (mostly tropical species) are predators of pest insects.

Lygaeidae are mostly herbivorous, but there are important predatory species in the genus *Geocoris*.

Especially well studied are predatory bugs occurring in cotton fields in the United States; the major species are bigeyed bugs *Geocoris pallens* Stål and *Geocoris punctipes* (Say), pirate bugs *Orius insidiosus* (Say) (Anthocoridae) and *Orius tristicolor* (White), and damsel bug *Nabis americoferus* Carayon (Nabidae). Studies on natural enemies of cabbage looper (*Trichoplusia ni* [Hübner], Lepidoptera: Noctuidae) in California cotton showed that predation on eggs and small larvae was the major mortality factor and that *G. pallens*, *O. tristicolor* and *N. americoferus* made up ca. 60–90 percent of the natural enemy fauna. Based on these findings Ehler & van den Bosch (1974) and Ehler (1977) argued for the idea of using general predators in classical biological control of pests of annual crops. In a study on natural control of three major lepidopteran pests in cotton, Ehler and Miller (1978) concluded that the predators are numerically dominant over parasitoids and primarily responsible for maintaining pest density below damaging levels. The natural predator complex accounted for 53–76 percent of the mortality at egg and nymphal stages of *Lygus hesperus* Knight (Heteroptera: Miridae); *G. pallens* was the most effective predator (Leigh & González 1976).

Predatory bugs can have very wide prey ranges. Twenty-six species of insects and mites are listed being preyed upon by *O. insidiosus* (Barber 1936), and 67 species of insects, mites and diplopods of size 0.25–8.7mm were preyed upon by *G. punctipes*, *Geocoris uliginosus* (Say) and *Geocoris bullatus* (Say) (Crocker & Whitcomb 1980). Most predatory bugs are capable of plant feeding (feeding on nectar, pollen and plant tissues), which allows maintenance of populations in absence of prey (Barber 1936; Crocker & Whitcomb 1980), but animal prey is usually needed for egg laying (e.g., Barber 1936). The optimal diet for the development, survival and egg production of *O. tristicolor* contained plant tissues, pollen and thrips, but even plant food sustained development. In spite of omnivory, the species shows clear preference to thrips over other prey (Salas-Aguilar 1977). *Geocoris pallens* preferred *Tetranychus* mites to *Frankliniella* thrips to cotton plant (requiring 5 times more mites than *O. tristicolor* to maintain its caloric level), and the mites provided the food base for the predators, ensuring better control of key pests (González & Wilson 1982).

Habitat range can be wide. For example, *Geocoris* spp. hunt and even feed on many annual and perennial plants in mesic or xeric habitats or microhabitats, and they commonly occur on the soil running about on exposed ground (Crocker & Whitcomb 1980; Bugg et al. 1987). *Orius insidiosus* occurs accross the United States on a wide variety of vegetation from trees to field crops., e.g., in Virginia it was the major insect enemy of corn earworm (*Helicoverpa zea*, Lepidoptera: Noctuidae), destroying on average 38 percent of the eggs (Barber 1936). *Orius amnesius* Ghauri and *Orius alpidipennis* (Reuter) were the dominant predators of cowpea flower thrips in Nigeria; others were rove beetles and ants (Matteson 1982). *Nabis* spp. were the most abundant predators on soybean in South Carolina, followed by spiders and *Geocoris* spp. (Shepard et al. 1974; see also Reed et al. 1984).

Predatory bugs can disperse by flying, e.g., *Geocoris* spp. (Crocker & Whitcomb 1980). They can reproduce relatively fast, e.g., *O. insidiosus* can lay eggs at a rate of 114/female under good conditions, takes 11–18 days from egg to adult, and has a minimum generation time of 24 days (Barber 1936). Being good colonizers, abundant, and having dietary plasticity that buffers them against food shortage, predatory bugs have high potential for biological pest control in certain annual field crops.

Ants

Ants occupy a special category among generalist insect predators. They are social insects that form large, stable colonies and have evolved strategies for food-finding and efficient use. Species diversity can be high, especially in the tropics. Ant communities exhibit a clear, sustained hierarchy, the dominant species being able to defend territories against other ant species (Way 1992). To some extent, some other social forms of stinging Hymenoptera, like vespid wasps, share the characteristics of ants, but their potential to control pests in field crops is not known.

Risch and Carroll (1982), Carroll and Risch (1990), and Way (1992) list several attributes of useful ant species. They are habitat generalists, can rapidly aggregate, showing numerical responses, and respond to prey populations in a density-dependent way. They attack low-density pests as well. The ability to store food delays satiation when prey is abundant. They can survive periods of prey scarcity by feeding on previously stored food and by cannibalizing their brood. They can both kill and deter pest species. Abundance and distribution of colonies can be managed. For example, in some species, polygyny allows establishment of new colonies from transferred fragments of old ones. If the species is dominant, it is not easily replaced by another species. On the other hand, negative traits are plant-feeding, and attendance of homopteran pests for collection of honeydew (Way 1963).

In a review on the role of ants in pest management Way (1992) gives examples of useful species in annual fields. For example, one imported and one indigenous *Solenopsis* species have proved to be valuable predators in crops like sugar cane and cotton in southern United States. However, tillage destroys nests of most species within annual fields, and this may often be the most difficult management problem in enhancing natural biological control by ants.

Spiders

Spiders are an important component of generalist predator complexes in both temperate and tropical crops (Riechert & Lockley 1984; Riechert & Bishop 1990). For example, in European cereals they often outnumber any other groups including carabids, and in the United Kingdom alone, at least 151 species from 12 families have been recorded (Sunderland 1987). Densities increase towards the end of season, and from 10–50/m^2 (Nyffeler & Benz 1981) up to ca. 200/m^2 (Sunderland 1987) have been reported in cereals. Some field spiders may survive the cultivation and overwinter in the field (Sunderland 1987). Colonization of the fields is by ballooning on silken threads (e.g., aeronautic Linyphiidae) or by crawling on the soil surface. Nyffeler and Benz (1988) suggested that the ballooning species, being the pioneers of early successional habitats, dominate the spider fauna in cereal fields because of their high dispersal ability. Sunderland (1987) mentioned a deposition rate of such immigrants of 30/m^2/hour in the United Kingdom. Ballooning linyphiids are the key predators of the major leafhopper and planthopper pests in Chinese and Japanese rice paddies (Riechert & Lockley 1984).

Analysis of faunal surveys of spiders in field crops in the United States by Young and Edwards (1990) indicates presence of at least 614 species. Highest number of species has been observed in cotton (308 spp.) and in soybean (262), which are multiple-branching dicotyledonous crops, while somewhat lower numbers have been seen in the monocotyledonous, architecturally less diverse crops, corn (136) and rice (75). Fifty-six percent of the species were wanderers and 44 percent web-spinners, the former being better represented proportionally in the field crop fauna than in all the United States fauna. Young and Edwards (1990) suggest the most frequently occurring three species, *Oxyopes salticus* Hentz (Oxyopidae), *Phidippus audax* (Hentz) (Salticidae), and *Tetragnatha laboriosa* Hentz (Araneidae), as prime candidates for augmentation.

Spiders are univoltine and thus incapable of a reproductive numerical response. Many species are territorial, extremely starvation-tolerant, sit-

and-wait predators with a very wide range of prey taxa. Although they may respond numerically to increasing prey densities by decreasing the territory, the density-dependent tracking of a prey population in an annual system is not possible. The size of the area defended against conspecifics is determined more by minima in prey availability and by availability of web and foraging sites than by occasional food abundance. However when prey is abundant, spiders will kill many times the number they consume, and the composite foraging activity of the assemblage of spider species serves as a buffer against establishment and initial growth of a pest population (Riechert & Lockley 1984; Riechert & Harp 1987).

Other Predators

Terrestrial predators can be found in most arthropod orders, with a wide range of habitat requirements, food specialization, etc. (for reviews see Hagen et al. 1976; Hagen 1987). Many minor taxa that may be locally or occasionally important generalist predators in annual fields are not discussed here. However, there are two common groups of predators that deserve attention in this context, namely 'aphid specialists' and predatory mites.

Many species of ladybird beetles (Coleoptera: Coccinellidae), larvae as well as adults, and many hoverfly (Diptera: Syrphidae) larvae are common predators of aphids or mites. In absence of prey some coccinellid species are able to switch to plant food for survival (Hodek 1973). The degree of food specificity varies, but coccinellids and syrphids are not true generalists: they are only able to grow, develop, and produce offspring on a narrow range of prey taxa. Most of the common Neuroptera, especially green lacewings (Chrysopidae) also specialize on aphids during their predaceous larval stages. Reviews on aphidophagous species are included in Minks and Harrewijn (1988).

Predatory mites are abundant predators of herbivorous mites and small insects like homopterans and thrips. The most important foliage inhabiting group are the phytoseiids (Acarina: Phytoseiidae), which are the main predators of spider mites. Some species are successfully used in biological control of tetranychid pests. The degree of food specificity varies, but the majority of species are probably generalists and can exploit even plant foods like pollen (McMurtry & Rodriguez 1987).

Diversity and Predation

This section discusses diversity in the context of enhancing predation. In principle, predation can be enhanced through increase in predator numbers and/or increase in success of individual predators in capturing prey.

Vegetational Diversity and Predator Population Densities

Reviews on effects of floral diversification on predators show that their numbers can be and often are increased through increased vegetational diversity. Of 16 published cases reviewed by Russell (1989), natural enemies were more abundant in polycultures than in monocultures in 10 cases, in 4 cases there was no effect, and the predator numbers were either lowered or the effect was not consistent in 2 cases. Andow's (1991a) data covered published studies on 90 species of predators and 40 parasitoids in annual and perennial systems. Of the predatory species, vegetational diversity was reported having increased population densities in 42.2 percent of the cases, in 12.2 percent of the cases the densities were lowered, in 30.0 percent of the cases the response was variable, and no change was found in 15.5 percent of the cases. Of the parasitoids, 75.0 percent had higher densities in diversified crops, while only one species (2.5 percent) had lowered numbers (Andow 1991a).

Sheehan (1986) discussed the responses of natural enemies on vegetational diversity. He extended the so-called resource concentration hypothesis to include natural enemies. Originally, this hypothesis was developed by Root (1973) to explain the effects of vegetational diversity on herbivorous arthropods. The hypothesis states that "herbivores are more likely to find and remain on hosts that are growing in dense or nearly pure stands; the most specialized species frequently attain higher relative densities in simple environments; and, as a result, biomass tends to become concentrated in a few species, causing a decrease in the diversity of herbivores in pure stands" (Root 1973; p. 95). Associational resistance mechanisms, as suggested by Tahvanainen and Root (1972), would then be the main cause to lowered herbivore numbers in diversified crops. As emphasized in many studies exploring these hypotheses, the processes most affected by resource concentration would then be the arthropod movement processes in approaching the crop and within it: immigration and emigration; host-finding, acceptance and tenure time; diffuse (non-migratory) movement, etc. (see Risch et al. 1983).

Sheehan (1986) pointed out that the mechanisms of resource concentration are likely to affect the specialized natural enemies in principally the same ways as they are likely to affect their specialized herbivore prey or host species. This suggestion is further supported by the theory of interactions among the three trophic levels (Price et al. 1980): e.g., increasing plant species diversity may "hide" a plant from a specialist herbivore, but associated plants may also "hide" the herbivore from its enemies if they use cues like plant volatiles in prey finding. Andow's (1991a) data summarized above do not clarify the issue; although parasitoids more often than predators were reported having increased densities in diverse crops, the effects on predation-rates (see next section) were not documented; the issues clearly warrant further refinement.

Unlike the case of specialized natural enemies, the resource concentration hypothesis does not, *a priori*, suggest any general pattern in effects of vegetational diversity on generalist arthropod predators. Increased vegetational diversity is likely to favour generalists by providing alternate food or prey. Architectural complexity may enhance tenure of adults and survival of immature stages by providing wider range of choices along environmental gradients (see, Sheehan 1986; Russell 1989). On the other hand, in diversified crops predators may aggregate on a hospitable plant species, in which case even increased overall densities do not necessarily result in increased predation on pests of other crop plants (Bugg et al. 1987; Bugg & Wilson 1989: examples on effects of weeds on abundance and efficiency of predators).

Vegetational Diversity and Predation Rates

Predation rate (numbers of prey-items eaten per predator) at a given prey density can be affected through changes in spatial distribution and location within vegetation, movement patterns and speed of movement of both the predator and the prey. Examples of such mechanisms are given here, but first, the relative importance of enemy impact and resource concentration in the context of pest control by diversification is discussed.

The response of monophagous vs. polyphagous herbivore species to plant species diversity clearly demonstrate the risk of increasing pest problems if the majority of pests are polyphagous (Table 2; Andow 1991a). The overall reduction in pest numbers through diversification was for monophagous herbivores twice that for polyphagous herbivores (p<0.001, logit-model, G^2–test; Payne 1987). The reduction in pest numbers was greater in perennial systems than in annual systems (0.05<p<0.1).

Reduction of monophagous pests was greater in perennial systems, where predators and parasitoids are conventionally thought to be relatively more important than in annual systems. This conventional wisdom derives from classical biological control, and is based on data that are biased towards perennial systems and specialized natural enemies. However, the reduction of polyphagous pest numbers was less in perennial systems than in annual systems (interaction term, system x food specialization:

Table 2. Percentage of cases with lowered numbers of arthropod herbivores in crops of increased vegetational diversity than in monocrops. Data from Andow (1991a).

	Monophagous Species	Polyphagous Species	All
Annual systems	53.5	33.3	48.5
Perennial systems	72.3	12.5	60.5
All	59.1	28.4	51.9[1]

[1]Out of the 287 species included in the data, 155 were monophagous in annual crops.

$0.001 < p < 0.01$) (Table 2). The data suggest that adding host plant species benefits polyphagous herbivores more in perennial systems than in annual systems. This would be explained by the fact that in perennial systems, vegetational diversification is likely to have a greater effect on resource availability. Annual systems have much less age or "successional" diversity and are more uniform in total plant density, irrespective of plant species diversity, than perennial systems.

Based on these data, Risch et al. (1983), Risch (1987), and Andow (1991a,b) suggested that mechanisms of resource concentration rather than natural enemies contribute more to lowered herbivore numbers in polycultures. The argument is that "if natural enemies were generally important in affecting herbivore response to vegetational diversity, then polyphagous herbivore populations should be lower in polycultures than monocultures, just like monophagous herbivore populations" (Andow 1991b, p. 267). The data are inconclusive in this respect, however. The pattern (Table 2) may as well emerge if (1) natural enemies were generally important *and* (2) the resource concentration mechanisms were generally less important for polyphagous herbivores than for monophagous herbivores (as was suggested by the same data, see above).

Further evidence can be sought from comparison of herbivore response to vegetational diversity in annual vs. perennial polycultures. By assuming that herbivores have greater difficulty finding annual plants than perennial plants (Feeny 1976; Rhoades & Gates 1976), and that enemy impact is greater in perennial systems than in annual systems (the "classical biocontrol paradigm"; see also Southwood & Way 1970), Andow (1991a) generated two predictions from the resource concentration hypothesis: (a) reduction of pest numbers in perennial polycultures would be less than in annual polycultures and (b) the relative reduction of monophagous herbivores compared with polyphagous herbivores, should be greater in annual systems than perennial systems. The empirical data on success rates are contradictory to these predictions (Table 2), and thus the data do not give evidence in favour of the resource concentration hypothesis. Omitting uncertain cases (David Andow, personal communication) that were tabulated as a variable effect or no change in herbivore numbers (Andow 1991a) does not change the result (the difference concerning point [a] is reversed but remains statistically non-significant).

From the re-analysis it is clear that data on published cases of increases or decreases in herbivore numbers in diversified crops cannot be reliably used to judge the relative importance of resource concentration and enemy impact. It is suggested here that rather than resource concentration alone, the interaction of habitat organization (*sensu* Liss et al. 1986), herbivore movement patterns and enemy impact may in many cases best explain the numerical response of herbivores to crop diversification.

Colonization, dispersal by diffusion, and foraging movements of predators can be influenced by habitat modifications. For example, increase in leaf area or decrease in relative density of host plants of the prey species may result in decreased searching efficiency (Sheehan 1986). In a greenhouse experiment, density but not diversity had an effect on the foraging rate of an individual predator, so that larger surface area resulted in more time spent searching (Risch et al. 1982). Even morphological differences between cultivars may affect predation. Leafless peas were easier for seven-spotted lady beetle (*Coccinella septempunctata* L., Coleoptera: Coccinellidae) to search than normal leafy peas, resulting in better aggregation of the predators to clusters of pea aphids on the leafless variety (Kareiva 1990b). Pilose leaves allowed easier capture of thrips by *O. tristicolor* than leaves with longer hairs because these interfered with the movement of thrips (Salas-Aquilar & Ehler 1977). The example suggests that diversification may affect prey rather than predator, but with a consequence on predation. In general, diversification that influences the foraging movements of predators or movements or distribution of prey may either interfere with or enhance the control of pest populations (Kareiva 1986a).

The diffusion rate of a coccinellid predator *Coleomegilla maculata* (DeGeer) was higher in polyculture than in an equal density monoculture of corn. The dispersion pattern was affected so that beetles were more clumped in the polycultures (Wetzler & Risch 1984). Colonization rate of anthocorid bug *O. tristicolor* was higher in polycultures than in monocultures, because of availability of alternative food (pollen) and because of more suitable microclimate (Letourneau & Altieri 1983). If diversification is done on a larger spatial scale, then diffusion rates and habitat preferences may play a key role in the distribution of predators across a field (Corbett 1993). Even some generalist feeders have specific habitat requirements (Nordlund et al. 1988). If synomones (chemicals from the prey's host plant) are involved in the host-finding, crop diversity may have an adverse effect on predation rate (Hagen 1987). However, much more is known about influences of allelochemicals on insect parasitoids than on predators (see Barbosa 1988).

Predator Diversity

How is the diversity of natural enemies of herbivores related to predation rates? Experience from multiple releases of exotic natural enemies in classical biological control suggests that competition between introduced species may in some cases impair control of the pest (Ehler & Hall 1982). However, in classical biological control most of the work has been done with specialist enemies, especially with parasitoids.

Natural biological control, e.g., the impact of indigenous natural enemies on an endemic pest, differs from classical biological control in that it involves a complex of native natural enemies rather than a limited number of imported enemies targeted at an exotic pest. Individual species of generalist predators may have little effect on the prey population, but the species together may significantly influence its density (e.g., Potts & Vickerman 1974; Luff 1983; Riechert & Lockley 1984; Liss et al. 1986). Rabb (1971) suggested that an important benefit of having a multiplicity of enemies is that they compensate for the variability across a broad range of predation rates, adding stability to the complex of natural enemies.

Management Considerations
Options and Decisions

Perrin (1980) defined the dimensions of agricultural diversity as crop/non-crop, spatial/temporal, and biotic/abiotic. The floral/faunal axis may be added as sub-dimensions of biotic diversity. Theoretically, the different components of diversity and their combinations offer an endless variety of management alternatives. However, the economic and technological production environment limits the options.

Practical possibilities in making a choice between environmentally acceptable and less acceptable management alternatives are usually limited. Often, it is necessary to compromise between conflicting needs. Ideally, a new control practice should be practically feasible, technically possible, economically desirable, environmentally acceptable, and politically advantageous (Norton 1991). Over-emphasis on technical and economic criteria has often led to adoption of unsustainable practices. The history of pesticide use offers examples of increasing public awareness transforming environmental problems into political ones.

In introducing alternative practices to replace commonly accepted ones, the problem may lie outside of research capabilities. Application of basic knowledge, via technology, is dependent on political, economic, and philosophical variables (Risch 1987). Norton (1991) pointed out that the major constraints consist of the socioeconomic infrastructures that are built around the already adopted technologies. These may "lock" the producer into a historically determined path, making it easier to adopt co-evolving practices rather than entirely new ones. Research and development workers should take such constraints into consideration, in design and in delivery (Norton 1991; see also Norton 1982; Mumford & Norton 1984). An example is given by Norton and Heong (1988), in their analysis of key factors in rice pest management in Malaysia.

Many reviews on aspects of vegetational diversity and pest control are available, covering intercropping, uses of multiline cultivars, undersowing, spatial patterning, strip cutting, multiple cropping, crop rotations, etc. (Table 3). Only pesticide use in relation to within-field biotic diversity and natural control of pest insects is discussed here.

Pesticides, Diversity, and Predation

An important management activity influencing within-field species diversity in present day agriculture is application of pesticides. Herbicides are used to maintain near-monocultures of crop plants, insecticides are used to dramatically reduce numbers of the target species, etc. Extensive discussion and review on effects of pesticides on natural enemies of arthropod pests is given by Croft (1990).

Natural enemies are affected not only by insecticides, but also by herbicides and fungicides. Predators are usually less susceptible to direct poisoning than parasitoids. Indirect effects via trophic links are superimposed on direct effects. For example, if a pesticide removes alternative prey, predators may emigrate. González and Wilson (1982) gave an example of how minor pests like *Tetranychus* spp. (Acari: Tetranychidae) may provide the food base for predators, in this case *Geocoris* spp. and *Orius* spp. in cotton, while the major pests, such as *Lygus* spp. and *Helicoverpa* spp. (Acari: Tetranychidae), are rarely sufficiently abundant to serve as a prime source of food for predatory arthropods. González and Wilson (1982) suggested assessment of pesticides for impact on arthropods that provide principal

Table 3. Reviews on vegetational diversity in pest control. (Reviews focusing on the role of weeds are not included.)

Authors	Year	Main Emphasis
Southwood & Way	1970	Ecology of pest management
Root	1973	Community ecology
Dempster & Coaker	1974	Applications for pest control
van Emden & Williams	1974	Agroecology
Norton	1975	Economic consideration
Litsinger & Moody	1976	Integrated pest management
Perrin	1977	Agroecology
Trenbath	1977	Modeling
Way	1977	Agroecology
Perrin & Phillips	1978	Population ecology
van Emden	1978	IPM strategy
Perrin	1980	Agroecology
Cromartie	1981	Pest management
Altieri & Letourneau	1982	Agroecology
Andow	1983	Agroecology
Altieri	1983	Habitat management
Kareiva	1983	Community ecology
Mumford & Baliddawa	1983	Cropping systems
Speight	1983	Agroecology
Risch	1983	Cultural pest control
Risch et al.	1983	Community ecology
Altieri & Letourneau	1984	Agroecology
Baliddawa	1985	Cropping systems
Altieri & Liebman	1986	Cropping systems
Herzog & Funderburk	1986	Ecology of cultural control
Sheehan	1986	Natural enemies
Risch	1987	Agroecology
Nordlund et al.	1988	Chemical ecology
Russell	1989	Natural enemies
Vandermeer	1989	Ecology of intercropping
Hokkanen	1990	Trap cropping
van Emden	1990	Natural enemies
Andow	1991a	Community ecology
Andow	1991b	Ecology of pest control
Brust & Stinner	1991	Crop rotation
Cromartie	1991	Pest management
Helenius	1991b	Intercropping and pest control

food for predators. Ables et al. (1978) give an opposite example: *Orius insidiosus* consumed fewer *Helicoverpa* eggs if aphids were present as alternative prey. On the other hand, if natural enemies of predators are suppressed, predators may become enhanced.

Following pesticide applications, recovery of predator communities is slow, as recolonization is possible only after the prey communities are recovered. Species richness and biomass of predators typically remain depressed for markedly longer time than richness and biomass of herbivores, but evenness (equality of species abundances) is less affected (Croft 1990).

An obvious way to increase within-field vegetational diversity is to reduce or exclude herbicide applications. The effects of increased weed diversity on natural control are discussed by Nentwig (this volume). Selective use of herbicides and exclusion of other pesticides from the outermost few meters of field edge of winter wheat increased vegetational diversity and resulted in higher densities of non-target arthropods, including polyphagous predators and their alternative prey (Chiverton & Sotherton 1991; see also S. Wratten, this volume). In case studies reviewed by Sunderland (1987), most (and all the most common) insecticides and herbicides resulted in depletion of spider populations in cereals.

In a four-year experiment comparing insecticide-treated and untreated fields of diverse crop rotation, insecticide use resulted in an average decline of 32 percent in abundance of carabids (Basedow 1990). Another long-term study assessed the effect of the continuous, increasing use of insecticides from 1971 to 1984. Compared to the control fields, carabid populations in the conventionally sprayed fields declined on average by 71 percent, one species disappeared and had not recolonized in 1989, and some key species in the genera *Bembidion* and *Pterostichus* declined by 90 percent or more (Basedow 1990). In a study by Duffield and Baker (1990), a single application of the insecticide dimethoate at the recommended rate resulted in a significant, immediate decline of carabid populations. Cumulative totals of carabids following the application were inversely related to the rate of change of pest aphid numbers, and the spatial pattern in abundance during the recovery (numbers increasing from the edges to the centre) was opposite to the pattern shown by aphids and Collembola (Duffield & Baker 1990). Metcalf (1986) reviewed several examples of development of resurgent and secondary pests in various cropping systems after insecticide applications that suppressed key predators.

Outbreak Frequencies, Economic Thresholds, and Enhancement

The main cause of pest outbreaks in annual systems is excessive colonization, typically a random event. Therefore, the primary goal in enhancement is to buffer the crop against all but rare outbreaks of colonists.

Enhancement of predators will reduce the probability of pests exceeding the economic threshold. Stable enhancement requires temporally stable guilds of generalist predator species that function as a "buffer" to outbreaks of pests. A theoretical model is presented that describes how the percentage of prey killed, $P(\%)$, and the value of the economic threshold (N_c) (expressed as a relative value against the distribution of mean pest colonist abundance), affect the probability of outbreak $p(N_c)$ (Figure 1). The graph shows that the change in $p(N_c)$, as a function of the change in proportion of pests killed, will depend on how low the threshold density is with respect to the mean of the pest abundance distribution. When the economic threshold is low, outbreak is generally more likely, and a small change in the proportion of pests killed by predation can make a large change in the outbreak probability. When the economic threshold is high, outbreak is generally unlikely and it takes a large change in the proportion of pests killed to cause the probability of outbreak to change. In general terms, changing the amount of pest mortality due to predation will have the largest effect in changing outbreak probability when the economic threshold is near or below the average pest abundance as calculated over time or across fields.

The modeling principle outlined above applies to all statistical distributions of pest densities. If the distribution is known and threshold densities are monitored in an IPM program, this approach could be used as a first step in assessing the possibilities for enhancement. If enhancement results in an increase in predation that acts before the monitoring takes place, there is no need to change the threshold. Densities would simply reach the threshold less frequently. In most situations, however, successful enhancement would decrease the frequency of outbreaks, as well as increase the threshold. Figure 1 represents a "minimum" model, as predation is assumed to have no effect on relative growth rate of the pest population while in reality, some suppression is to be expected. The model assumes that there is no negative density dependence in the rate of growth of the remaining pest population that would undermine the effect of predation. Theoretically, intense intraspecific competition would produce such a response. However, intraspecific competition among herbivores is rare (Lawton & Strong 1981) and is likely to occur only at densities far beyond the damage threshold.

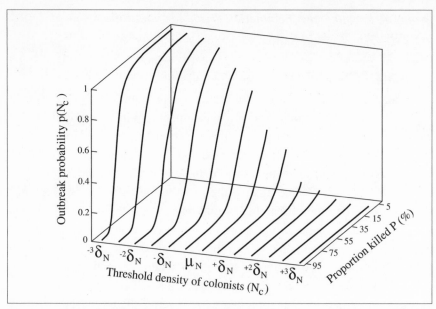

Figure 1. Outbreak probability (p(N_c)) as a function of proportion of colonists killed (P) and relative threshold density of colonists (N_c). The relative threshold density is expressed on a standardized scale of normally distributed colonist densities (N). (m = mean; d = standard deviation). The equation was

$$p(N_c) = f(N_c/(1-P)), \text{ where}$$

f(N_c/(1-P)) = normal probability integral from N_c/(1-P) to + ∞.

Note that for P = 0, p(N_c) = f(N_c) and, for example, when $N_c = m_N$, the probability of outbreak or outbreak frequency is 50%.

Case Study: Aphid Predation in Cereals in Finland

Background

In Finland, spring barley and spring oats are grown near their northern limit. The thermal growing season is ca. 150 days in which time a total of ca. 1200 day-degrees (°C) is accumulated. The major pest of the cereals is bird cherry–oats aphid (*Rhopalosiphum padi* [L.], Homoptera: Aphididae). Due to freezing winter temperatures, the aphid is exclusively holocyclic, i.e., it overwinters as an egg on its winter host, bird cherry. In May through early June, from three-leaf stage to early tillering of spring cereals, alate migrants leave the winter host and colonize wild and cultivated plants of the grass family. From wild grasses the aphids transmit barley yellow dwarf virus (BYDV) to the cereals.

The aphid populations reach outbreak levels at 5–10 year intervals, and since 1928 severe damage has occurred during nine growing seasons. BYDV frequently caused crop losses. Thirty years of intensive research into *R. padi* has provided detailed information on geographic distribution, damage effects, biology, prediction, and natural enemies. Research for developing integrated control of the pest in Finland is summarized in Helenius (1991a). Currently, predictions and recommendations to farmers are based on a sampling scheme that covers the whole cereal-growing area. This centrally operated system has been successful in determining the need for aphicide treatments (Kurppa 1989). However, control of the pest still relies entirely on pesticides, and the monitoring system does not include natural enemies.

Among the specialist enemies, the dominating groups are coccinellids, especially *C. septempunctata*, and syrphid larvae. Hymenopterous parasitoids (Aphidiidae and Aphelinidae) and entomopthoraceous fungi use *R. padi* as a host. The seasonal abundances of the specialists generally follow the within-season development of the host's populations (Rautapää 1976; Leather & Lehti 1982; Helenius 1990a). There is a reproductive numerical response in coccinellids, but with an unavoidable time lag that reduces the control value. However, *C. septempunctata* alone can control *R. padi*, provided that the abundance (predator/prey ratio) and weather-driven activity are high enough during the early establishment phase of the prey population (see, Helenius 1991a). Too little is known about population dynamics of coccinellids to predict their abundance for an IPM scheme, and the unpredictability of weather during the critical phase further complicates the use of coccinellids into an IPM monitoring-system.

The major groups of generalist predators are the epigeal carabids, staphylinids and spiders (Helenius 1990b; Holopainen & Helenius 1992). The species lists (Raatikainen & Huhta 1968; Varis et al. 1984; Niemelä et al. 1987; Helenius et al., in preparation) closely resemble those given for other northern European countries (see above). By experimentally manipulating densities of epigeal predators, aphid infestations were reduced and yields improved. In these studies, *Bembidion* spp. (Carabidae) were indicated as the key group, because of the highest activity-abundance during the colonization and establishment of *R. padi* (Helenius 1990b; see also, Ekbom & Wiktelius 1985; Chiverton 1986, 1987).

The most abundant *Bembidion* species in arable land in Finland are *B. properans* (Stephens), *B. lampros*, *B. guttula* (Fabricius), and *B. quadrimaculatum* (Linnaeus) (e.g., Varis et al. 1984; Helenius et al. in preparation). All these are univoltine spring breeders. Overwintered adults disperse into fields from field margins in spring (Wallin 1987). In May when spring cereals are shown, their densities can be high and peak in early June. Eggs are deposited and the predatory larvae develop in the soil (Mitchell 1963). The first newly emerged adults ("tenerals") are found in July. By September, the new generation has completed its development, and the adults migrate to field margins in search of overwintering sites.

Experiment for Enhancement of *Bembidion*

An experiment was run in 1989–1990 in which the goal was to increase numbers of *Bembidion* spp. by improving their reproductive success through undersowing. In undersowing, either red clover (*Trifolium pratense*), or annual ryegrass (*Lolium multiflorum*) was sown with the spring barley (*Hordeum vulgare*). After harvest of the barley, the undersown crop was allowed to grow until late plowing in October. For details, see Helenius et al. (1995) and Helenius (1995).

A one-hectare block of barley (Viikki Experimental Farm, 61°12¢ N) in the middle of a larger field was used for the experiment. Treatments were assigned to 25m square plots laid out in a 4x4 Latin square design. These were (1) a monocrop of barley (= control); (2) barley undersown with clover and fertilized with supplemental nitrogen at the same rate as the barley monocrop (90kg N/ha); (3) barley undersown with clover and fertilized with reduced amounts of nitrogen (70kg N/ha); and (4) barley undersown with Italian ryegrass (90kg N/ha).

The emergence rates of teneral *Bembidion* spp. were measured by traps of the type described by Helenius (1991a). In both years, the trapping period covered the eclosion period of the beetles (i.e., the first and the last catches from the pitfalls did not contain tenerals).

For statistical analysis of trap data (beetle frequencies), generalized linear modeling (Nelder & Wedderburn 1972) was applied. The data were overdispersed, and the negative-binomial distribution was simulated by scaling (see, Baker et al. 1987: p. 112) in models that were run with Poisson errors (GLIM-package: Payne 1987). For significance testing, chi-square apporoximation was used (see, Baker et al. 1987).

The catch of *Bembidion* spp. from the emergence traps was 88 former generation adults and 96 tenerals in 1989, and 133 adults, and 321 tenerals in 1990. *Bembidion guttula* was the most abundant species (64.8 percent of the adults and 57.3 percent of the tenerals in 1989, and 93.2 percent of the adults, and 98.4 percent of the tenerals in 1990), the other species being *B. lampros*, *B. properans*, *B. quadrimaculatum*, and *B. gilvipes* Sturm.

In both years, the undersowing regime significantly affected the catches of teneral beetles (p<0.01, X^2=13.51, d.f.=3). In the best treatments, the recruitment rates calculated from mean weekly catch rates from the emergence traps were 2.0–2.4 times higher than in the monocrop, and the highest rate was in the ryegrass treatment in 1990, in which 897,000 tenerals/ha were produced (Figure 2). Habitat choice (non-random dispersal among experimental plots) of the parent generation, and thus, rates of egg laying, may explain the differences between treatments in recruitment rates. Alternatively, the patterns in recruitment rates may reflect effects of undersowing on larval survival, possibly due to increased cover or increased food availability, or both.

The undersowing system producing highest numbers of new generation beetles varied markedly between the two years. In 1989, undersowing with clover combined with reduced nitrogen application gave significantly higher recruitment rate than the other treatments (p<0.01, X^2=7.46, d.f.=1). The failure in this treatment the next year may have been due to excessive snail damage to clover combined with poor growth of clover due to drought. In 1990, the other two undersowing regimes, i.e., clover with higher nitrogen level and ryegrass, together gave significantly higher rates than the monocrop of barley (p<0.001, X^2=12.61, d.f.=1). Confounded with the undersowing effect, there was a significant within-field spatial pattern in the emergence rates (Helenius 1995). The seemingly homogenous crop field had a high microsite diversity.

Positive effects of undersowing or weed cover on carabids and predation have been reported earlier (Luff 1987). Our study paralleled the work on increasing predator numbers by increased availability of overwintering sites (see S. Wratten this volume). In our study, however, the availability and distribution of overwintering sites was not considered a limiting factor for the predators' densities. Finnish field sizes are relatively small, with a higher perimeter-to-area ratio than in many other regions of Europe.

The data indicate that numbers of *Bembidion* spp. can be increased by 100 percent or more by undersowing spring cereals with clover or ryegrass. There may be some (but not necessarily any) loss in grain yield due to competition with undersown crop (Kauppila & Helenius unpublished). However, this and the extra costs of seed and labor may be compensated for by reduced rate of soil erosion and nitrogen leaching in autumn, after the harvest of the main crop, and improved soil structure. Further experimentation is needed to evaluate the value of undersowing to pest control. Even if the recruitment is improved, increased predator densities in autumn may not translate into increased densities and/or higher predation rates the next sea-

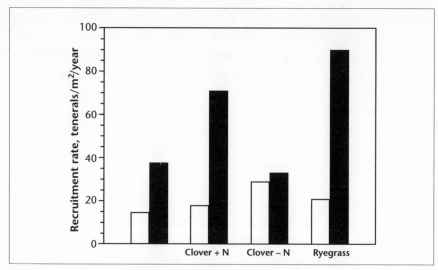

Figure 2. Emergence rates of *Bembidion* spp. beetles as numbers of teneral beetles per m² per season in barley monocultures and in crops of barley undersown with red clover (two nitrogen fertilization levels) or with annual ryegrass. (Values are calculated from mean weekly catch rates). Open bars: 1989; black bars: 1990.

son. In conclusion, the undersowing method shows promise as a way to decrease the outbreak frequency of *R. padi* in spring cereals through enhancement of the natural enemy buffer.

Summary and Conclusions

Stable equilibrium population densities for pest-enemy interactions in annual crops is in most cases unattainable. The spatial and temporal variability and unpredictability of the crop habitats and associated arthropods disrupt the synchrony of predator-prey populations. Perennial systems have experienced a large number of pest reductions through importation of exotic natural enemies, most likely due to the continuity in predator/parasitoid–prey interactions. However, the largely unexplored diversity of generalist predators offers possibilities for use of naturally occurring biological control. In virtually all annual cropping systems large numbers of species of predatory beetles, true bugs, and spiders can be found, and in the tropics, also ants.

In their life cycles or in their population dynamics, generalist predators are not coupled with the pest species the way specialists are. Enhancement of predation should aim at increasing the *predator buffer*. An ideal buffer would be created by stable, high density guilds of predators, capable of rapid aggregation to high density "hot spots" of colonizing pests, and capable of switching to pest species from alternative food. Because of the lack of true regulation of the pest population in such a buffered system, the objective would be to minimize the pests' numbers and even to drive their within-field populations to extinction. Such a predator buffer would reduce the frequency of outbreaks, rather than eliminate the possibility of one.

Management of within-field species diversity (floral and faunal) and spatio-temporal patterning can be used to influence predator numbers and predation rates. Exactly how the diversity should be manipulated depends on the cropping system and crop in question. For example, increasing veg-etational diversity *per se* is not useful and may be counterproductive. Tai-loring to each system is required, but this does not mean that only manage-ment-intensive and skill-demanding options remain. Simple solutions may well be found, as shown by the example of restoration of the natural enemy buffer in wheat by reducing pesticide applications (Chiverton and Sotherton 1991). The case study presented here on epigeal predators of cereal aphids illustrates the possibility of enhancing natural control by a simple intercrop-ping system.

In present day agriculture, motivation to enhance natural control is based primarily on environmental and, increasingly, on political arguments. In many cases, the practical aim is to reduce the need for insecticide appli-cations. Along with other options, increasing within-field diversity may in many cases prove useful. However, future work on natural control should include the technological and socioeconomic constraints to implementation of new control tactics, in addition to understanding agroecological relation-ships.

Acknowledgments

I thank the editors of this book, and also Mr. Steve Schoenig and Professor Michael Way for constructive criticism on the manuscript. The study was funded by the Finnish National Research Council for Agriculture and For-estry (grants no. 1011093 and no. 1011619). I wrote the first draft during my visit to Silwood Centre for Pest Management, Imperial College at Silwood Park in England, and the final version during my stay at South Savo Research Station, Agricultural Research Centre in Mikkeli, Finland.

References

Ables, J. R., S. L. Jones & D. W. McCommas, Jr. 1978. Response of selected predator species to different densities of *Aphis gossypii* and *Heliothis virescens* eggs. Environ. Entomol. 7:402-404.

Allen, R. T. 1979. The occurrence and importance of ground beetles in agricultural and surrounding habitats, pp. 485-507. *In* T. L. Ervin, G. E. Ball & D. R. Whitehead [eds.], Carabid beetles: their evolution, natural history and classification. Dr. W. Junk, The Hague.

Altieri, M. A. 1983. Vegetational designs for insect-habitat management. Environ. Management 7:3-7.

Altieri, M. A. & D. K. Letourneau. 1982. Vegetation management and biological control in agroecosystems. Crop Protection 1:405-430.

1984. Vegetation diversity and insect pest outbreaks. CRC Critical Rev. Plant Sci. 2:131-169.

Altieri, M. A & M. Liebman. 1986. Insect, weed, and plant disease management in multiple cropping systems, pp. 183-218. *In* C. A. Francis [ed.], Multiple cropping systems. MacMillan, New York.

Andow, D. 1983. The effect of agricultural diversity on insect populations, pp. 91-116. *In* W. Lockeretz [ed.], Environmentally sound agriculture. Proceedings, the Fourth IFOAM International Scientific Conference. Praeger, New York.

Andow, D. A. 1991a. Vegetational diversity and arthropod population response. Annu. Rev. Entomol. 36:561-586.

1991b. Control of arthropods using crop diversity, pp. 257-284. *In* D. Pimentel [ed.], CRC handbook of pest management in agriculture, 2nd ed. CRC Press, Boca Raton.

Anon. 1991. FAO Yearbook 1990, 44. FAO Statistics Series 99. Food and Agriculture Organization, Rome.

Baker, R. J., M. R. B. Clarke, B. Francis, M. Green, J. A. Nelder, C. D. Payne, R. A. Reese & J. Webb. 1987. User's guide. *In* C. D. Payne [ed.], The GLIM system release 3.77 manual, 2nd ed. Royal Statistical Society, Oxford, UK.

Baliddawa, C. W. 1985. Plant species diversity and crop pest control. An analytical review. Insect Sci. Appl. 6:479-487.

Barber, G. W. 1936. *Orius insidiosus* (Say), an important natural enemy of the corn earworm. United States Department of Agriculture Technical Bulletin 504.

Barbosa, P. 1988. Natural enemies and herbivore-plant interactions: influence of plant allelochemicals and host specificity, pp. 201-229. *In* P. Barbosa & D. K. Letourneau [eds.], Novel aspects of insect-plant interactions. Wiley, New York.

Barnes, H. F. 1956. Gall midges of cereal crops. Gall midges of economic importance, vol. 7. Crosby Lockwood & Son, London.

Basedow, T. 1973. Der Einfluss epigäischer Raubarthropoden auf die Abundanz phytophagen Insekten in der Agrarlandschaft. Pedobiologia 13:410-422.

1989. Polyphagous predators (mainly Col., Carabidae) controlling cereal aphids (Hom., Aphididae) on winter barley during summer. SROP/WPRS Bulletin 12:54-62.

1990. Effects of insecticides on Carabidae and the significance of these effects for agriculture and species number, pp. 115-125. *In* N. E. Stork [ed.], The role of ground beetles in ecological and environmental studies. Intercept, Andover, U.K.

Beddington, J. R., C. A. Free & J. H. Lawton. 1978. Characteristics of successful natural enemies in models of biological control of insect pests. Nature 273:513-519.

Breene, R. G., W. L. Sterling & D. A. Dean. 1989. Predators of the cotton fleahopper on cotton. Southw. Entomol. 14:159-166.

Brust, G. E. & B. R. Stinner. 1991. Crop rotation for insect, plant pathogen, and weed control, pp. 217-236. *In* D. Pimentel [ed.], CRC handbook of pest management in agriculture, 2nd ed. CRC Press, Boca Raton, Florida.

Bryan, K. M. & S. D. Wratten. 1984. The responses of polyphagous predators to prey spatial heterogeneity: aggregation by carabid and staphylinid beetles to their cereal aphid prey. Ecol. Entomol. 9:251-259.

Bugg, R. L., L. E. Ehler & L. T. Wilson. 1987. Effect of common knotweed (*Polygonum aviculare*) on abundance and efficiency of insect predators of crop pests. Hilgardia 55:(7):1-52.

Bugg, R. L. & L. T. Wilson. 1989. *Amni visnaga* (L.) Lamarck (Apiaceae): associated beneficial insects and implications for biological control, with emphasis on the bell-pepper agroecosystem. Biol. Agric. Hortic. 6:241-268.

Carroll, C. R. & S. Risch. 1990. An evaluation of ants as possible candidates for biological control in tropical annual agroecosystems, pp. 30-46. *In* S. R. Gliessman [ed.], Agroecology: researching the ecological basis for sustainable agriculture. Springer Verlag, New York.

Chiverton, P. A. 1986. Predator density manipulation and its effects on populations of *Rhopalosiphum padi* (Hom.: Aphididae) in spring barley. Ann. Appl. Biol. 109:49-60.

1987. Predation of *Rhopalosiphum padi* (Homoptera: Aphididae) by polyphagous predatory arthropods during the aphids' pre-peak period in spring barley. Ann. Appl. Biol. 111:257-269.

Chiverton, P. A. & N. W. Sotherton. 1991. The effects on beneficial arthropods of the exclusion of herbicides from cereal crop edges. J. Appl. Ecol. 28:1027-1039.

Coaker, T. H. & D. A. Williams. 1963. The importance of some Carabidae and Staphylinidae as predators of the cabbage root fly, *Erioischia brassicae* (Bouche). Entomol. Exp. Appl. 6:156-164.

Coombes, D. S. & N. W. Sotherton. 1986. The dispersal and distribution of polyphagous predatory Coleoptera in cereals. Ann. Appl. Biol. 108:461-474.

Corbett, A. 1993. The role of movement in the response of natural enemies to agroecosystem diversification: a theoretical evaluation. Environ. Entomol. 22:519-531.

Crawley, M. J. 1975. The numerical responses of insect predators to changes in prey density. J. Animal Ecol. 44:877-892.

Crocker, R. L. & W. H. Whitcomb. 1980. Feeding niches of the big-eyed bugs *Geocoris bullatus, G. punctipes,* and *G. uliginosus* (Hemiptera: Lygaeidae: Geocorinae). Environ. Entomol. 9:508-513.

Croft, B. A. 1990. Arthropod biological control agents and pesticides. Wiley, New York.

Cromartie, W. J., Jr. 1981. The environmental control of insects using crop diversity, pp. 223-251. *In* D. Pimentel [ed.], CRC handbook of pest management in agriculture. Volume 1. CRC Press, Boca Raton, Florida.

1991. The environmental control of insects using crop diversity, pp. 183-216. *In* D. Pimentel [ed.], CRC handbook of pest management in agriculture, 2nd ed. Volume 1. CRC Press, Boca Raton, Florida.

Dempster, J. P. & T. H. Coaker. 1974. Diversification of crop ecosystems as a means of controlling pests, pp. 106- 114. *In* D. Price Jones & M. E. Solomon [eds.], Biology in pest and disease control. Oxford Press. London and New York.

Den Boer, P. J. 1981. On the survival of populations in a heterogeneous and variable environment. Oecologia 67:322-330.

1990. The survival value of dispersal in terrestrial arthropods. Biol. Conservation 54:175-192.

Duffield, S. J. & S. E. Baker. 1990. Spatial and temporal effects of dimethoate use on populations of Carabidae and their prey in winter wheat, pp. 95-104. *In* N. E. Stork [ed.], The role of ground beetles in ecological and environmental studies. Intercept, Andover, U.K.

Edwards, C. A., K. D. Sunderland & K. S. George 1979. Studies on polyphagous predators of cereal aphids. J. Appl. Ecol. 16:811-823.

Ehler, L. E. 1977. Natural enemies of cabbage looper on cotton in the San Joaquin Valley. Hilgardia 45(3):73-106.

Ehler, L. & R. Hall. 1982. Evidence for competitive exclusion of introduced natural enemies in biological control. Environ. Entomol. 11:1-4.

Ehler, L. E. & J. C. Miller. 1978. Biological control in temporary agroecosystems. Entomophaga 23:207-212.

Ehler, L. E. & R. van den Bosch. 1974. An analysis of the natural biological control of *Trichoplusia ni* (Lepidoptera: Noctuidae) on cotton in California. Can. Entomol. 106:1067-1073.

Ekbom, B. S. & S. Wiktelius. 1985. Polyphagous arthropod predators in cereal crops in central Sweden 1979-1982. Zeitschrift für Angewandte Entomologie 99:433-442.

Feeny, P. P. 1976. Plant apparency and chemical defense. Recent Adv. Phytochem. 10:1-40.

Finch, S. 1988. Entomology of crucifers and agriculture—Diversification of the agroecosystem in relation to pest damage in cruciferous crops, pp. 39-71. *In* M. K. Harris & C. E. Rogers [eds.], The entomology of indigenous and naturalized systems in agriculture. Westview Press, Boulder & London.

Godfret, K. E., W. H. Whitcomb & J. L. Stimal. 1989. Arthropod predators of velvetbean caterpillar, *Anticarsia gemmatalis* Hübner (Lepidoptera: Noctuidae) eggs and larvae. Environ. Entomol. 18:118-123.

González, D. & L. T. Wilson 1982. A food-web approach to economic thresholds: a sequence of pests/predaceous arthropods on California cotton. Entomophaga 27:31-43.

Gravesen, E. & S. Toft. 1987. Grass fields as reservoirs for polyphagous predators (Arthropoda) of aphids (Homopt., Aphididae). J. Appl. Entomol. 104:461-473.

Hagen, K. S. 1987. Nutritional ecology of terrestrial insect predators, pp. 533-577. *In* F. Slansky & J. G. Rodriguez [eds.], Nutritional ecology of insects, mites, spiders and related invertebrates. Wiley, New York.

Hagen, K. S., S. Bombosch & J. A. McMurtry. 1976. The biology and impact of predators, pp. 93-142. *In* C. B. Huffaker & P. S. Messenger [eds.], Theory and practice of biological control. Academic Press, New York.

Harrison, S. 1991. Local extinction in a metapopulation context: an empirical evaluation. Biol. J. Linnean Soc. 42:73-88.

Hassell, M. P. 1978. The dynamics of arthropod predator-prey systems. Princeton University Press, Princeton.

Helenius, J. 1990a. Incidence of specialist natural enemies of *Rhopalosiphum padi* (L.) (Hom., Aphididae) on oats in the monocrops and mixed intercrops with faba bean. J. Appl. Entomol. 109:136-143.

1990b. Effect of epigeal predators on infestation by the aphid *Rhopalosiphum padi* and on grain yield of oats in monocrops and mixed intercrops. Entomol. Exp. Appl. 54:225-236.

1991a. Integrated control of *Rhopalosiphum padi,* and the role of epigeal predators in Finland. IOBC/WPRS Bulletin 14:123-130.

1991b. Insect numbers and pest damage in intercrops vs. monocrops: concepts and evidence from a systems of faba bean, oats and *Rhopalosiphum padi* (Homoptera, Aphididae). J. Sustainable Agric. 1:57-80.

Helenius, J. 1995. Rate and local scale spatial pattern of adult emergence of the generalist predator *Bembidion guttula* in an agricultural field. Acta Jutlandica 70:101-111.

Helenius, J., J. Holopainen, M. Muhojoki, P. Pokki, T. Tolonen & A. Venäläinen. 1995. Effect of undersowing and green manuring on abundance of ground beetles (Coleoptera, Carabidae) in cereals. Acta Zool. Fennica 196:156-159.

Herzog, D. C. & J. E. Funderburk. 1986. Ecological basis for habitat management and pest cultural control, pp. 217-250. *In* M. Kogan [ed.], Ecological theory and integrated pest management practice. Wiley, New York.

Hodek, I. 1973. Biology of Coccinellidae. Junk Publishers, The Hague.

Hokkanen, H. M. T. 1990. Trap cropping in pest management. Annu. Rev. Entomol. 36:119-138.

Holopainen, J. K. & J. Helenius. 1992. Pitfall catch and diet of ground beetles (Coleoptera, Carabidae) during an outbreak of *Rhopalosiphum padi* (Hom., Aphididae) in a spring barley field in Finland. Acta Agricultura Scandinavica 42:57-61.

Kagan, M. 1943. The Araneida found on cotton in central Texas. Ann. Entomol. Soc. Am. 36:257-258.

Kareiva, P. 1983. Influence of vegetation texture on herbivore populations: resource concentration and herbivore movement, pp. 259-289. *In* R. F. Denno & M. S. McClure [eds.], Variable plants and herbivores in natural and managed ecosystems. Academic Press, New York.

Kareiva, P. M. 1984. Predator-prey dynamics in spatially structured populations: manipulating dispersal in a coccinellid-aphid interaction. Lecture Notes in Biomathematics 54:368-389.

Kareiva, P. 1986a. Trivial movement and foraging by crop colonizers, pp. 59-82. In M. Kogan [ed.], Ecological theory and integrated pest management practice. Wiley, New York.

1986b. Patchiness, dispersal, and species interactions: consequences for communities of herbivorous insects, pp. 192-206. In J. Diamond & T. J. Case [eds.], Community ecology. Harper & Row, New York.

1990a. Establishing a foothold for theory in biocontrol practice: using models to guide experimental design and release protocols, pp. 65-81. In R. R. Baker & P. E. Dunn [eds.], New directions in biological control: alternatives for suppressing agricultural pests and diseases. Alan R. Liss, Inc., New York.

1990b. The spatial dimension in pest-enemy interactions, pp. 213-227. In M. Mackauer, L. E. Ehler & J. Roland [eds.], Critical issues in biological control. Intercept, Andover, U.K.

Kiritani, K., S. Kawahara, T. Sasaba & F. Nakasuji. 1972. Quantitative evaluation of predation by spiders on the green rice leafhopper, Nephotettix cincticeps Uhler, by a sight-count method. Res. Pop. Ecol., Kyoto University 13:187-200.

Kiritani, K. & N. Kakiya. 1975. An analysis of the predator-prey system in the paddy field. Res. Pop. Ecol., Kyoto University 17:29-38.

Knutson, A. E. & F. E. Gilstrap. 1989. Direct evaluation of natural enemies of the southwestern corn borer (Lepidoptera: Pyralidae) in Texas corn. Environ. Entomol. 18:732-739.

Kurppa, S. 1989. Predicting outbreaks of Rhopalosiphum padi in Finland. Ann. Agric. Fenniae 28:333-347.

Lawton, J. H. & D. R. Strong. 1981. Community patterns and competition in folivorous insects. Am. Nat. 118:317-338.

Leather, S. R. & J. P. Lehti. 1982. Field studies on the factors affecting the population dynamics of the bird cherry–oat aphid, Rhopalosiphum padi (L.) in Finland. Annales Agriculturae Fenniae 21:20-31.

Leigh, T. F. & D. González. 1976. Field cage evaluation of predators for control of Lygus hesperus Knight on cotton. Environ. Entomol. 5:948-952.

Letourneau, D. K. 1990. Two examples of natural enemy augmentation: a consequence of crop diversification, pp. 11-29. In S. R. Gliessman [ed.], Agroecology. Researching the ecological basis for sustainable agriculture. Springer-Verlag, New York.

Letourneau, D. K. & M. A. Altieri. 1983. Abundance patterns of a predator, Orius tristicolor (Hemiptera: Anthocoridae), and its prey, Frankliniella occidentalis (Thysanoptera: Thripidae): habitat attraction in polycultures versus monocultures. Environ. Entomol. 12:1464-1469.

Levins, R. 1970. Extinction. Lectures Math. Life Sci. 2:75-107.

Liss, W. J., L. J. Gut, P. H. Westigard & C. E. Warren. 1986. Perspectives on arthropod community structure, organization, and development in agricultural crops. Annu. Rev. Entomol. 31:455-478.

HELENIUS 155

Litsinger, J. A. & K. Moody. 1976. Integrated pest management in multiple cropping systems, pp. 293-316. *In* R. I. Papendick, P. A. Sanchez & G. P. Triplett [eds.], Multiple cropping. American Society for Agronomy Special Publications 27.

Lövei, G.L. & K.D. Sunderland. 1996. Ecology and behavior of ground beetles (Coleoptera: Carabidae). Annu. Rev. Entomol. 41:231-256.

Luff, M. L. 1983. The potential of predators for pest control. Agric. Ecosyst. Environ. 10:159-181.

1987. Biology of polyphagous ground beetles in agriculture. Agric. Zool. Rev. 2:237-278. Intercept, Andover, U.K.

1990. Spatial and temporal stability of carabid communities in a grass/arable mosaic, pp. 191-200. *In* N. E. Stork [ed.], The role of ground beetles in ecological and environmental studies. Intercept, Andover.

Mader, H. J. [ed.] 1990. Survival and dispersal of animals in cultivated landscapes. Biol. Conservation 54:167-290.

May, R. M. & M. P. Hassell. 1988. Population dynamics and biological control. Philos. Trans. R. Soc., London 318:129-169.

Matteson, P. C. 1982. The effects of intercropping with cereals and minimal permethrin applications on insect pests of cowpea and their natural enemies in Nigeria. Trop. Pest Manage. 28:372-380.

McMurtry, J. A. & J. G. Rodriguez. 1987. Nutritional ecology of phytoseiid mites, pp. 609-644. *In* F. Slansky & J. G. Rodriguez [eds.], Nutritional ecology of insects, mites, spiders and related invertebrates. Wiley, New York.

Metcalf, R. L. 1986. The ecology of insecticides and the chemical control of insects, pp. 251-297. *In* M. Kogan [ed.], Ecological theory and integrated pest management practice. Wiley, New York.

Minks, A. K. & P. Harrewijn [ed.] 1988. Aphids, their biology, natural enemies and control, 2B. Elsevier, Amsterdam.

Mitchell, B. 1963. Ecology of two carabid beetles, *Bembidion lampros* (Herbst) and *Trechus quadristriatus* (Schrank), I. Life cycles and feeding behaviour. J. Animal Ecol. 32:289-299.

Mols, P. J. M. 1987. Hunger in relation to searching behaviour, predation and egg production of the carabid beetle *Pterostichus coerulescens* L.: results of simulation. Acta Phytopathologica et Entomologica Hungarica 22:187-296.

Mowat, D. J. & S. J. Martin. 1981. The contribution of predatory beetles (Coleoptera: Carabidae and Staphylinidae) and seed-bed-applied insecticide to the control of cabbage root fly, *Delia brassicae* (Wied.), in transplanted cauliflowers. Hortic. Res. 21:127-136.

Mumford, J. D. & C. W. Baliddawa. 1983. Factors affecting pest occurrence in various cropping systems. J. Insect Sci. Appl. 4:59-64.

Mumford, J. D. & G. A. Norton. 1984. Economics of decision making in pest management. Annu. Rev. Entomol. 29:157-174.

Murdoch, W. W. 1969. Switching in general predators: experiments on predator specificity and stability of prey populations. Ecol. Monogr. 39:335-354.

1975. Diversity, complexity, stability, and pest control. J. Appl. Ecol. 12:795-807.

1990. Ecological theory and biological control, pp. 49-55. *In* R. R. Baker & P. E. Dunn [eds.] New directions in biological control: alternatives for suppressing agricultural pests and diseases. Alan R. Liss, Inc., New York.

Murdoch, W. W., J. Chesson & P. L. Chesson. 1985. Biological control in theory and practice. Am. Nat. 125:344-366.

Murdoch, W. W. & A. Oaten. 1975. Predation and population stability. Adv. Ecol. Res. 9:1-131.

Nelder, J. A. & R. W. M. Wedderburn. 1972. Generalized linear models. J. R. Statistical Soc., Series A, 135:370-384.

Niemelä, J., Y. Haila & E. Halme. 1987. Carabid assemblages in Southern Finland, a forest-field comparison. Acta Phytopathologica et Entomologica Hungarica 22:417-424.

Nordlund, D. A., W. J. Lewis & M. A. Altieri. 1988. Influences of plant-produced allelochemicals on the host/prey selection behaviour of entomophagous insects, pp. 65-90. *In* P. Barbosa & D. K. Letourneau [eds.], Novel aspects of insect-plant interactions. Wiley, New York.

Norton, G. A. 1975. Multiple cropping and pest control. An economic perspective. Mededelingen van de Faculteit Landbouwwetenschappen, Rijksuniversiteit Gent, 40:219-228.

1982. A decision-analysis approach to integrated pest control. Crop Protection 1:147-164.

1993. Philosophy, concepts and techniques, pp 1-21. *In* G.A. Norton & J.D. Mumford [eds.], Decision tools for pest management. Wallingford, Oxon, UK.

Norton, G. A. & K. L. Heong. 1988. An approach to improving pest management: rice in Malaysia. Crop Protection 7:84-90.

Nyffeler, M. & G. Benz. 1981. Ökologische Bedeutung der Spinnen als Insekten-prädatoren in Wiesen und Getreidefeldern. Mitteilungen der Deutschen Gesellschaft für Allgemeine und Angewandte Entomologie 3:33-35.

1988. Prey and predatory importance of micryphantid spiders in winter wheat fields and hay meadows. J. Appl. Entomol. 105:190-197.

Nyffeler, M., D. A. Dean & W. L. Sterling. 1988. Prey records of the web-building spiders *Dictyna segregata* (Dictynidae), *Theridion australe* (Theridiidae), *Tidarren haemorrhoidale* (Theridiidae), and *Frontinella pyramitela* (Linyphiidae) in a cotton agroecosystem. Southw. Nat. 33:215-218.

O'Neill, R. J. 1989. Comparison of laboratory and field measurements of the functional response of *Podisus maculiventris* (Heteroptera: Pentatomidae). J. Kansas Entomol. Soc. 62:148-155.

Payne, C. D. [ed.] 1987. The GLIM system release 3.77 manual, 2nd ed. Royal Statistical Society, Oxford.

Perrin, R. M. 1977. Pest management in multiple cropping systems. Agro-Ecosyst. 3:93-118.

1980. The role of environmental diversity in crop protection. Protection Ecol. 2:77-114.

Perrin, R. M. & M. L. Phillips. 1978. Some effects of mixed cropping on the population dynamics of insect pests. Entomol. Exp. Appl. 24:585-593.

Peter, C. 1988. New records of natural enemies associated with the brown plant-hopper, *Nilaparvata lugens* (Stal). Curr. Sci. 57:1087-1088.

Pimm, S. C. 1984. The complexity and stability of ecosystems. Nature 307:321-326.

Potts, G. R. & G. P. Vickerman. 1974. Studies on the cereal ecosystem. Adv. Ecol. Res. 8:107-197.

Price, P. W. 1976. Colonization of crops by arthropods: non-equilibrium communities in soybean fields. Environ. Entomol. 5:605-611.

1984. Insect ecology. Wiley, New York.

1987. The role of natural enemies in insect populations, pp. 287-312. In P. Barbosa & J. C. Shultz [eds.], Insect outbreaks. Academic Press, San Diego.

Price, P. W., C. E. Bouton, P. Gross, B. A. McPherson, J. N. Thompson & A. E. Weise. 1980. Interactions among three trophic levels: influence of plants on interactions between insect herbivores and natural enemies. Annu. Rev. Ecol. Syst. 11:4-65.

Raatikainen, M. & V. Huhta. 1968. On the spider fauna of Finnish oat fields. Annales Zoologici Fennici 5:254-261.

Rabb, R. L. 1971. Naturally-occurring biological control in the Eastern United States, with particular reference to tobacco insects, pp. 294-311. In C. B. Huffaker [ed.], Biological control. Plenum, New York.

Rautapää, J. 1976. Population dynamics of cereal aphids and method of predicting population trends. Annales Agriculturae Fenniae 11:424-436.

Reed, T., M. Shepard & S. G. Turnipseed. 1984. Assessment of the impact of arthropod predators on noctuid larvae in cages in soybean fields. Environ. Entomol. 13:954-961.

Rhoades, D. F. & R. G. Cates. 1976. Toward a general theory of plant antiherbivore chemistry. Recent Adv. Phytochem. 10:168-213.

Riechert, S. E. & L. Bishop. 1990. Prey control by an assemblage of generalist predators: spiders in garden test systems. Ecology 71:1441-1450.

Riechert, S. E. & J. M. Harp. 1987. Nutritional ecology of spiders, pp. 645-672. In F. Slansky & J. G. Rodriguez [eds.], Nutritional ecology of insects, mites, spiders and related invertebrates. Wiley, New York.

Riechert, S. E. & T. Lockley. 1984. Spiders as biological control agents. Annu. Rev. Entomol. 29:299-320.

Risch, S. J. 1983. Intercropping as cultural pest control: prospects and limitations. Environ. Management 7:9-14.

1987. Agricultural ecology and insect outbreaks, pp. 217-238. In P. Barbosa & J. C. Schultz [eds.], Insect outbreaks. Academic Press, San Diego.

Risch, S. J., D. Andow & M. A. Altieri. 1983. Agroecosystem diversity and pest control: data, tentative conclusions, and new research directions. Environ. Entomol. 12:625-629.

Risch, S. J. & C. R. Carroll. 1982. The ecological role of ants in two Mexican agroecosystems. Oecologia 55:114-119.

Risch, S. J., R. Wrubel & D. Andow. 1982. Foraging by a predaceous beetle, Coleomegilla maculata (Coleoptera: Coccinellidae), in a polyculture: effects of plant density and diversity. Environ. Entomol. 11:949-950.

Root, R. B. 1973. Organization of a plant-arthropod association in simple and diverse habitats: the fauna of collards (*Brassica oleracea*). Ecol. Monogr. 43:95-124.

Russell, E. P. 1989. Enemies hypothesis: a review of the effect of vegetational diversity on predatory insects and parasitoids. Environ. Entomol. 18:590-599.

Salas-Aguilar, J. & L. E. Ehler. 1977. Feeding habits of *Orius tristicolor*. Ann. Entomol. Soc. Am. 70:60- 62.

Samal, P. & B. C. Misra. 1975. Spiders: the most effective natural enemies of the brown planthopper in rice. Rice Entomol. Newsletter 3:31.

1978. *Casnoidea indica* (Thunb.) a carabid ground beetle predating on brown plant hopper, *Nilaparvata lugens* (Stal) of rice. Curr. Sci. 47:688-689.

Sasaba, T., K. Kiritani & T. Urabe. 1973. A preliminary model to simulate the effect of insecticides on a spider-leafhopper system in the paddy field. Res. Pop. Ecol., Kyoto University 15:9-22.

Scheller, H. V. 1984. The role of ground beetles (Carabidae) as predators on early populations of cereal aphids in spring barley. Zeitschrift für Angewandte Entomologie 97:451-463.

Scherney, F. 1959. Der biologische Wirkungseffekt von Carabiden der Gattung Carabus auf Kartoffelkäferlarven, Verhandlungen 4. Internationale Pflanzenschutz Kongress, Hamburg 1957, 1:1035-1038.

1960. Beiträge zur Biologie und ökonomischen Bedeutung räuberish lebender Käferarten. Untersuchungen über das Auftreten von Laufkäfern (Carabidae) in Feldkulturen (Teil II). Zeitschrift für Angewandte Entomologie 47:231-255.

Sharma, V. K. & P. Sarup. 1980. Predatory role of spiders in the integrated control of the maize stalk borer, *Chilo partellus* (Swinhoe). Indian J. Entomol. 42:229-231.

Sheehan, W. 1986. Response by specialist and generalist natural enemies to agroecosystem diversification: a selective review. Environ. Entomol. 15:456-461.

Shepard, M., G. R. Carner & S. G. Turnipseed. 1974. Seasonal abundance of predaceous arthropods in soybeans. Environ. Entomol. 3:985-988.

Singh, B., G.S. Battu & A.S. Atwal. 1975. Studies on the spider predators of the maize borer *Chilo partellus* (Swinhoe) in the Punjab. Indian J. Entomol. 37:72-76.

Sotherton, N. W. 1984. The distribution and abundance of predatory arthropods overwintering on farmland. Ann. Appl. Biol. 105:423-429.

1985. The distribution and abundance of predatory arthropods overwintering in field boundaries. Ann. Appl. Biol. 106:17-21.

Southwood, T. R. E. 1962. Migration of terrestrial arthropods in relation to habitat. Biol. Rev. 37:171-214.

Southwood, T. R. E & H. N. Comins. 1976. A synoptic population model. J. Animal Ecol. 45:949-965.

Southwood, T. R. E. & M. J. Way. 1970. Ecological background to pest management, pp. 6-29. *In* R. L. Rabb & F. E. Guthrie [eds.], Concepts of pest management. Proceedings of a conference held March 25-27 1970 at North Carolina State University, Raleigh. North Carolina State University print shop, Raleigh, NC.

Speight, M. R. 1983. The potential of ecosystem management for pest control. Agric. Ecosyst. Environ. 10:183-199.

Stork, N. E. 1990. The Role of ground beetles in ecological and environmental studies. Intercept, Andover.

Sturm, M. M. & W. L. Sterling. 1990. Geographical patterns of boll weevil mortality: observations and hypothesis. Environ. Entomol. 19:59-65.

Sunderland, K. D. 1975. The diet of some predatory arthropods in cereal crops. J. Appl. Ecol. 12:507-515.

1987. Spiders and cereal aphids in Europe. Section Regionale Ouest Palearctique (SROP)/West Palaearctic Regional Section (WPRS) Bulletin 10:82-102.

1988. Quantitative methods for detecting invertebrate predation occurring in the field. Ann. Appl. Biol. 112:201-224.

Sunderland, K. D., N. E. Crook, D. L. Stacey & B. J. Fuller. 1987. A study of feeding by polyphagous predators on cereal aphids using ELISA and gut dissection. J. Appl. Ecol. 24:907-933.

Sunderland, K. D., A. M. Fraser & A. F. G. Dixon. 1986. Field and laboratory studies on money spiders (Linyphiidae) as predators of cereal aphids. J. Appl. Ecol. 23:433-447.

Tahvanainen, J. O. & R. B. Root. 1972. The influence of vegetational diversity on the population ecology of a specialized herbivore *Phyllotreta cruciferae* (Col.: Chrysomelidae). Oecologia 10:321-346.

Thomas, M. J., K. Balakrishna Pillai, K. V. Mammen & N. R. Nair. 1979. Spiders check plant hopper population. Int. Rice Res. Newsletter 4:18-19.

Thiele, H. U. 1977. Carabid beetles in their environments. Springer Verlag, Berlin.

Trenbath, B. R. 1977. Interactions among diverse hosts and diverse parasites. Ann. New York Acad. Sci. 287:124-150.

Turnbull, A. L. & D. A. Chant. 1961. The practice and theory of biological control of insects in Canada. Can. J. Zool. 39:697-753.

Van den Bosch, R. & K. S. Hagen. 1966. Predaceous and parasitic arthropods in California cotton fields. Bull. Univ. of Calif. Agric. Exp. Stn. 820:1-27.

Vandermeer, J. 1989. The ecology of intercropping. Cambridge University Press, Cambridge.

van Emden, H. F. 1978. Insect pest management in multiple cropping systems—a strategy, pp. 325-343. *In* Symposium, Cropping Systems Research and Development for the Asian Rice Farmer. International Rice Research Institute, Los Banos, Laguna, Philippines.

1990. Plant diversity and natural enemy efficiency in agroecosystems, pp. 63-80. *In* M. Mackauer, L. E. Ehler & J. Roland [eds.], Critical issues in biological control. Intercept, Andover, U.K.

van Emden, H. F. & G. F. Williams. 1974. Insect stability and diversity in agro-ecosystems. Annu. Rev. Entomol. 19:455-475.

van Vreden, G. & A. L. Ahmadzabidi. 1986. Pests of rice and their natural enemies in peninsular Malaysia. Pudoc, Wageningen.

Varis, A.-L., J. K. Holopainen & M. Koponen. 1984. Abundance and seasonal occurrence of adult Carabidae (Coleoptera) in cabbage, sugar beet and timothy fields in southern Finland. Zeitschrift für Angewandte Entomol. 98:62-73.

Vickerman, G. P. & K. D. Sunderland. 1975. Arthropods in cereal crops: nocturnal activity, vertical distribution and aphid predation. J. Appl. Ecol. 12:755-766.

Wallin, H. 1987. Dispersal and migration of carabid beetles inhabiting cereal fields. Acta Phytopathologica et Entomologica Hungarica 22:449-453.

Wallin, H. & B. S. Ekbom. 1988. Movements of carabid beetles (Coleoptera, Carabidae) inhabiting cereal fields: a field tracing study. Oecologia 77:39-43.

Way, M. J. 1963. Mutualism between ants and honeydew producing Homoptera. Annu. Rev. Entomol. 8:307-344.

1977. Pest and disease status in mixed stands vs. monocultures; the relevance of ecosystem stability, pp. 127-138. In J. M. Cherrett & G. R. Sagar [eds.], Origins of pest, parasite, disease and weed problems. Blackwell, Oxford.

1992. Role of ants in pest management. Annu. Rev. Entomol. 37:479-503.

Wetzler, R. E. & S. J. Risch. 1984. Experimental studies of beetle diffusion in simple and complex crop habitats. J. Ani. Ecol. 53:1-19.

Whitcomb, W. A. & K. Bell. 1964. Predaceous insects, spiders and mites of Arkansas cotton fields. Bull. Univ. of Arkansas Agric. Exp. Stn. 690.

Whitcomb, W. H., H. Exline & R. C. Hunter. 1963. Spiders of the Arkansas cotton field. Ann. Entomol. Soc. Am. 56:653-660.

Winder, L. 1990. Predation of the cereal aphid *Sitobion avenae* by polyphagous predators on the ground. Ecol. Entomol. 15:105-110.

Wratten, S. D. 1987. The effectiveness of native natural enemies, pp. 88-112. In A. V. Burn, T. H. Coaker & P. C. Jepson [eds.], Integrated pest management. Academic, London.

Wu, Y., Y. Li & D. Jiang. 1981. Integrated control of cotton pests in Nanyang region. Acta Entomologica Sinica 24:34-41.

Young, O. P. & G. B. Edwards. 1990. Spiders in United States field crops and their potential effect on crop pests. J. Arach. 18:1-27.

Natural Regulation at the Farm Level

Fritz J. Häni—Swiss College of Agriculture, CH-3052 Zollikofen, Switzerland
Ernst F. Boller—Swiss Federal Research Station for Arboriculture, Viticulture and Horticulture, CH-8820 Wädenswil, Switzerland
Siegfried Keller—Swiss Federal Research Station of Agronomy, CH-8046 Zürich-Reckenholz, Switzerland

Introduction

Other chapters in this volume emphasize individual aspects and possibilities of enhancing biological control of arthropod pests. Here we consider the farming system as a whole. A farmer cannot focus on one individual aspect of natural regulation, e.g., the control of a single pest species, because of the need to integrate many different economic and ecological goals. For example, soil tillage has to be taken into account at the same time as seed bed preparation; control of weeds, pests, and diseases; mineralization and leaching of nitrate; influence on soil organisms, plant residues, etc., but at least as important are economic and technical aspects like labor and machinery. Even well-established natural control methods often fail if economic feasibility is neglected in the framework of the whole farming system.

The logical response to this situation is a holistic approach to farming that focuses on agroecosystems and their capacity for natural regulation (Baggiolini 1978, 1990; Delucchi 1979; Altieri 1987; El Titi 1989; Häni 1989; Gliessman 1990; Häni et al. 1990, 1991). Natural regulation in this context is not restricted to the control of pests by natural enemies, but includes many interactions between crop and environment, such as nitrogen fixation by nodule bacteria on legumes, resistance to stress due to mycorrhizae, disease resistance due to mixtures of varieties, etc.

Our proposed concepts are based mainly on our experiences in Western Europe, especially Switzerland. We will discuss Integrated Farming Systems as well as Organic Farming, but will emphasize the former, because with this concept, all important links between farming activities and natural regulation can be discussed (e.g., use of selective pesticides).

Although we discuss specific tactics for enhancing natural regulation, we emphasize the necessity of holism and integration at the farm level.

Farming Systems

From Conventional Farming to Integrated Pest Management—
The First Type of Solution

Disadvantages of the conventional "high input" strategy are becoming increasingly obvious, and include economic and environmental problems. These entail increased production that was at first encouraging, but has since led to problems of surpluses; deterioration of agroecosystems through reduction of biodiversity and, therefore, of natural regulation; pesticide resistance; deficiencies in the environment and in foodstuffs; and interference with the landscape. Agriculture may be viewed as both a cause and a victim of environmental damage.

Technologies that deal with pest and soil fertility issues in isolation and on a one-by-one basis we term collectively "the first type of solution." By contrast, we apply the term "second type of solution" to farming systems that imply a holistic perspective, taking into account specific functions of agriculture (e.g., production of food and fiber) and its general functions (e.g., decentralized colonization, landscape preservation).

The cases of Integrated Pest Control (IPC) and Integrated Pest Management (IPM) typify and illustrate problems with the first type of solution. IPC was initially developed by entomologists as an answer to the undesirable side-effects of insecticides, "hoping to escape from the cycles of pest resurgence, secondary pests, and insect resistance to pesticides" (Zadoks 1989). Shortly after the conception of IPC in the fifties (Stern et al. 1959; Steiner 1965), it was recognized worldwide as the leading principle of modern crop protection, aiming at minimal economic and environmental costs. The original concept stressed a combination of biological and chemical control measures. This concept was soon broadened to IPM, which emphasized the combination of all suitable methods in as compatible a manner as possible. However, investment in research, extension, and education could not prevent the evolution of a primarily chemical form of crop protection with all its adverse consequences. There are two main reasons for the limited success of IPM, especially in arable crops: (1) major causes of pest problems were not addressed: insufficient crop rotation, vulnerable varieties, and high inputs of N-fertilizer; and (2) IPM deals with individual pest species, and it is difficult to simultaneously use many different specific control strategies for individual pests, diseases, and weeds of several crops.

Beyond IPM: Holistic Approaches—The Second Type of Solution

In Europe, the evolution from conventional to Integrated Pest Management

Table 1. World market–oriented and ecosystem-oriented farming. Priorities of the two extreme farming visions (after Vereijken, 1990b, modified.

Values	World market–oriented	Ecosystem-oriented
1. Food supply	+	+++
2. Employment	+	+++
3. Income and profit	+++	+
4. Abiotic environment	+	+++
5. Nature and landscape	+	+++
6. Health and well-being	+	+++

and to Integrated Farming (or Low Input Sustainable Agriculture) has been extensively reviewed (e.g., Delucchi 1987; Baggiolini 1990; Boller 1990; Schäfermeyer & Dickler 1991). The decisive step from the older and more limited idea of Integrated Pest Management to Integrated Farming Systems is the transition from individual, single-tactic solutions to multiple-tactic solutions that recognize complex biological interactions.

Ecologically oriented agricultural movements arose almost simultaneously in Europe and the United States (Edens et al. 1985; Altieri 1987; National Research Council 1989; Edwards et al. 1990; Gliessman 1990). Many variations of this second type of solution have been suggested, such as organic (Lampkin 1990), ecological (Herrmann & Placolm 1991), low-input and sustainable—"LISA" (Edens et al. 1985; Edwards et al. 1990; USDA 1990, 1991), alternative (National Research Council 1989), and integrated (Anonymous 1977; Vereijken and Royle 1989; Diercks and Heitefuss 1990; Häni 1990a; Häni and Vereijken 1990). The existence of so many different terms may be confusing, because they are often insufficiently defined, developed, or evaluated. For an evaluation of these terms, it may be helpful to distinguish between two extreme visions of agriculture after Vereijken (1990b) (Table 1): (1) a world market–oriented vision aimed at maximum profits and minimally regulated or subsidized; and (2) an ecosystem-oriented vision emphasizing management of agroecosystems in such a manner that sufficient food production is sustainable, of high quality, and takes place with minimal environmental disturbance (Colby 1990). The latter form of agriculture would preserve nature for its own sake, be based on respect and responsibility for the biosphere, and be supported by national and international laws and agreements. It is obvious that for enhancing biological control through habitat management, only farming systems approximating vision (2) are suitable.

Integrated Farming Compared with Organic and Conventional Farming

For practical reasons, it is helpful to reduce the theoretically infinite number of agricultural systems to 3 different approaches: (1) Conventional Agriculture; (2) Integrated Farming Systems (English-speaking European countries), Integrated Production (German- and French-speaking European countries), or Low Input Sustainable Agriculture "LISA" (USA); and (3) Organic Farming. Conventional Agriculture may be considered close to vision (1), i.e., world market–oriented, and Organic Farming close to vision (2), i.e., ecosystem-oriented, whereas Integrated Farming (IF) is an intermediate approach. Table 2 presents important characteristics, along with documentary references (Steinmann 1983; Keller 1985; Diercks 1986; Keller & Weisskopf 1987; Häni 1989, 1990a; Vereijken & Royle 1989; Zadoks et al. 1989; Anonymous 1990; Häni et al. 1990; Häni & Vereijken 1990; Anonymous 1991a; El Titi et al. 1993).

On conventional farms, the legal limits for the use of chemicals, for animal husbandry, and for the use of fertilizers are often reached (and sometimes exceeded), whereas farmers using integrated and organic methods collectively adopt stricter guidelines, with some individuals often imposing their own voluntary restraints. This demonstrates an ethos reflecting a greater value and respect for nature. Enhancement of natural regulation, through mechanisms such as habitat management is emphasized and plays a central role in Organic as well as in Integrated Farming, whereas in conventional agriculture it is often neglected.

Some forms of Organic Farming may be considered the most radical and consistent approaches to the principles and objectives of Integrated Farming. However, an important difference between Integrated and Organic Farming lies in the means of achieving the objectives. Organic Farming excludes synthetic pesticides; Integrated Farming maximizes natural regulation in order to minimize the use of pesticides. To "minimize" expresses a continuum of use scenario. The ultimate minimization would be "no pesticides," but this would be achieved only over time (El Titi 1991). Organic Farming and Integrated Farming approaches have strong and weak points. In Organic Farming, the exclusion of pesticides creates strong pressures for enhancing natural regulation processes. Whereas Organic Farming has severe conversion difficulties, Integrated Farming may be considered as a step-by-step process by which a large number of farmers can start immediately and progress continuously to higher levels of integration. Prescriptive guidelines applied by some certifying organic organizations may lead to rote adherence by farmers and may require less thorough understanding by the farmer than the stepwise approach used in Integrated Farming. Weak points of Integrated Farming might include the continued use of potentially dangerous or environmentally damaging pesticides.

The initial step from IPM to Integrated Production was taken in Europe for fruit production (Anonymous 1977), as reviewed by Baggiolini (1990) and Schäfermeyer and Dickler (1991). In arable farming, the first on-farm system research was begun in Germany in 1977 (El Titi 1990a,b), in the Netherlands in 1979 (Vereijken 1990a) and in Switzerland (including animal husbandry) in 1985, with bookkeeping comparisons having been maintained since 1981 (Häni 1990a). These steps were soon followed in other countries. In 1981, the International Organization for Biological Control of Noxious Animals and Plants/West Palaearctic Regional Section (IOBC-WPRS) founded the group of "Integrated Arable Farming Systems." Presently, 12 European Union (EU) countries and 4 countries outside EU participate (Vereijken 1995) participate. Results obtained in different countries have been reviewed (Vereijken & Royle 1989; Häni & Vereijken 1990). The data generally show economic improvement, reduction of inputs (especially of pesticides), and enhancement of natural regulation (El Titi 1992). Examples of such results are detailed later in this chapter.

Implementation, Certification, and Economic Evaluation of Ecologically Oriented Farming Systems

In Organic Farming, well-established standards and guidelines exist on national and international levels (Anonymous 1991a; Lampkin 1990). However, until recently, Integrated Farming guidelines in Europe existed only for a few crops (mainly fruits) and only on a regional basis (Baggiolini 1990). During 1992, a panel of the IOBC/WPRS approved in Wädenswil, Switzerland, a definition, objectives, and principles for Integrated Farming. The following is a close paraphrase of the definition:

1. Integrated Farming (Integrated Production) aims at the production of high-quality food and other outputs, using environmentally sound methods. These methods maximize integration of natural resources and regulating mechanisms into farming systems to replace polluting inputs and to secure sustainable farming.
2. A holistic systems approach is emphasized, involving the entire farm as the basic unit, on the central role of agroecosystems, on balanced nutrient cycles, and on the welfare of all species in animal

Table 2. Comparison of farming systems (after Häni et al., 1990, modified)

General Goals	Conventional Agriculture	Integrated Farming	Organic Farming
Production cycles	High off-farm inputs	Production cycles are more or less closed (sometimes through interfarm exchange)	Production cycles closed at the farm level as much as possible. Basic cyclic principle
Soil utilization	Optimized according to economic criteria	Optimized according to economic criteria while respecting ecological criteria	Economic optimization only when compatible with a high ecological performance
Use of auxiliary inputs (including energy)	Optimized according to economic criteria	Limited. Ecological criteria are respected	Very limited. No synthetic products
Environmental impacts	Tolerated when they are economically justified and within legal limits	Limited	Strongly limited
Animal husbandry	Optimized economically	Specific animal welfare requirements are met as are ethical and ecological requirements	Species welfare, ethical and ecological requirements are sometimes farther reaching than in Integrated Farming
Ethics of the individual farmer	Economically oriented management	Economic criteria are woven into an ecological management structure. There is a consciousness of the interrelations involved in production techniques and off-farm inputs	Similar to Integrated Farming

Table 2. *continued*

Farm Structure	Conventional Agriculture	Integrated Farming	Organic Farming
Degree of specialization	Low to extremely high	Low to high	Low. Farm activities are adapted to the agro-habitat
Production intensity (inputs and energy for each unit of production)	Generally high	Medium to high when ecologically acceptable	Low to medium and only occasionally high (e.g., for vegetable production). Harvest levels are sustainable in the long term
Use of the agro-habitat	Crops ill adapted to the agro-habitat are sometimes grown (e.g. corn on sloping terrain)	Growing crops adapted to the agro-habitat is a basic rule for an Integrated Farming System	Crops must be adapted to the agro-habitat
Degree of mechanization	Low to high	Low to high	Low to high
Degree of fodder self-sufficiency	0–100%	+/-100% (groups of farms)	+/-100% (farm level)
Animal density per hectare of agricultural land	Legal limit	Maximum of 2.5 Large Animal Units (LU)	Maximum of 2.5 LU
Relative labor requirements	100%	105–115%	115–130%
Marketing	Mostly indirect	Mostly indirect/Partly direct	Often direct

continued on next page

Table 2. *continued*

Cultivation Techniques	Conventional Agriculture	Integrated Farming	Organic Farming
Rotation	Monoculture to balanced	Multicropping, balanced	Multicropping, balanced
Intercropping	Only partly used	Application of the evergreen method (soil covered by plants as much as possible)	Systematic application of the evergreen method
Mixed cropping, undersowing	Seldom used	Strongly used in some crops	Regularly practiced
Soil protection, soil activity	Sometimes neglected leading to lower soil fertility	Important soil protection measures are taken in order to preserve soil fertility	Maintaining soil fertility is a central goal of farm management
Tillage	Sometimes negative for soil structure	Sometimes superficial. Favors good soil structure	Sometimes superficial. Favors good soil structure
Fertilizer	Sometimes more fertilizer is applied than is absorbed by plants. High proportion of mineral fertilizers. Organic and green manure are not always optimized	A balanced fertilization with no surplus. Priority is given to organic and green manure	Mainly organic fertilizers that activate soil biology. No highly soluble fertilizers. Legumes are widely used in the rotation

Table 2. *continued*

Plant Protection	Conventional Agriculture	Integrated Farming	Organic Farming
According to threshold levels	Seldom used	Used when known	Only exceptionally (as there are few intervention possibilities)
Through preventive cultivation techniques	Of little importance	Very important	Basically the exclusive regulation method (with thermal weed control)
Through biological control methods	Seldom (price dependent)	Frequently used when available. Enhancement of natural enemies	Frequently used when available. Enhancement of natural enemies
Mechanical weed control	Seldom (labor costs)	Frequent, sometimes in combination with other techniques	Basically the exclusive method (with thermal weed control)
Chemical weed control	Often	Reduced chemical weed control. Only when other methods are unsuccessful	Excluded

Ecological Landscape Mgmt.	Conventional Agriculture	Integrated Farming	Organic Farming
Biological diversity	Not enhanced, or only when economic use permits	The biodiversity has to be actively increased at all 3 levels (genetic, species, ecosystem). Zones for ecological compensation (ZEC) to cover at least 5% of farm land (to be increased over time to 10%) The lateral dimension of an individual field in annual crops may not exceed 100 m. Otherwise fields divided by ZECs.	As for Integrated Farming

husbandry. The preservation and improvement of soil fertility and of a diversified environment are essential components.
3. Biological, technical, and chemical methods are balanced carefully, taking into account the protection of the environment, profitability, and social requirements.

The panel also approved minimum requirements for an IOBC label (El Titi et al. 1993).

Existing Guidelines

Three categories of guidelines are distinguished: (1) strict rules (orders and permissions), which are easy to supervise but do not encourage the farmer to explore new possibilities; (2) a strict level of minimum requirements supplemented by a mixture of rules and recommendations; and (3) rating systems: strict prohibitions, i.e., malus points, and options that are ranked according to their ecological, ethical, or economic impact.

Strict rules as in category 1 include the "organic" requirements applied by various governmental and non-governmental certifying agencies. The "Third Way Project" (Häni 1990a) exemplifies category 2; for marketing under the label of "Integrated Produce" (Hofer 1990), minimal criteria must be met; beyond this, farmers may choose from a set of recommendations and rules (Anonymous 1990).

A category 3–type approach is found in the guidelines for grape growers developed by the Swiss Federal Research Station at Wädenswil ("Wädenswil Model," Boller et al. 1990; Boller 1989). These constitute technical guidelines, an ecological bookkeeping method, and a rating system. Strict prohibitions (malus points) define the line between good and bad practices whereby farmers are either qualified or disqualified for certification. Farmers are also awarded bonus points according to the number of environmentally beneficial practices employed and the degree to which these are used.

Participants in the Integrated Farming programs for viticulture in northern Switzerland follow a set of rules that were established in discussions between research staff of Wädenswil and grape growers and were tested in field trials conducted during 1986–1988. The performance of the grape growers is evaluated annually. Some 50 individual aspects subjected to program evaluation are grouped into 7 compartments as follows: (1) soil and fertilizers; (2) soil cultivation and weed management; (3) disease control; (4) arthropod pest control; (5) yield and quality; (6) farm records, training, and equipment; and (7) wine-making. Each compartment contains a list of defined problems with major environmental impacts that are rated and evaluated according to the solution taken by the grower. Bonus points are given to solutions and actions calculated to reduce the use of pesticides, fertilizers, and fuel; diversify the vineyard agroecosystem; and improve grape quality. More bonus points are given to solutions with minimal negative impact on the environment. No points are given to solutions that follow traditional agricultural practices ("conventional viticulture") and do not make use of available ecologically oriented alternatives. Finally, there are some 42 well-defined criteria built into the scheme that specify serious violations of the objectives of Integrated Farming as defined by IOBC. Any one of these malus points observed on a farm will cause disqualification. They include such items as N-input in excess of 50 kg/ha/year, copper input exceeding 3 kg/ha/year, application of more than 7 fungicide treatments per season against downy mildew, etc.

Such a rating system has definite advantages over other systems operating with sets of rules and recommendations. In particular, the farmer can choose the ecological level of farm management according to the prevailing local conditions and his or her willingness to accept higher risks. The competitive aspect of the system allows a direct comparison between farms and stimulates the farmer to match or exceed the achievements of fellow farmers. The system is flexible as it allows the incorporation of new scientific findings and practical experience, the allocation of greater weight (bonus points) to activities with higher environmental impact, and the consideration of economic aspects. The program stimulates experimentation and self-evaluation by the farmer. This system also provides insight and background to the customer who is evaluating the real value of, and achievements behind, the farmer's certificates and labelled products. However, compared with other systems, the controlling organization must invest more time and care in field inspections and in the evaluation of the farm records.

Economic and Agrotechnical Effects of Farming Systems

In Switzerland, the "Third Way Project" involves commercial farms (called "pilot-farms") that are operated as integrated systems. When compared to conventional farms as well as to reference areas on the integrated pilot-farms free of pesticides, integrated farms gave very promising economic results (Figure 1). Investigations in other European countries yielded similar findings (Vereijken et al. 1986; Vereijken & Royle 1989; Häni & Vereijken 1990).

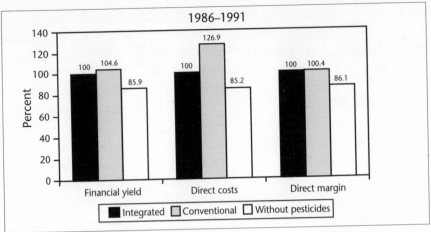

Figure 1. Direct comparisons between "Integrated," "Conventional" and "Without pesticides" on the 3 pilot-farms. Financial results of winter wheat ("Integrated "= 100%). (After Häni 1990a and unpublished.)

Various studies have shown that as the degree of adaptation to ecological requirements increases, the labor requirements increase. This is due to the need for additional monitoring and analyses, care of zones of ecological compensation, increased reliance on cultural control of pests, etc. (Häni 1993, 1994). The higher labor costs and the slightly lower yields are only partly offset by the lower direct costs. However, with greater experience in the use of this new farming technology, the disparity in labor costs will probably be reduced. The remaining gap in profitability could be made up by state subsidies giving direct payments linked to the ecological performance of a farming system. Such a state effort would recognize the ecological benefits for the whole of society. In fact, in Switzerland, such direct payments were introduced in 1993 for organic and integrated farms that fulfill specific requirements (farm bill, article 31b), and also for individual on-farm accomplishments and improvements, e.g., the creation of zones of ecological compensation (Wolfe et al. 1993).

Farming Systems and Bioindicators

Farmers and the general public alike want to know how farming systems affect the surrounding environment; such an evaluation, though important, is difficult. Clear criteria include quantity of fertilizers and pesticides and the resulting residues in soil, water, or plants. In the "Third Way Project" in Switzerland, where pilot-farms were operated as completely integrated systems from 1986 through 1992, chemicals were used to protect the plants 1.3 times per year on average for all crops, whereas they were used 3.6 times

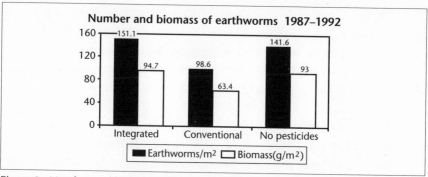

Figure 2. Number and biomass of earthworms (Lumbricidae). Averages of 3 pilot-farms, 1987–92. Thirteen trials (5 winter wheat, 3 maize, 3 potatoes, 2 sugar beet), 4 replicates each. The differences are significant (P<0.01) between *Integrated* and *Conventional*, and between *No pesticides* and *Conventional*, but not between *Integrated* and *No pesticides*.

per year in conventional production (Häni 1993, 1994). By comparison, a similar test in The Netherlands showed 3.6 treatments per year in integrated farming, compared with 8.5 in conventional agriculture (Vereijken 1990a).

The direct ecological effects of farming systems can be judged by indicator plants and animals. Frequently used bioindicators of the soil system are earthworms (El Titi 1990a; Häni 1990a, 1990b) (Figure 2). Other possible bioindicators for the soil system are Collembola and mites (El Titi 1984), cellulose or filter paper degradation (Jäggi 1989; Häni 1990b), and CO_2 release. Despite methodological progress, it is still doubtful whether mycorrhizae are suitable bioindicators (Schüepp et al. 1991, 1992).

Ground-breeding birds can be used as indicators of the use of pesticides and fertilizer in arable crops or of grassland management, respectively. The breeding success of the partridge (*Perdix perdix* L., Galliformes: Phasianidae) is better where pesticides were not used on the headlands (Sotherton 1990). The skylark (*Alauda arvensis* L., Passeriformes: Alaudidae) and the whinchat (*Saxicola rubetra* [L.], Passeriformes: Muscicapidae) cannot breed successfully when the meadows are cut early. In grain fields, the skylark can breed successfully when an undersowing or weed between the rows is present (Jenny 1990).

Another possibility is to investigate the general faunistic and botanical diversity in an agroecosystem, as is now done in vineyards in Switzerland (Remund et al. 1989, 1992). Grassland management may be judged by botanical diversity alone (Thomet et al. 1989) or in combination with the arthropod diversity. Van Wingerden et al. (1992) showed that diversity of grasshopper species decreased with increasing fertilization and mowing frequency.

Farming systems may also affect the abundance and activity of polypha-gous and specialized predators, which in turn may affect the regulatory capacity for pests. In the following sections, we discuss the structure and specific elements of agroecosystems and their documented effects on natu-ral enemies and the regulation of agricultural pests and diseases.

Structure of Agroecosystems

To demostrate the components of complex agroecosystems and of their interactions, a series of diagrams was developed. Figure 3 depicts a network of components and their qualitative interactions for a typical Swiss vine-yard. Interactions among elements relevant for plant protection are rated as positive or negative with respect to the interest of the farmer or crop plant, respectively (Boller 1988).

A diagram resembling Figure 3 might facilitate the analysis and plan-ning of scientific experiments, but be unsuitable for daily use by a farmer. Therefore, we have rearranged the main elements, reduced the complex interactions to a few facts (Figure 4), and condensed the information into a crop-specific diagram for viticulture (Figure 5). The latter is used to train Swiss grape growers in integrated production. Figure 6 is a similar diagram for maize production in Europe (Bigler 1988).

With these tools, farmers can check the concrete, individual aspects on the concentric circles around the biocoenotic center, and thereby verify the impact of various activities on agroecosystem stability.

The Elements of Agroecosystems and Examples
of Natural Regulation of Pests and Diseases

The division between pest and beneficial organisms is arbitrary, because no organism is exclusively noxious. Every species has a role to play in the ecosystem through interactions with neighbors and environment while be-ing itself influenced by both. It is estimated that there are 40,000 animal species in central Europe, of which 30,000 are insect species. The finely-tuned balance of natural wildlife systems depends upon this diversity.

Individual organisms do not cause significant harvest losses; such losses are caused rather by large populations of noxious organisms. In general, population explosions of pests should be regarded as a consequence and not a cause of a disrupted natural balance. Humans have always tried to ma-nipulate this natural balance in their favor, and in so doing have, con-sciously or unconsciously, reduced or eliminated some species while in-creasing others. The methods of intervention have never been as diverse or

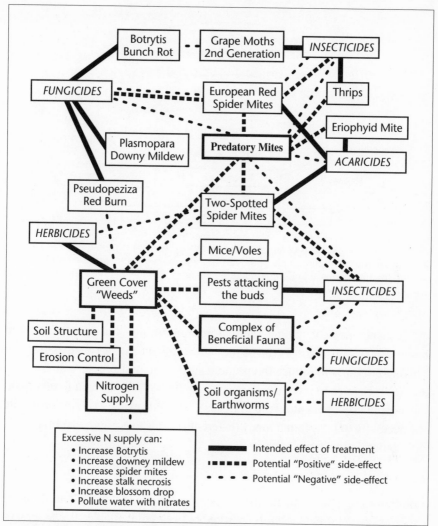

Figure 3. Interactions between major components of the agrosystem "Vineyard" in Eastern Switzerland. Cover plants emerge in this graphic as an ecological turntable for enhancement of natural regulation (after Boller 1988).

as well developed as those of today. Such powerful technology requires awareness of the responsibilities implied. Clearly, human ability to regulate ecosystems has limits. Moreover, the more extreme types of intervention require justification. This is true because little is known about the role played by each species and the interactions within an ecosystem. In addition, interventions often undermine natural diversity, making agroecosystems less stable and more susceptible to stress.

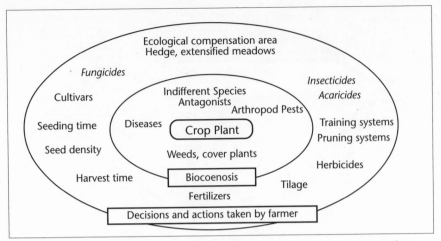

Figure 4. The generalized components of an agroecosystem can be rearranged to meet the practical requirements of the farmer. The farmer's activities are grouped around the biocoenosis, and the impact of specified activities on the system can be visualized. (Boller et al. 1998a)

Enhancement of natural regulation aims to alter the ratio of pests to beneficial organisms in favor of the latter. Enhancement schemes may involve: (1) efforts to reduce the pest density or the pest pressure on the crop and (2) measures to increase abundance and/or activity of beneficial organisms (predators, parasitoids, pathogens). The following examples indicate the necessity of integrating many different aspects to achieve maximal natural regulation in a farming system.

Crop Rotation

Rotation of annual crops has developed over centuries and is based on the experience of farmers. Before pesticides became available, it was practically the only method to control soil-borne pests and diseases. In the middle of this century, a well-developed rotation consisting of 6–8 crops in sequence was considered a good agricultural practice (Koblet 1965).

The wide use of pesticides and increasing economic pressure drastically changed the structure of agriculture. Farmers specialized and industrialized their enterprise, land use was maximized, and ecologically valuable zones disappeared. Rotation was reduced to a few species of crop or even down to a monoculture. Yields increased, often through excessive application of fertilizers and pesticides. These measures favored the pest species and impaired the beneficial organisms, damaging natural regulation (Basedow 1987a). Beneficial organisims were probably not driven to local extinction,

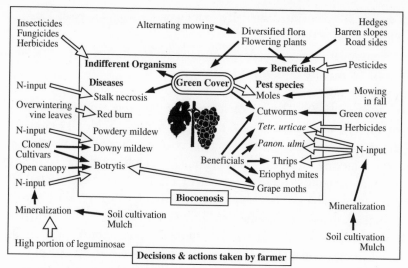

Figure 5. Structure of a typical perennial agroecosystem "Vineyard" in Eastern Switzerland and resulting plant protection measures in the framework of Integrated Production. The most important positive external and internal influences acting upon and between the individual components of the system are indicated by black arrows (➤) and demonstrate clearly the complexity of the entire network of relationships. Only major pests and diseases are displayed in this graph. Operations with negative impact on the agro-ecosystem are marked by outline arrows (⇨). The cover plants function as an ecological turntable. (After Boller 1988).

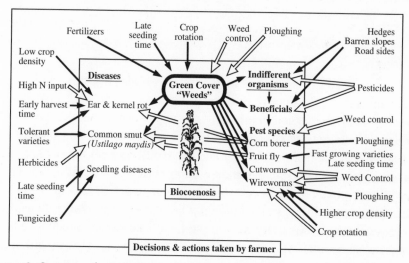

Figure 6. Structure of a typical annual agroecosystem (maize) in Switzerland and resulting plant protection measures in the framework of Integrated Production. The most important external and internal influences acting upon and between the individual components of the system are indicated by arrows indicating positive (➤) or negative (⇨) interactions as seen by the farmer. The cover plants function as an ecological turntable. (After Bigler 1988).

but rather were separated spatially or temporally from the pests they attack (Keller & Duelli 1990).

The modern holistic concept of agriculture aims at a reversal of this process. One remedy is the reestablishment of a "sound" crop rotation, limiting the proportion of any specific crop and including sown meadows (Anonymous 1990). The principles remain the same: suppression of soil-borne diseases and pests by a time-tested rotation (Table 3, Figure 7). In addition to the traditional crop rotation, covering the soil by short-duration or overwintering crops or green manures is emphasized. These measures favor abundant and diverse soil biota (El Titi 1984; Troxler & Zettel 1987) and reduce erosion and leaching of mineralized fertilizer, mainly nitrogen.

Special resistant intermediate crops, like resistant oil seed radish (*Raphanus sativus* var. *oleiformis*), can be used to suppress plant-parasitic nematodes. Intermediate crops can also prevent unspecialized phytophagous or saprophagous arthropods from damaging subsequent crops (Heitefuss & Garbe 1986).

In a crop rotation, fields may be allocated in two different ways: "shifting" and "jumping." Shifting means that a crop is cultivated adjacent to the field where it was cultivated the previous season. Jumping implies no common borderlines with fields of the same crop of the previous season. In production of peas (*Pisum sativum*), jumping rotation can be used to avoid problems with thrips (Thysanoptera) and to reduce incidence of *Contarinia pisi* Winn. (Diptera: Cecidomyiidae) (Keller unpublished.).

Tillage

As summarized in Table 4, soil tillage may be divided into three basic types: (1) conventional destructive or turning tillage (e.g., with moldboard plows or disk harrows); (2) conservation tillage, which leaves the soil layers more or less intact (this can employ chisels or rippers to loosen and spring-tooth or spike-tooth harrows to loosen and mix soil); and (3) non-tillage (with crops planted directly into standing or mowed residue). The first is the traditional ploughing practice, which turns the upper 20–25 cm of the soil, thus placing the biologically active top layer into lower soil strata and the biologically less-active subsoil at the surface. This method produces a compact and sometimes impermeable layer at plough-depth, where plant debris may accumulate.

Conventional tillage is being abandoned in favor of conservation and non-tillage, mainly for reasons of soil conservation. However, reduced tillage can also reduce some pests (Corbett & Webb 1970; Graber & Suter 1985; El Titi 1987) and diseases (Palti 1981; El Titi 1987) (Table 4).

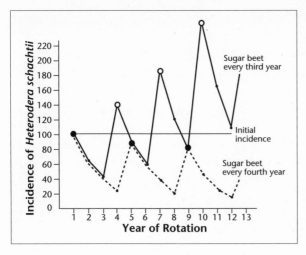

Figure 7. Influence of the crop rotation on propagation of the plant-parasitic nematode *Heterodera schachtii*. (After Schäufele, from Diercks, 1986.)

Ground beetles and spiders are good indicators for soil tillage (Basedow 1987a, 1990; Welling 1990; Völkl & Keller 1991). Reduced tillage favors many species of ground beetles (Coleoptera: Carabidae), spiders (Araneae), predatory mites (Acari: Gamasidae), Collembola, and anecic earthworms (e.g., *Lumbricus terrestris* L., Oligochaeta: Lumbricidae) as well as entomopathogenic fungi, e.g., Entomophthorales attacking aphids, which can survive only on the soil surface (Latteur 1977). By contrast, ploughing or intensive soil cultivation is helpful in reducing European corn borer (*Ostrinia nubilalis* Hbn., Lepidoptera: Pyralidae), larval cockchafer (*Melolontha melolontha* L., Coleoptera: Scarabaeidae) and various other scarabaeids, Hessian fly (*Mayetiola destructor* Say, Diptera: Cecido-myiidae), and the incidence of some diseases (Yarham 1979; Palti 1981; Yarris 1982). Specific cases include reduction of tan spot (causal agent *Drechslera tritici-repentis*, a fungal pathogen of wheat [*Triticum aestivum*]), (Heitefuss et al. 1984) and of *Fusarium* spp. (Cassini 1969, 1973; Häni 1980, 1981). Because tillage influences many factors, specific pests or diseases can be either increased or decreased by the same tillage method applied at different times of the year. Take-all fungus (*Gaeumannomyces graminis* (Sacc.) Arx et Olivier, Sphaeriales: Diaporthaceae), for example, can be decreased through reduced tillage by increased activity of antagonists and higher concentration of phosphorus, but lower concentration of nitrogen in the top soil. By contrast, reduced tillage may also favor take-all fungus by leaving inoculum-carrying debris in the upper soil strata (Palti 1981). The field slug, (*Deroceras reticulatum* O.F. Müller,

Table 3. Pests and diseases that can be favored by a high percentage of host plants in rotations of Integrated Farming Systems. After Baeumer, in Diercks and Heitefuss, 1990, modified. Main host plants +; possible host plants (+).

	Wheat	Barley	Rye	Oats	Maize	Cruciferae[1]	Legumes	Beta[1] beets	Potatoes
Insects									
Collembola								+	
Atomaria linearis								+	
Haplodiplosis equestris	+	+	+						
Ostrinia nubilalis					+				
Psylliodes chrysocephala						+			
Dasineura brassicae						+			
Nematodes									
Heterodera avenae	+	+		+					
Heterodera schachtii						+		+	
Globodera spp.									+
Ditylenchus dipsaci			+	+	+	+		+	
Pratylenchus spp.	+	+	+	+					
Diseases									
Streptomyces scabies									+
Pseudocercosporella herpotrichoides (& other pathogens)	+[2]	+	+						
Gaeumannomyces graminis	+[2]	+	+						
Fusarium spp. in cereals	+	+	+	(+)					
Drechslera tritici-repentis		(+)	(+)		+				
Fusarium spp. in legumes							+		
Sclerotinia sclerotiorum						+	+		+
Phoma lingam						+			
Maximum percentage for Integrated Farming									
Favorable site	33	40	50	25	40	33	25	33	33
Unfavorable site	25	33	33	25	25	25	20	25	25

[1] The total of Brassicaceae and Beta beets should not exceed 25% or 33%, respectively (risk of nematodes).

Stylommatophora: Limacidae) is favored by ploughed soils during the oviposition period, because such tillage permits slugs to lay their eggs in deeper soil layers with high moisture content, improving egg survival (El Titi 1987). By contrast, during other periods of the year, slugs may be favored by retention of plant debris on the soil surface as a result of reduced tillage.

There are many other indirect and direct effects of tillage methods. Many soil fauna are increased in diversity and densities by reduced tillage (Baeumer 1990). This implies stabilization, i.e., natural regulation, but sometimes outbreaks of Collembola occur (related to a high amount of organic matter), damaging young sugar beet (*Beta vulgaris*). Tillage method influences the sowing date and relates to nutrient availability to crop plants and to weed flora. These in turn may influence arthropod fauna.

Crop Varieties and Species, Including Mixtures

Many varieties of annual crops are available, with characteristics influencing both pest and beneficial species. To reduce pest and disease attack, a variety with resistance characteristics best adapted to a given situation is chosen. However, no single variety is resistant to all pests and diseases. Therefore, it is best to cultivate two or more varieties, multilines (mixtures of strains of a crop differing at only one or a few genetic loci), or different species with complementary resistance characteristics. This agronomic tool is not so readily available in perennial crops, where external and internal quality of the final product is of paramount importance, e.g., wine made from *Vitis vinifera* grape, or apple (*malus pumila*) varieties with specific gustatory and visual attributes. Therefore, breeding programs for disease and pest resistance in fruit and grape production are still in their early stages.

For annual crops, the developmental rate and the time from sowing or emergence to harvest are important varietal characteristics, influencing pests and beneficials. For example, early-growing oilseed rape (*Brassica napus* var. *oleifera*) varieties are more attractive for stem borers (*Ceutorhynchus* spp., Coleoptera: Curculionidae) (Büchi 1989). Attractive, early-growing plants can be used as "baiting crops" as proposed by Büchi et al. (1987) and Büchi (1990) (the term "trap crop" is more commonly used in the United States). Turnip rape or bird rape (*Brassica rapa* var. silvestris) can be seeded in strips at least 9m wide along the borders of fields with oilseed rape. The faster-growing, brighter-green, and earlier-flowering turnip rape is more attractive to several oilseed rape pests, including *Meligethes* spp., than is oilseed rape itself. These pests therefore concentrate on turnip rape and

Table 4. Differences of basic soil tillage methods and examples of beneficial organisms of pests, weeds and diseases that are relatively favored (+) or reduced (-), compared to other methods. Drawing after Buchner & Köller, 1990.

	Conventional tillage Loosening and turning	Conservation tillage Loosening and mixing	Loosening	No tillage Direct drilling
Beneficial Organisms				
Earthworms	–	+	+	+
Carabidae	–	+	+	+
Araneae	–	+	+	+
Gamasidae	–	+	+	+
Entomophthorales	–	+	+	+
Pests				
Ostrinia nubilalis	–	+	+	+
Scarabaeidae	–	+	+	+
Mayetiola destructor	–	+	+	+
Slugs	+/–	+/–	+/–	+/–
Heterodera avenae	+	–	–	–
Delia coarctata	+	–	–	–
Perennial Weeds[1]	–	+	+	+
Diseases				
Erysiphe graminis	+	–	–	–
Puccinia recondita	+	–	–	–
Drechslera tritici-repentis	–	+	+	+
Fusarium spp. in maize[2]	–	+	+	+
Kabatiella zeae	–	+	+	+
Drechslera turcica	–	+	+	+
Gaeumannomyces graminis	+/–	+/–	+/–	+/–
Pseudocercosporella herpotrichoides[3]	+/–	+/–	+/–	+/–

[1] Conventional tillage (plowing) can also reduce some annual weeds, e.g., *Alopecurus myosuroides, Apera spica-venti, Bromus sterilis.*

[2] Main effect on *F. graminearum* (rarely chlamydospores), less on *F. culmorum.*

[3] New results show a disease increase by conventional tillage (plowing). Anken et al. 1997.

damage this species. The system, however, works only at relatively low pest pressure. If despite this technique threshold values are exceeded, it is often possible to restrict an insecticide application to the border.

Sowing

Late-sown maize (*zea mays*) develops faster than that maize that is early sown and may escape the attack of the frit fly *Oscinella frit* L. (Diptera: Chloropidae). Late-sown sugar beet also may escape some pest species, but by contrast may suffer more from fungal attack. Attacks by the gout fly (*Chlorops pumilionis* Bjerk, Diptera: Chloropidae) can be avoided by early-sown spring cereal or relatively late-sown autumn cereal. A method to avoid infections by the barley yellow dwarf virus is timely (relatively late) sowing of the crop, so that it emerges after the autumnal flight of the bird cherry–oat aphid (*Rhopalosiphum padi* L. Homoptera: Aphididae), the main vector. Early- and late-sown peas are both attacked by the pea midge (*Contarinia pisi* Winn., Diptera: Cecidomyiidae). But for the early-sown peas, attack occurs at a later developing stage and does not affect yield (Keller 1989).

Weed Management and Undersowing

In both annual and perennial commercial crops, cover crops (e.g., clovers or grasses) may be undersown or some weeds tolerated. These techniques can convert monocultures into more biologically diverse agroecosystems. Besides the prevention of erosion and leaching of nutrients, cover crops are an important tool for enhancing the abundance and activity of beneficial arthropods.

Examples of such enhancement using a green cover (resident vegetation) can be found in orchards and especially in cool climate vineyards (Remund et al. 1989). The procedure consists of the permanent presence of a diversified resident flora. An alternate mowing regime, whereby only every second alley is mowed at one time, allows plants to flower throughout the season. Alternate mowing of the green cover of a vineyard at Walenstadt increased the species diversity of flowering plants (Figure 8). The number of potential pest species remained about the same, but these fluctuated at significantly lower population densities. Flower-visiting antagonists such as hymenopterous parasitoids, Syrphidae (Diptera), Chrysopidae (Neuroptera), and Anthocoridae (Hemiptera) increased in abundance (Figure 9). Similar results were obtained in a 4-year analysis of 21 different vineyards (Figure 10) (Remund et al. 1992). In large vineyards without green cover, Remund and Boller (1991) found no parasitization of the eggs of grape berry moth

Figure 8. Traditional European vineyard with open soil due to mechanical and chemical weed control (A) and modern vineyard with permanent natural green cover with flowering plants (B). The plant diversity and permanent flower supply are achieved by alternating mowing.

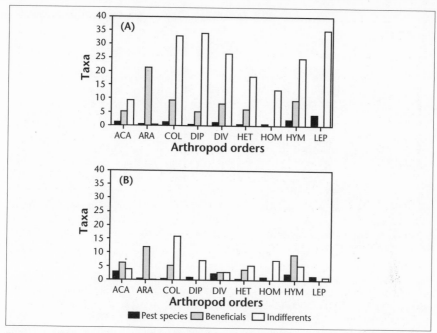

Figure 9. Abundance of arthropods (taxa) in an experimental vineyard at Walenstadt with flowering cover plants (A) and without coverplants (B). ACA = Acarina (mites), ARA = Araneida (spiders), COL = Coleoptera (beetles), DIP = Diptera, DIV = different orders, HET = Heteroptera (bugs), HOM = Homoptera (leafhoppers, aphids), HYM = Hymenoptera (wasps and parasitoids), LEP = Lepidoptera.

(*Eupoecilia ambiguella* Hb., Lepidoptera: Tortricidae). By contrast, in vineyards with green cover, *Trichogramma* spp. (Hymenoptera: Trichogrammatidae) egg parasitoids parasitized up to 87%; this higher level probably resulted from alternate lepidopteran hosts associated with the green cover (Remund & Boller 1991).

Flowering vegetation between the grape (*vitis vinifera*) vines also provides pollen, which serves as an alternate food to the predatory mite *Typhlodromus pyri* Scheuten (Acari: Phytoseiidae). In early spring until June, most pollen originates from trees and shrubs outside the vineyard. During the summer generations of the predator, pollen is provided mainly by cover plants within the vineyards (e.g., grasses, *Urtica, Plantago*) (Figure 11) (Wiedmer & Boller 1990). Nutritional quality of different pollens for the survival and reproduction of the predator was assessed in the laboratory by several workers (Boller & Frey 1990; Engel 1991; Remund & Boller 1992). Among the predominant grasses (Poaceae), greatest survival and fecundity were achieved in descending order using the pollens of Kentucky bluegrass (*Poa pratensis*) > ryegrass (*Lolium multiflorum*) > orchardgrass

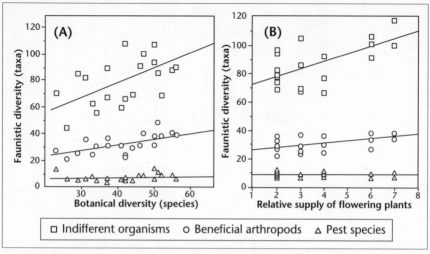

Figure 10. Correlation between botanical and faunistic diversity in 21 vineyards of Eastern Switzerland in the year 1990. A) Increasing botanical diversity, expressed in number of plant species, correlates significantly with increasing number of indifferent taxa ($r = 0.65$; $p < 0.05$) and of benefical organisms ($r = 0.67$; $p < 0.05$). Diversity of pest species is not influenced. B) Increasing supply of flowering plants (relative scale between 1 and 10) enhances significantly number of indifferent arthropods ($r = 0.68$; $p < 0.05$) and beneficial arthropods ($r = 0.60$; $p < 0.05$). (Remund et al., 1992)

(*Dactylis glomerata*). Low fecundity was achieved with quackgrass (*Agropyron repens*), *Arrhenaterum elatius*, and red fescue (*Festuca rubra*), whereas no eggs were produced with poverty brome (*Bromus sterilis*) (Remund & Boller 1992).

In hop (*Humulus lupulus*) fields, crop plants in plots with soil-covering weeds had lower densities of spider mite (*Tetranychus urticae* Koch, Acari: Tetranychidae) than did those in weed-free plots. This is ascribed to the preference of spider mites for some weeds (e.g., *Convolvulus arvensis*, *Galinsoga parviflora*, *Solanum nigrum*, *Stellaria media*, over hop plants. Under average pest pressure, the weed cover was sufficient to prevent mite damage on hop plants. Under heavy pest pressure, however, it was insufficient, and an acaricide was required (Schweizer 1992).

Experimental sugar beet plots without weeds, when compared to those with weeds, had larger and longer-lasting aphid populations, more plants infected with the sugar beet yellow virus, and, in most cases, fewer ground beetles and fungus-infected aphids (Häni et al. 1990). Yield, however, was not affected. Similarly, herbicide-treated cereals contained more aphids than did untreated ones (Storck-Weyhermüller 1990).

Flowering weeds attract pollen- and nectar-feeding beneficial insects, such as adult syrphids and parasitoids. Hoverflies are more abundant where

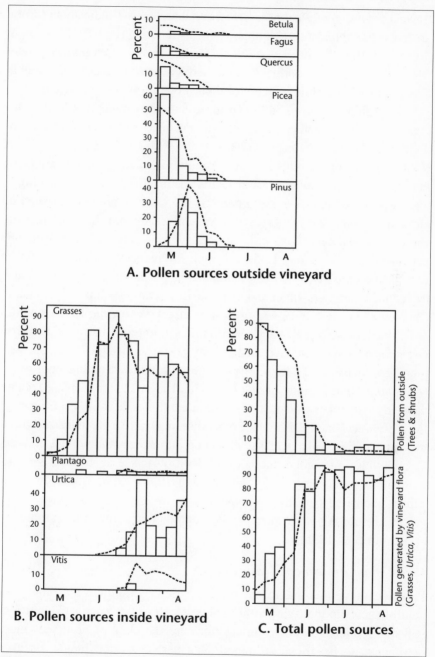

Figure 11. Composition of pollen observed between May and August 1987 in a vineyard near Wädenswil, Switzerland. Lines indicate relative pollen densities found on the leaf surface of grapevine, bars indicate pollen caught in vineyard on pollen capture devices (from Wiedmer & Boller 1990).

flowers are present in and near fields, and therefore are indicators of floral diversity and abundance (Molthan & Ruppert 1988; Häni 1989; Weiss & Stettmer 1991). Densities of adult syrphids were twice as high in winter wheat not treated with herbicides as in herbicide-treated wheat, resulting in about a three- to sixfold increase of their progeny (Figure 12) (Häni 1989, 1990a). Aphids occurred earlier in those untreated fields, but, due to increased syrphid abundance, populations declined earlier and before economic thresholds were reached (Häni 1989, 1990a).

Undersowing (seeding a second crop into a standing crop) is a type of intercropping that may increase parasitism (Powell 1986). Undersowing is mainly practiced in maize, using ryegrass (*Lolium perenne*) or a mixture of clover (*Trifolium* spp.) and ryegrass. Recently, the more shade-tolerant orchard grass (*Dactylis glomerata*) has been substituted for ryegrass (Ammon, personal communication). Undersowing led to reduced egg mass densities. European corn borer (*Ostrinia nubilalis* Hbn., Lepidoptera: Pyralidae) and to reduced numbers of cereal leaf beetles (*Oulema* spp., Coleoptera: Chrysomelidae), but there was little or no effect on *Ostrinia* larvae, aphids, or frit flies. Sowing maize amid a meadow, led to pronounced reduction of *Ostrinia* and aphids and increased activity of soil surface predators, but also increased density of slugs (Bigler, pers. comm.). Sowing sunflower (*Helianthus annuus*) into maize (2–3 plants/100 m²) led to increased abundance of beneficials over that in maize alone (Suter 1988). Flowering sunflowers attract beneficials into the otherwise unattractive green maize, where they find aphids. This increased attraction into maize is important because Coccinellidae and other entomophaga feed on maize aphids and thereby build nutritional reserves before hibernating (Bigler 1986). Undersowing in cereals increased parasitism of aphids (Powell 1983) and activity of the ground beetle *Agonum dorsalis* Pontoppidan (Coleoptera: Carabidae) (Wichtrup et al. 1985).

Direct Plant Protection Measures

Pesticides have to some extent been replaced by cultural and biological control agents in managing arthropod pests. However, chemical control is still the dominant measure. Insecticides are the class of pesticides most detrimental to beneficial arthropods, but herbicides and fungicides may affect ground-dwelling predators or entomopathogenic fungi, respectively (Delorme & Fritz 1978; Wilding & Brobyn 1980; Zimmermann & Basedow 1980; Keller & Schweizer 1991). The working group on pesticides and beneficial organisms of the IOBC/WPRS has tested some 200 compounds

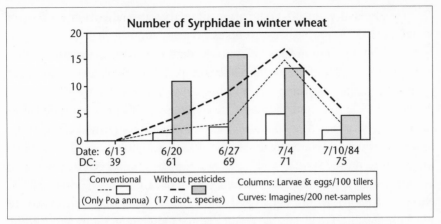

Figure 12. Five species of Syrphidae, with *Sphaerophoria scripta* (L.) the most frequent, in 2 neighboring fields of winter wheat. Significant differences at every date (Wilcoxon Rank Sum Test). (After Häni 1990a.)

for activity against some 35 antagonist species (Hassan 1985, 1991). Thorough investigation and documentation of the ecological side effects of pesticides is a prerequisite for the planning of IPM programs in Integrated Farming Systems (Boller et al. 1989).

When considering a pesticide application the following points should be observed:

1. The use of pesticides should be limited and that of preventive methods maximized.

2. Pesticide application should be localized; parts of the field should remain untreated to promote recolonization of the treated area by beneficials. Localized application can be through strip spraying along drill lines or by appropriate product formulation. Options include seed treating, baiting for snails, granular pesticides placed on the drill line, and restricting application to heavily-infested parts of the field.

3. Timely application and use of pesticides with short residual activities can help avoid massive destruction of beneficials and the development of pesticide-resistant pests.

4. Reduced pesticide doses permit low pest densities to remain, which in turn permit beneficials to survive. Reduced doses are important for insecticides and herbicides, but their use requires knowledge of the pest and evaluation of the field infestation.

5. Selective products should be given priority over broad-spectrum products that affect beneficials.

Timely and reduced application of active ingredients may protect beneficials. Most published examples concern the effects on natural enemies of aphids, ground beetles, and spiders (e.g., Storck-Weyhermüller 1987; Pöhling 1987; Frings 1988; Ulber et al. 1990). By contrast, improper insecticide applications may provoke pest outbreaks by destroying beneficials. For example, the use of insecticides for the control of the European corn borer often results in aphid outbreaks through reduction of the aphid antagonists (Naïbo & Foulgocq 1985). Chemical aphid control in sugar beet may favor a leaf miner, *Pegomyia betae* Curt. (Diptera: Anthomyiidae), probably by reducing its natural enemies (Frings 1988). More and improved biological and biotechnical pest control methods are becoming available, as has been detailed for the United States and Europe (Schönbeck et al. 1988; Krieg & Franz 1989), for Western European countries (Bigler 1986, 1987; Aeschlimann & Carl 1987; Staal 1987; Van Lenteren 1987a, b 1988) and for Eastern countries (Hluchý 1987, 1988, 1989; Anonymous 1991b; Bigler & Cerutti 1991; Hluchý & Pospísil 1991).

Predatory mites, especially *Typhlodromus pyri* and *Amblyseius* spp., play major roles in the biological control of spider mites, eriophyid mites, and thrips in Swiss orchards and vineyards. Their effectiveness as natural control agents justifies protection and augmentation in Integrated Farming Systems. The bacterium *Bacillus thuringiensis* Berliner is widely used in vineyards against grape moth (*Eupoecilia ambiguella* Hb., Lepidoptera: Tortricidae). Addition of 1% sugar to the commercial bacterial products enhances the effectiveness of this biological control method to > 90%. *Bacillus thuringiensis* is also applied with varying degree of success against other lepidopteran pests in fruit orchards and vegetable crops. Specific granulosis viruses are commercially available in Switzerland and other European countries against two major apple pests, codling moth (*Cydia pomonella* [L.], Lepidoptera: Tortricidae) and summer fruit tortrix moth (*Adoxophyes orana* Fischer von Röslerstamm, Lepidoptera: Tortricidae).

Few biological control methods are applied in arable crops in western Europe. The most widely used is the release of *Trichogramma evanescens* Westw. (Hymenoptera: Trichogrammatidae) for control of European corn borer. In Switzerland, *Bacillus thuringiensis* var. *tenebrionis* is being increasingly used for the control of Colorado potato beetle (*Leptinotarsa decemlineata* [Say], Coleoptera: Chrysomelidae). For the control of larvae of cockchafer (*Melolontha melolontha* L., Coleoptera: Scarabaeidae) on grassland and in orchards, the white muscardine fungus (*Beauveria brongniartii*) is mainly used (Keller et al. 1989). The green muscardine fungus (*Metarhizium anisopliae*) is being tested for possible use in wireworm (Coleoptera: Elateridae) control.

In the field of biotechnical methods, the use of pheromones as a mating disruptant has gained a prominent place in viticulture where it is sucessfully used on a large scale against the grape moth (*Eupoecilia ambiguella* Hb., Lepidoptera: Tortricidae). The use of pheromones in combination with traps or applied as a mating confusant may be deleterious to antagonists that use pheromones for host-seeking.

Soil Fertilization

It is well known that nutrients affect the relations between crop and pathogen in many different ways, but mainly through modifying the host vigor and resistance (Palti 1981). Nutrient insufficiency, especially of P and K, may favor certain diseases and pests (Diercks 1986). Calcium influences soil pH, which in turn may affect the plant growth, soil pests (El Titi 1987), and diseases (Garrett 1970). For example, liming before a potato (*Solanum tuberosum*) crop can increase common scab (*Streptomyces scabies*). Calcium also affects cell-wall development. The resistance of lucerne (*Medicago sativa*) to the plant-parasitic nematode *Ditylenchus dipsaci* Kuehn (Tylenchida: Tylenchidae) is greatest when calcium is sufficient (Fuchs & Grossmann 1972). Nitrogen is of paramount importance for pests and diseases. High nitrogen input generally reduces plant resistance and favours plant diseases and pest insects (Hansen 1986). For example, populations of two-spotted spider mite *Tetranychus urticae* Koch (Acari: Tetranychidae) on apple leaves grew faster when trees had a higher N supply and foliar N content (Wermelinger et al. 1991). Organic fertilizers may increase insect species diversity and densities (Troxler and Zettel 1987), whereas composts inhibit some diseases of seedlings (Tränkner & Liesenfeld 1990). Abstaining from both pesticides and N fertilizers in the field margins may enhance species richness of weeds and the associated beneficial organisms (Welling 1990).

Measures Outside the Crop

Only a few useful or noxious arthropods can survive for long periods in a cultivated field. This is especially true for annual crops that are temporary monocultures with limited ecological diversity. Most animals are either repelled or eliminated by the absence of vegetation after harvest, by tillage, and by pesticide applications. Most agriculturally-important arthropods persist in places with permanent vegetation. Several pest species, e.g., aphids, migrate in spring from their hibernation sites to distant fields. Entomophagous arthropods seek refuge in various biotopes, mainly near the relevant agroecosystems. Therefore, it has been recommended that field size be limited to a few hectares and that field margins, hedges, and marginal biotopes be planted with appropriate flora and maintained. Such areas have been termed "zones of ecological compensation" (ZECs). These measures favor faunistic diversity for improved natural regulation (Basedow 1987b, 1990; Duelli et al. 1991; Storck-Weyhermüller 1987; Welling et al. 1987).

In general, beneficial organisms are less adapted to the special environment of a crop than are pest species. This is particularly true for species with few generations per year, like many Carabidae and Syrphidae. Natural enemies with several generations per year, e.g., parasitoids of aphids, must spend at least one generation in alternative hosts or in original hosts outside the crop. *Aphidius ervi* Hal. (Hymenoptera: Aphidiidae), the most important parasitoid of pea aphid (*Acyrthosiphon pisum* Harr., Homoptera: Aphididae) can develop in at least 14 aphid species, although with different reproductive success (Suter 1977). The same was demonstrated for other Aphidiidae (Basedow 1987b, Table 5), and is probably true for many other parasitoids of pests. For a discusssion of the relative merits of generalist and specialist natural enemies in arable crops see Helenius (this volume).

Various beneficial organisms migrate seasonally (Hemptinne & Naisse 1988) (Figure 13). After hibernation in forests, Coccinellidae disperse to hedges, nettle patches, and orchards, and then to the annual crop. Before migration to hibernation sites, they concentrate in maize fields or on plants in marginal biotopes, such as aphid-infested thistles (Figure 14) (Hemptinne 1988; Völkl 1988).

As mentioned earlier, adult aphidophagous syrphids respond to the presence of nectar and pollen produced by flowering weeds, and are therefore indicators for the intensity of weed control and for field border management (Molthan and Ruppert 1988; Häni 1989; Weiss & Stettmer 1991). Parasitoids, especially multivoltine species such as Aphidiidae, require flowers or honeydew and alternative hosts to survive fallow periods or to overwinter (Table 5). In southern Switzerland, *Anagrus atomus* Hal. (Hymenoptera: Mymaridae) is an important egg parasitoid of grape leafhopper (*Empoasca vitis* Goethe, Homoptera: Cicadellidae). Because the *E. vitis* overwinters as an adult, *A. atomus* attacks other leafhoppers that overwinter in the egg stage near the vineyard (Cerutti 1989). Further examples were discussed by Völkl and Keller (1991), demonstrating the use of parasitoids as indicators of landscape diversity.

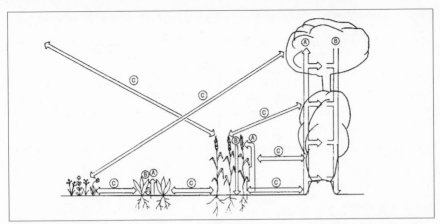

Figure 13. Migration paths of field fauna (schematic). Pests are a prerequisite for the development of beneficial arthropods. (Enhancement of the latter requires low levels of pest densities in the crop.) Some pests migrate to crops from hedges, etc. But they are also followed by their predators, which could develop in the hedge due to the presence of the pests. From Knauer, 1990, modified. A = Ascension from the soil surface into vegetation, B = Migration to protected places inside the same biotope, C = Migration to other biotopes.

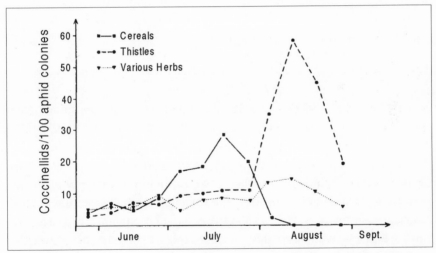

Figure 14. Incidence of lady beetles (Coleoptera: Coccinellidae) on various plant types from June to August (average of three areas; Völkl 1988).

A network or mosaic of ZECs can be introduced outside the crop area to provide resources needed by beneficial organisms (Duelli 1992). Such a mosaic might include marginal biotopes, hedges with herbaceous borders, field borders with 3–4 meters width (Welling 1990), and the subdivision of large fields by herbaceous strips ("artificial borders") (Nentwig 1989;

Table 5. The host range of five aphid parasitoids within and outside the crop (after Basedow 1987b)

Aphids on		Parasitized by				
		Aphidius picipes	*Aphidius ervi*	*Aphidius urticae*	*Praon volucre*	*Diaeretiella rapae*
Crop	Cereals	•	•			
	Beet	•	•			
	Peas	•	•	•		
	Brassica				•	•
	Potatoes		•			•
Plants in hedges						
	Rubus spp.		•	•	•	
	Rosa spp.	•			•	
	Urtica dioica		•	•	•	
Weeds						
	Galium spp.				•	•
	Atriplex spp.					•
	Chenopodium spp.					•
Plants at field margins						
	Galium spp.				•	•
	Trigonella sp.			•		
	Malva spp.					•
	Grass	•	•			•

Heitzmann et al. 1992) (Figure 15). Field borders with appropriate plant communities can also be attractive for pests and reduce the attack of the crop (Röser 1988). This was shown for field brome grass (*Bromus arvensis*), which is highly attractive to the wheat stem sawfly (*Cephus cinctus* Nort. Hymenoptera: Cephidae) and when sown on field borders may divert the pest from wheat (Van Emden 1981). To enhance a broad spectrum of entomophagous arthropods, borders and strips should be botanically diverse and contain either herbaceous plants, bushes or trees. Plants should be attractive to arthropods during both flowering and vegetative stages (see chapters by Bugg et al., Nentwig, and Wratten et al. this volume). The value of such zones in enhancing natural regulation has been demonstrated repeatedly (Stechmann & Zwölfer 1988). Hedges, borders of forests, and marginal biotopes contain many entomophagous arthropods, some of which disperse into the crop and act as beneficials. For example, Staphylinidae (Coleoptera) migrate between external biotopes and field margins (Figure 16). Use of external biotopes by various predators was reviewed by Röser (1988). Field-

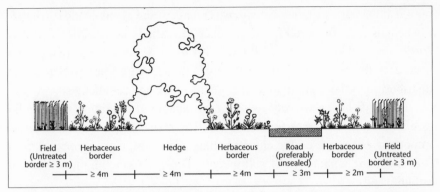

Figure 15. Layout of a hedge (schematic). For enhancement of natural regulation the herbaceous border of hedges and fields is of paramount importance. If roads are paved, they form a barrier for some predators such as Carabidae (Mader et al., 1988). (After Miess, 1987, modified.)

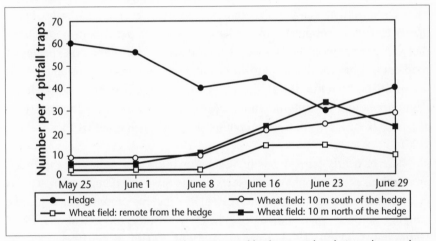

Figure 16. Influence of a hedge (with unmanured herbaceous border) on the number of rove beetles (Coleoptera: Staphylinidae), polyphagous predators, in an adjacent winter wheat field (Ruf & Häni unpublished).

margins also afford hibernation sites for entomopathogenic fungi that attack pests (Keller 1987; Keller, unpublished).

Predatory mites of orchards and vineyards may disperse seasonally from border vegetation. Mites are blown by the wind into perennial crops from the surroundings and thereafter spread through the agroecosystems. A well-documented case is the dispersal of *Typhlodromus pyri* from hedges and shrubs into Swiss vineyards. An irregular distributional pattern of the predatory mite within a vineyard in Northern Switzerland suggested an investigation as to the sources of *T. pyri* (Boller et al. 1988b). The predator's distribution at the border of the vineyard was mapped, and plants of the surrounding hedges were surveyed for its presence, revealing that blackberry (*Rubus fruticosus*) and less frequently dogwood (*Cornus sanguinea*) were the main sources of the predator. Predator density in the vineyard was correlated with distribution of these alternative host plants in the adjacent hedges.

Wetlands host relatively few species capable of migration and regulation of agricultural pests within fields (Röser 1988), but are nevertheless important as reservoirs for entomopathogenic fungi, e.g., *Entomophthora brevinucleata,* a pathogen of gall midges (Diptera: Cecidomyiidae) (Keller unpublished).

The natural regulation of aphid populations can be enhanced by artificial meadows, mainly pure lucerne (*Medicago sativa*) or a mixture of lucerne and red clover (*Trifolium pratense*) with one or more grass species (e.g., *Dactylis glomerata, Phleum pratense,* or *Lolium* spp.) (Suter & Keller 1977). This environment enables the aphids, mainly pea aphid (*Acyrthosiphon pisum* Harris, Homoptera: Aphididae), to establish permanent populations that serve throughout the season as a food source for numerous species of predators, some species of parasitoids, and the two most important species of aphid-pathogenic fungi, i.e., *Erynia neoaphidis* Rem. and Henn. (Entomophthorales: Entomophthoraceae) and *Conidiobolus obscurus* (Hall & Dunn) Rem and Kell. (Entomophthorales: Ancylistaceae) (Keller & Suter 1980).

Pure or nearly pure stands of vegetation are better suited for the "mass production" of aphidophagous insects than is diversified vegetation. This is true whether the vegetation is cultivated, like lucerne or maize, or wild, like stinging nettle (*Urtica dioica*) or blackberry (Stary 1983; Kokobu 1986; Hartmann & Duelli 1988; Katz et al. 1990; Duelli et al. 1991). This is probably due to the action of polyphagous natural enemies that are able to control the aphid development on single plants. In pure stands, the aphids escape their generalist natural enemies and can attain high densities until their specific antagonists arrive, encountering favourable conditions for their own reproduction. This hypothesis is supported by the results from weed management and undersowing (treated earlier in this chapter).

Permanent meadows with low inputs of fertilizer and high botanical diversity are of prime importance for beneficials, because the botanical richness meets the food requirements of a multitude of phytophagous insects. These insects have little or no importance as crop pests, but serve as food for beneficials in the meadows, as well as hosts to entomopathogens, enabling these meadows to serve as permanent reservoirs of biological control agents. An example illustrating this function of meadows is the aphid prediction system. Using densities of aphids and beneficials measured at the beginning of a growing season in the grassland communities adjacent to the crop, it is possible to predict subsequent changes in the aphid population. An early high density of aphids in the meadows leads to a multiplication of natural enemies that later migrate into the crop where they will help reduce aphids throughout the season. However, if the meadows harbor low aphid densities in the spring, natural enemies will be insufficient to limit an eventual aphid population explosion in the crop (Suter & Keller 1977).

Overall Farmland and Landscape Management

Diversification of agricultural landscapes plays an important role in pest control (e.g., Altieri 1985; El Titi 1987). Biological diversity can be reestablished in several ways, including cultivation methods, diversification of flora in the fields, rotations, mixed variety cultivation, permanent plant cover (intercropping), organic manure, undercropping, "rest weeds" (weedy crop, i.e., remnant weeds in the crops that remain below the economic threshold), untreated edge strips (weedy headlands), and subdivision of large fields using herbaceous strips, etc. (Wratten et al. this volume).

Landscape planning can also be important, as it can include the introduction and maintenance of ZECs, such as forest edges, hedges, permanent meadows, orchards with green undercover, river banks, wetlands, unpaved roads and roadsides that harbor weeds, farm courtyards, and fallow land. ZECs have different positive effects such as wind and erosion reduction, pollution mitigation, shelter for rare animals and plants, and recreation areas for contact with nature and rural life. Their most important function in agricultural production is as reservoirs for beneficial organisms.

For conservation of species and ecosystems, it may be reasonable to separate fully protected areas from the farmland (the "segregation" concept, per Hampicke 1988). For enhancement of natural regulation, only ZECs close to the fields are suitable (the "integration" concept, per Thomet & Thomet 1991). ZECs should be arranged in a chain-like network. Conservationists have repeatedly requested that about 10% of the farmland be allocated to ZECs (Röser 1988; Broggi & Schlegel 1989), and that there be a maximal mesh distance of about 400 meters with an optimum of 75–125 meters (Knauer 1990). Such a network would also be suitable for enhancement of beneficials, but this approach must be supplemented and supported by the previously discussed measures inside and outside the crop. ZEC methodology has been reviewed, including choice of varieties, arrangement, planting, and subsequent maintenance (Miess 1987; Röser 1988; Amstutz et al. 1990; Knauer 1990; Häni 1992). Effects of different ZECs on individual beneficial species have also been reviewed (Röser 1988). Plant composition should be determined carefully to optimize the enhancement of beneficials and to prevent outbreaks of pests or diseases due to introduced host plants (Häni 1992).

Conclusion

As of 1996, the results of the GATT (General Agreement on Tariffs and Trade) negotiations were leading to reduced public subsidies in the form of price supports for agricultural products, and more market-oriented production strategies. Some farmers counteracted decreasing prices by further intensifying farm operations. At the same time, there was increasing public and political pressure to shift from policies aimed at destruction of surpluses to a strategy subsidizing farmers who employ sustainable practices. One approach to this is direct payments to farmers, linked to ecological performance. Such payments can be used to encourage increased use of natural farm resources and regulation. An example is the implementation of a policy of direct payments to farmers practicing Organic Agriculture or Integrated Farming and for zones of ecological compensation in Switzerland (farm bill, article 316, 1993). For the first time, taxpayers honor ecologically oriented forms of agriculture that strive to increase biological diversity at varietal, species, and ecosystem levels. Compensation for lost farm income to subsidize an ecologically oriented, sustainable agriculture will increase public awareness and motivation of farmers to reintroduce, protect, and enhance those natural regulating factors that are the main theme of this book. At the printing of this book (1998) these visions have also been realized in the EU by ecologically motivated direct payments (directive 2078/92).

Integrated Farming and related concepts may be an attractive option for Eastern European and other countries lacking capital but having a good basis for biological control (cf. Hluchý 1987; Edwards et al. 1990). It is hoped that those countries will not be lured into short-term gain using high inputs of chemical pesticides and fertilizers. Whatever the political and economic framework might be, agroecosystem-oriented, sustainable farming systems that protect natural resources and enhance natural regulation should be considered worldwide as the only reasonable and responsible approach.

Rapid progress can be made if on-farm system research is intensified, transfer of knowledge improved, and coordination of research activities facilitated through formation of national and international networks of pilot- and experimental farms. Such a research network was started in 1991 by six countries of the European Community: Germany, Netherlands, Denmark, France, Italy, and Great Britain. Research networks have also been started on national levels. For example, as of 1990, Switzerland had some 200 pilot- and experimental farms involved in integrated or organic production, with guidance from local and regional advisors.

We anticipate that scientists in Europe and worldwide will give greater focus to critical aspects of ecosystem management. We also foresee a faster rate of increase in scientific knowledge and its implementation. The new approaches to agriculture will require increased protection of the soil, e.g., through extended green cover. Other priorities will include the protection and expansion of existing zones of ecological compensation, and a general reduction in broad-spectrum and persistent pesticides. These approaches will enhance the abundance and impact of beneficial and indifferent organisms alike and stabilize pest populations at lower densities.

Acknowledgment

We would like to thank H. Ramseier, J. Pitt (Swiss College of Agriculture, Zollikofen), the late U. Remund (Swiss Federal Research Station for Arboriculture, Viticulture & Horticulture, Wädenswil) and the editors R. L. Bugg and C. H. Pickett for their helpful suggestions and comments.

References

Aeschlimann, J. P. & K. P. Carl. 1987. Les tactiques curatives à l'aide d'ennemis natureles, pp. 167-192. In Delucchi V. (Ed.) Protection intégrée—quo vadis? Parasitis, Geneva, Switzerland.

Altieri, M. A. 1985. Diversification of agricultural landscapes—a vital element for pest control in sustainable agriculture, pp. 166-184. In T. C. Edens, C. Fridgen, S. L. Battenfield [eds.]. Sustainable agriculture and integrated farming systems. Michigan State University Press, East Lansing.

Altieri, M. A. 1987. Agroecology: the scientific basis of alternative agriculture. Westview Press.

Amstutz, M., M. Dick, & N. Hufschmid. 1990. Natur aus Bauernhand, ein Leitfaden zur ökologischen Landschaftsgestaltung. Forschungsinstitut für Biologischen Landbau, Basel-Oberwil.

Anken, T., J. Heusser, P. Weisskopf, U. Zihlmann, H. R. Forrer, C. Hoegger, C. Scherrer, A. Mozafar & W. G. Sturny 1997. Bodenbearbeitungssysteme: Die Direktsaat stellt hoechste Anforderungen. Eidegenoessische Forschungsanstalt für Agrarwirtschaft und Landtechnik-Berichte Nr. 501, pp.1-14.

Anonymous. 1977. Vers la production agricole intégrée / (an approach towards integrated agricultural production through integrated plant protection). Bulletin 4 of the International Organization for Biological Control of Noxious Animals and Plants / West Palaearctic Regional Section.

Anonymous. 1990. Anbauempfehlungen für integrierten Ackerbau. Schweizerischer Verband der Ingenieur-Agronomen, Landwirtschaftliche Lehrmittelzentrale Zollikofen 42 pp., 2nd edition.

Anonymous. 1991a. Richtlinien über Verkaufsprodukte aus biologischem/ökologischem Landbau. Vereinigung schweizerischer biologischer Landbau-Organisationen, 8th edition.

Anonymous. 1991b. Biopesticidy V Zemedelstvi. Ministerstvo Zemedelstvi CR, Praha.

Baeumer, K. 1990. Verfahren und Wirkungen der Bodenbearbeitung, pp. 68-87. In R. Diercks & R. Heitefuss [eds.], Integrierter Landbau. Verlagsunion Agrar, München.

Baggiolini, M. 1978. La valorisation qualitative de la production agricole. Revue Suisse Viticulture, Arboriculture, Horticulture 10:51-57.

Baggiolini, M. 1990. Production Intégrée en Suisse. Aperçu historique de la "production agricole intégrée." Mitteilungen Schweizerische Entomologische Gesellschaft 63:493-500.

Basedow, T. 1987a. Der Einfluss gesteigerter Bewirtschaftungsintensität im Getreidebau auf die Laufkäfer (Coleoptera, Carabidae). Mitteilungen Biologische Bundesanstalt für Land- und Forstwirtschaft 235.

Basedow, T. 1987b. Die Bedeutung von Hecken, Feldrainen und pflanzenschutzmittelfreien Ackerrandstreifen für die Tierwelt der Äcker. Gesunde Pflanzen 39:421-429.

Basedow, T. 1990. Zum Einfluss von Feldrainen und Hecken auf Blattlausräuber, Blattlausbefall und die Notwendigkeit von Insektizideinsätzen im Zuckerrübenanbau. Gesunde Pflanzen 42:241-245.

Bigler, F. 1986. Möglichkeiten und Grenzen der biologischen Schädlingsbekämpfung im integrierten Pflanzenschutz in der Schweiz. Schweizerische Landwirtschaftliche Forschung 25:325-339.

Bigler, F. 1987. Gegenwärtiger Stand und Zukunftsaussichten der biologischen Schädlingsbekämpfung in der Schweiz. Schriftenreihe "Biologischer Pflanzenschutz" des Bundesministers für Ernährung, Landwirtschaft und Forsten (Germany), Heft 344.

Bigler, F. 1988. Qualitative Vernetzungen im Agro-Oekosystem des Mais unter besonderer Berücksichtigung des Pflanzenschutzes. Schweizerische Landwirtschaftliche Forschung 27:63-71.

Bigler, F. & F. Cerutti. 1991. Bericht über den Besuch am "Union Institut für Biologische Methoden im Pflanzenschutz" in Kishinev, Moldau, USSR. Unpublished report, Swiss Federal Research Station for Agronomy, Zurich-Reckenholz.

Boller, E. 1988. Das mehrjährige Agro-Oekosystem "Rebberg." Schweizerische Landwirtschaftliche Forschung 27:55-61.

Boller, E. 1989. The ecosystem approach to plan and implement integrated plant protection in viticulture of Eastern Switzerland, pp. 607-617. *In* R. Cavalloro (ed.). Plant-protection problems and prospects of integrated control in viticulture. Proceedings of the Commission of the European Community / International Organization for Biological and Integrated Control of Noxious Animals and Plants (IOBC) International Symposium, Lissabon, 6-9 June 1988.

Boller, E. 1990. Integrierte Produktion in der Schweiz. II. Weiterentwicklung der integrierten Produktion in den achtziger Jahren, heutiger Stand und künftige Entwicklungen. Mitteilungen Schweizerische Entomologische Gesellschaft 63:501-505.

Boller, E., P. Basler & W. Koblet. 1990. Integrated production in viticulture of Eastern Switzerland: concepts and organization (the "Wädenswil Model"). Schweizerische Landwirtschaftliche Forschung 29:287-291.

Boller, E., F. Bigler, M. Bieri, F. Häni & A. Stäubli. 1989. Nebenwirkungen von Pestiziden auf die Nützlingsfauna landwirtschaftlicher Kulturen. Schweizerische Landwirtschaftliche Forschung 28:3-40.

Boller, E., F. Bigler, J. O. Derron, H. R. Forrer & P. M. Fried. 1988a. Allgemeiner Aufbau eines Agro-Oekosystems aus phytomedizinischer Sicht und mögliche Anwendung in der Praxis. Schweizerische Landwirtschaftliche Forschung 27:49-53.

Boller, E. & B. Frey. 1990. Blühende Rebberge in der Ostschweiz: 1. Zur Bedeutung des Pollens für die Raubmilben. Schweizerische Zeitschrift für Obst- und Weinbau 126:401-405.

Boller, E., U. Remund & M.P. Candolfi. 1988b. Hedges as potential sources of *Typhlodromus pyri*—the most important predatory mite in vineyards of Northern Switzerland. Entomophaga 33:15-22.

Broggi, M. F. & H. Schlegel. 1989. Mindestbedarf an naturnahen Flächen in der Kulturlandschaft. Bericht 31 des Nationalen Forschungsprogramms Boden, Bern.

Büchi, R. 1989. Unterschiedlicher Befall von 4 verschiedenen 00-Sorten durch den Stengelrüssler (*C. napi*) in der Schweiz im Jahre 1988. Bulletin of the Groupe Consultatif International de Recherche sur le Colza (GCIRC) 5:66.

Büchi, R. 1990. Investigations on the use of turnip rape as trap plant to control oilseed rape pests. Bulletin of the International Organization for Biological Control of Noxious Animals and Plants / West Palaearctic Regional Section 13: 32-39.

Büchi, R., F. Häni, B. Schenk, P. Frei. 1987. Rübsen in Raps als Fangpflanzen für Rapsschädlinge. Mitteilungen Schweizerische Landwirtschaft 35 (1/2):34-40.

Buchner, W. & K. Köller. 1990. Integrierte Bodenbearbeitung. Ulmer, Stuttgart.

Cassini, R. 1969. Les fusarioses du blé. Bulletin Technique d'Information 244:819-822.

Cassini, R. 1973. Influences des techniques de culture sur le développement des maladies des céréales. pp. 17-34. *In* Proceedings of Journées d'études sur la lutte contre les maladies des céréales. Versailles.

Cerutti, F. 1989. Modellizzazione della dinamica delle popolazioni di *Empoasca vitis* Goethe (Hom., Cicadellidae) nei vigneti del Cantone Ticino e influsso della flora circostante sulla presenza del parassitoide *Anagrus atomus* Haliday (Hym., Mymaridae). Dissertation. Eidgenössische Technische Hochschule Zürich, Nr. 9019.

Colby, M. E. 1990. Environmental management in development—the evolution of paradigms. World Bank Discussion Papers. No. 80. Washington, D.C.

Corbett, D. C. & R. M. Webb. 1970. Plant and soil nematode population changes in wheat grown continuously in ploughed and in unploughed soil. Ann. Appl. Biol. 65:327-335.

Delorme, R. & R. Fritz. 1978. Action de divers fongicides sur le développement d'une mycose à *Entomophthora aphidis*. Entomophaga 23:389-401.

Delucchi, V. 1979. Le paradigme "intégré". Recherche Agronomique en Suisse 18:213-223.

Delucchi, V. (ed.). 1987. Integrated Pest Management—quo vadis? Parasitis, Geneva, Switzerland.

Diercks, R. 1986. Alternativen im Landbau. Ulmer, Stuttgart. 379 pp., 2. Aufl.

Diercks, R. & R. Heitefuss, (eds.). 1990. Integrierter Landbau. Verlagsunion Agrar, München.

Duelli, P. 1992. Mosaikkonzept und Inseltheorie in der Kulturlandschaft. Verhandlungen der Gesellschaft für Ökologie. Freising-Weihenstephan 21:379-383.

Duelli, P., E. Blank & M. Frech. 1991. The contributions of seminatural habitats to arthropod diversity in agricultural areas. Proceedings of the European Congress of Entomology, Budapest (Gödöllö).

Edens, T. C., C. Fridgen & S. L. Battenfield. 1985. Sustainable agriculture and integrated farming systems. 1984 Conference Proceedings. Michigan St. Univ. Press, East Lansing.

Edwards C. A., R. Lal., P. Madden, R. H. Miller & G. House. 1990. Sustainable agricultural systems. Soil and Water Conservation Society, Ankeny, Iowa.

El Titi, A. 1984. Auswirkungen der Bodenbearbeitungsart auf die edaphischen Raubmilben (Mesostigmata: Acarina). Pedobiologia 27:79-88.

El Titi, A. 1987. Environmental manipulation detrimental to pests, pp. 105-121. *In* V. Delucchi [ed.]. Integrated Pest Management—quo vadis? Parasitis, Geneva.

El Titi, A. 1989. Farming systems research at Lautenbach. Bulletin of the International Organization for Biological Control of Noxious Animals and Plants / West Palaearctic Regional Section 12:21-35.

El Titi, A. 1990a. Modellvorhaben "Lautenbacher Hof," pp. 316-329. *In* R. Diercks & R. Heitefuss [eds.]. Integrierter Landbau. Verlagsunion Agrar, München.

El Titi, A. 1990b. Farming system research at Lautenbach, Germany. Schweizerische Landwirtschaftliche Forschung 29:237-247.

El Titi, A. 1991. Integrated farming system. Motivations/definition & principles/framework/guidelines. Unpublished report.

El Titi, A. 1992. Integrated farming: an ecological farming approach in European agriculture. Outlook on Agriculture 21:33-39.

El Titi, A., E. Boller & J.P. Gendrier, eds. 1993. Integrated production: principles and technical guidelines. Bulletin of the International Organization for Biological Control of Noxious Animals and Plants / West Palaearctic Regional Section 16(1).

Engel, R. 1991. Die Raubmilbe *Typhlodromus pyri*—mehr als nur ein Spinnmilbenvertilger. Deutsches Weinbau-Jahrbuch 42:217-224.

Frings, B. 1988. Untersuchungen über die Möglichkeiten der Erhaltung und Förderung von Nützlingen im Zuckerrübenanbau. Dissertation, Universtität Bonn.

Fuchs, W. H. & F. Grossmann. 1972. Ernährung und Resistenz von Kulturpflanzen gegenüber Krankheitserregern und Schädlingen, pp. 1006-1107. *In* H. Linser (ed.). Handbuch der Pflanzenernährung und Düngung. Springer Verlag, Berlin.

Garrett, S. D. 1970. Pathogenic root infecting fungi. Cambridge Univiversity Press, UK.

Gliessmann, S. R. 1990. Agroecology: researching the ecological basis for sustainable agriculture. Springer, New York.

Graber, C. & H. Suter. 1985. Schnecken-Regulierung. Forschungs-Institut für biologischer Landbau, Basel-Oberwil.

Hampicke, U. 1988. Extensivierung der Landwirtschaft für den Naturschutz: Ziele, Rahmenbedingungen und Massnahmen. Schriftenreihe Bayer, Landesamt für Umweltschutz, München, 84:9-35.

Häni, A., H.U. Ammon & S. Keller. 1990. Vom Nutzen der Unkräuter. Landwirtschaft Schweiz 3:217-221.

Häni, F. 1980. Fusarium diseases of cereals in Switzerland: prevalence on cereal seed, on wheat ears, and in soil. Zeitschift für Pflanzenkrankeiten und Pflanzenschutz. 87:257-280.

Häni, F. 1981. Zur Biologie unde Bekämpfung von Fusariosen bei Weizen und Roggen. Phytopathologische Zeitschrift 100:44-87.

Häni, F. 1989. Third Way, a research project in ecologically oriented farming systems in Switzerland. Bulletin of the International Organization for Biological Control of Noxious Animals and Plants/West Palaearctic Regional Section 12(5):51-66.

Häni, F. 1990a. Farming systems research at Ipsach, Switzerland—the "Third Way" project. Schweizerische Landwirtschaftliche Forschung 29(4):257-271.

Häni, F. 1990b. Oekologische und ökonomische Auswirkungen des Bewirtschaftungssystems. Schweizerische Landwirtschaftliche Forschung 29:83-97.

Häni, F. 1992. Nützlinge und Landschaftsgestaltung, pp. 305-366. *In* F. Häni,G. Popow, H. Reinhard, A. Schwarz, & K. Tanner. Pflanzenschutz im Integrierten Ackerbau. 3. Ergänzte Auflage, landwirtschaftliche Lehrmittel-zentrale, Zollikofen.

Häni, F. 1993. Weiterentwicklung umweltschonender Bewirtschaftungssysteme: Projekt "Dritter Weg." Schweizerische Landwirtschaftliche Forschung 32:341-364.

Häni, F. 1994. Entwicklung ökologisch ausgerichteter Bewirtschaftungssysteme in der Schweiz, pp. 329-338, 426-427. *In* R. Diercks & R. Heitefuß [eds.]. Integrierter Landbau. Verlagsunion Agrar, München. 2nd edtion.

Häni, F., E. Boller & F. Bigler. 1990. Integrierte Produktion—ein ökologisch ausgerichtetes Bewirtschaftungssystem. Schweizerische Landwirtschaftliche Forschung 29:101-115.

Häni, F., E. Boller & F. Bigler. 1991. Integrated production—a way to achieve ecologically sound agriculture. World Farmer's Times 1:15-18.

Häni, F. & P. Vereijken, eds. 1990. Development of ecosystem-oriented farming. Schweizerische Landwirtschaftliche Forschung 29:221-436.

Hansen, W. 1986. Die Populationsdynamik von Blattläusen an Weizen in Abhängigkeit von der Qualität des Phloemsaftes bei unterschiedlicher N-Düngung. Dissertation. Universität Göttingen.

Hartmann, K. & P. Duelli. 1988. Flächenbezogene Angaben zur Erfassung der Produktion (Schlüpfrate) von adulten Aphidophagen in naturnahen Gebieten. Mitteilungen der Deutschen Gesellschaft für allgemeine und angewandte Entomologie 6:182-187.

Hassan, S. A. (ed.). 1985. Standard methods to test the side-effects of pesticides on natural enemies of insects and mites developed by the IOBC/WPRS-Working Group on "Pesticides and Beneficial Organisms." European and Mediterranean Plant Protection Organization (EPPO) Bulletin 15:214-255.

Hassan, S. A. (ed.). 1991. Results of the fifth joint pesticide testing programme carried out by the International Organization For Biological and Integrated Control of Noxious Animals and Plants / West Palaearctic Regional Section—Working Group on "Pesticides and Beneficial Organisms." Entomophaga 36:55-67.

Heitefuss, R. & V. Garbe. 1986. Pflügen oder nicht pflügen-Konsequenzen für den Pflanzenschutz. Gesunde Pfanzen 38:529-533.

Heitefuss, R., K. König, A. Obst, & M. Reschke. 1984. Pflanzenkrankheiten und Schädlinge im Ackerbau. DLG-Verlag. Frankfurt (Main).

Heitzmann, A., G. A. Lys & W. Nentwig. 1992. Nützlingsförderung am Rand-oder: vom Sinn des Unkrauts. Landwirtschaft Schweiz 5:25-36.

Hemptinne, J. L. 1988. Ecological requirements for hibernating *Propylea quatuor-decimpunctata* (L.) and *Coccinella septempunctata* (Col.: Coccinellidae). Entomophaga 33:505-515.

Hemptinne, J. L. & J. Naisse. 1988. Life cycle strategy of *Adalia bipunctata* (L.) (Col., Coccinellidae) in a temperate country, pp. 71-77. *In* E. Niemczyk & A.F.G. Dixon [eds.]. Ecology and effectiveness of Aphidophaga. SPB Academic Publishers, The Hague, Netherlands.

Herrmann, G. & G. Plakolm. 1991. Oekologischer Landbau. Verlagsunion Agrar, Vienna, Austria.

Hluchý, M. 1987. Soucasny stav v CSSR pouzivanych biologickych metod ochrany rostlin (Present state of the biological methods of plant protection currently used in the CSSR. In Czech, English summary). Agrochémia 26:113-115.

Hluchý, M. 1988. Hodnoceni rozshau uplatneni metod biologické ochrany rostlin v CSSR v roce 1987. Agrochemicky Zpravodaj, 1-3.

Hluchý, M. 1989. Erfahrungen mit dem Einsatz der Raubmilbe *Typhlodromus pyri* zum biologischen Schutz der Weinrebe in der Tschechoslowakei. Der Pflanzenarzt 11-12:12-15.

Hluchý, M. & Z. Pospísil. 1991. Use of the predatory mite *Typhlodromus pyri* Scheuten (Acari, Phytoseiidae) for biological protection of grape vines from phytophagous mites, pp. 655-660. *In* F. Dusbabek & V. Bukva V. [eds.]. Modern acarology, Vol. 2. Academia, Prague and SPB Academic Publishing, The Hague, Netherlands.

Hofer, H. 1990. "Agri natura" guarantees agricultural production which respects the environment and animals, pp. 313-323. *In* F. Häni & P. Vereijken [eds.]. Development of ecosystem-oriented farming. Schweizerische Landwirtschaftliche Forschung 29.

Jäggi, W. 1989. Bestimmung des Zellulose-Abbaus im Boden im Freilandversuch. Unpublished report, Swiss Federal Research Station Zurich-Reckenholz.

Jenny, M. 1990. Die Feldlerche, ein Charaktervogel des Wies- und Ackerlandes in Gefahr. Schweizerische Vogelwarte Sempach.

Katz, E., P. Duelli & P. Wiedemeier. 1990. Der Einfluss der Nachbarschaft naturnaher Biotope auf Phänologie und Produktion von entomophagen Arthropoden in Intensivkulturen. Mitteilungen der Deutschen Gesellschaft für Allgemeine und Angewandte Entomologie 7:306-310.

Keller, E. R. 1985. Integrierte Pflanzenproduktion—Konzept für die Erzeugung gesunder Nahrungs- und Futtermittel. Schweizerische Landwirtschaftliche Monatshefte 63:233-258.

Keller, E. R. & P. Weisskopf. 1987. Integrierte Pflanzenproduktion. Landwirtschaftliche Lehrmittelzentrale, Zollikofen.

Keller, S. 1987. Observations on the overwintering of *Entomophthora planchoniana*. J. Invert. Pathol. 50:333-335.

Keller, S. 1989. Auftreten der Erbsengallmücke *Contarinia pisi* Winn. bei Konservenerbsen und Möglichkeiten ihrer Bekämpfung. Landwirtschaftliches Jahrbuch der Schweiz 2:57-62.

Keller, S. 1991. Nützlinge und ihre praktische Bedeutung. Die Grüne 40:20-24.

Keller, S. & P. Duelli. 1990. Ökologische Ausgleichsflächen und ihr Einfluss auf die Regulierung von Schädlingspopulationen. Mitteilungen Schweizerische Entomologische Gesellschaft 63:431-437.

Keller, S., E. Keller, C. Schweizer, J. A. L. Auden & A. Smith. 1989. Two large field trials to control the cockchafer *Melolontha melolontha* L. with the fungus *Beau-veria brongniartii* (Sacc.) Petch, pp. 183-190. *In* Progress and Prospects in Insect Control. British Crop Protection Council Monogr. No. 43.

Keller, S. & C. Schweizer. 1991. Die Wirkung von Herbiziden auf das Sporulierungsvermögen des blattlauspathogenen Pilzes *Erynia neoaphidis*. Anzeiger Für Schädlingskunde, Pflanzenschutz, Umweltschutz 64:134-136.

Keller, S. & H. Suter. 1980. Epizootiologische Untersuchungen über das *Entomophthora*-Auftreten bei feldbaulich wichtigen Blattlausarten. Acta Oecologica, Oecologica Applicata 1:63-81.

Knauer, N. 1990. Agroökosysteme im konventionellen und im integrierten Landbau, pp. 19-31. *In* R. Diercks & R. Heitefuss [eds.]. Integrierter Landbau. Verlagsunion Agrar, München.

Koblet, R. 1965. Der Landwirtschaftliche Pflanzenbau. Birkhäuser Basel.

Kokobu, H. 1986. Migration rates, *in situ* reproduction and flight characteristics of aphidophagous insects in cornfields. Dissertation. Universität Basel, Switzerland.

Krieg, A. & J. M. Franz. 1989. Lehrbuch der biologischen Schädlingsbekämpfung. Paul Parey, Berlin.

Lampkin, N. 1990. Organic Farming. Farming Press. Ipswich, U.K.

Latteur, G. 1977. Sur la possibilité d'infection directe d'aphides par *Entomophthora* à partir de sols hébergeants un inoculum naturel. Comptes Rendus de Séances de l'Académie de Sciences, Série D, Paris, Sciences Naturelles 284:2253-2256.

Mader, H.-J., C. Schell, & P. Kornacker. 1988. Feldwege-Lebensraum und Barriere. Natur und Landschaft 63:251-256.

Miess, B. 1987. Landschaft als Lebensraum-Biotopvernetzung in der Flur. Baden-Württemberg (Germany), Ministerium für ländlichen Raum, Ernährung, Landwirtschaft und Forsten.

Molthan, J. & V. Ruppert. 1988. Zur Bedeutung blühender Wildkräuter in Feldrainen und Äckern für blütenbesuchende Nutzinsekten. Mitteilungen aus der Biologischen Bundesanstalt für Land- und Forstwirtschaft, Berlin-Dahlem 247:85-99.

Naïbo, B. & L. Foulgocq. 1985. Mise en évidence d'un effet favorisant le puceron *Rhopalosiphum padi* à la suite d'un traitement par pyrethrinoides sous formulation liquide dirigé contre la Pyrale du maïs, *Ostrinia nubilalis* Hbn, pp. 345-354. *In* J. Thiault, A. Fougeroux & N. Beyt [eds.]. Faune et flore auxiliaire en agriculture. ACTA, Paris.

National Research Council. 1989. Alternative agriculture. Natural Academy Press, Washington D.C.

Nentwig, W. 1989. Augmentation of beneficial arthropods by strip management. II. Successional strips in a winter wheat field. Zeitschrift für Pflanzenkrankheiten und Pflanzenschutz 96:89-99.

Palti, J. 1981. Cultural practices and infectious crop diseases. Springer Verlag, Berlin.

Pöhling, H.M. 1987. Effects of reduced dose rates of pirimicarb and fenvalerate on aphids and beneficials arthropods in winter wheat. Bulletin of the International Organization for Biological Control of Noxious Animals and Plants / West Palaearctic Regional Section 10(1):184-193.

Powell, W. 1983. The role of parasitoids in limiting cereal aphid populations, pp. 50-56. *In* R. Cavalloro (ed.). Aphid antagonists. Balkema, Rotterdam.

Powell, W. 1986. Enhancing parasitoid activity in crops, pp. 319-340. *In* J. Waage & D. Greathead [eds.]. Insect parasitoids. Academic Press, London.

Remund, U. & E. Boller. 1991. Möglichkeiten und Grenzen von Eiparasiten zur Traubenwicklerbekämpfung. Schweizerische Zeitschrift für Obst-Weinbau 127:535-540.

Remund, U. & E. Boller. 1992. Blühende Rebberge in der Ostschweiz. 3. Ergänzende Pollenfrassversuche mit Raubmilben. Schweizerische Zeitschrift für Obst- und Weinbau 128:237-240.

Remund, U., D. Gut & E. Boller. 1992. Beziehungen zwischen Begleitflora und Arthropodenfauna in Ostschweizer Rebbergen.—Einfluss der botanischen Vielfalt auf die ökologische Stabilität. Schweizerische Zeitschrift für Obst- und Weinbau 128:527-540.

Remund, U., U. Niggli & E. Boller. 1989. Faunistische und botanische Erhebungen in einem Rebberg der Ostschweiz—Einfluss der Unterwuchsbewirtschaftung auf des Oekosystem Rebberg. Landwirtschaft Schweiz 2:393-408.

Röser, B. 1988. Saum- und Kleinbiotope. Oekologische Funktion, wirtschaftliche Bedeutung und Schutzwürdigkeit in Agrarlandschaften. Ecomed Verlag, Landsberg (Germany).

Schäfermeyer, S. & E. Dickler. 1991. Vergleichende Untersuchungen zu Richtlinien für die integrierte Kernobstproduktion in Europa. Mitteilungen aus der Biologischen Bundesanstalt für Land- und Forstwirtschaft, Berlin-Dahlem.

Schönbeck, F., F. Klingauf, and P. Kraus. 1988. Situation, Aufgaben und Perspektiven des Biologischen Pflanzenschutzes. Gesunde Pflanzen 40:86-96.

Schüepp, H. & Bodmer M. 1991. Complex response of VA-Mycorrhizae to xenobiotic substances. Toxic. Environ. Chem. 30:193-199.

Schüepp, H., M. Bodmer & D.D. Miller. 1992. A cuvette system designed to enable monitoring of hyphae of vesicular-arbuscular mycorrhizal fungi external to plant roots. Methods Microbiol. 24:67-76.

Schweizer, C. 1992. Einfluss von Unkraut auf Spinnmilben in Hopfenkulturen. Landwirtschaft Schweiz 5:597-599.

Sotherton, N. W. 1990. The environmental benefits of conservation headlands in cereal fields. Pesticide Outlook 1:14-18.

Staal, G. B. 1987. Juvenoids and antijuvenile hormone agents as insect growth regulators, pp. 277-292. In Delucchi V. (ed.). Integrated Pest Management—quo vadis? Parasitis, CH-Geneva.

Stary, P. 1983. The perennial stinging nettle (Urtica dioica) as a reservoir of aphid parasitoids (Hymenoptera, Aphidiidae). Acta Entomologica Bohemoslovaca 80:81-86.

Stechmann, D.H. & H. Zwölfer. 1988. Die Bedeutung von Hecken für Nutzarthropoden in Agrarökosystemen. Landwirtschaft-Angewandte Wissenschaft 365:31-55.

Steiner, H. 1965. Zwölf Jahre Arbeit am integrierten Pflanzenschutz im Obstbau. Gesunde Pflanzen 17:5-13.

Steinmann, R. 1983. Der biologische Landbau—ein betriebswirtschaftlicher Vergleich. Eidgenössische Forschungsanstalt für Betriebswirtschaft und Landtechnik, Tänikon, Schrift Nr. 19.

Stern, V.M., R. Van den Bosch & K.S. Hagen. 1959. The integrated control concept. Hilgardia 29(2):81-101.

Storck-Weyhermüller, S. 1987. Untersuchungen zum Einfluss natürlicher Feinde auf die Populationsdynamik der Getreideblattläuse sowie über die Wirkung niedriger Dosierungen selektiver Insektizide auf die Aphiden und deren spezifische Predatoren. Dissertation. Universität Giessen.

Storck-Weyhermüller, S. 1990. Herbizidfreie Ackerrandstreifen—faunistische Untersuchungen zum hessischen Ackerschonstreifenprogramm. Mitteilungen aus der biologischen Bundesanstalt für Land- und Forstwirtschaft, Berlin-Dahlem 266:80.

Suter, H. 1977. Populationsdynamik der Erbsenblattlaus (*Acyrthosiphon pisum* Harris) und ihrer Antagonisten. Eidgenössische Technische Hochschule Zürich Dissertation. Nr. 5932.

Suter, H. 1988. Mais: vom Öko-Saulus zum Öko-Paulus. Mais 2:42-44.

Suter, H. & S. Keller. 1977. Ökologische Untersuchungen an feldbaulich wichtigen Blattlausarten als Grundlage für eine Befallsprognose. Zeitschrift für angewandte Entomologie 83:371-393.

Thomet, P., W. Schmid & R. Daccord. 1989. Erhaltung von artenreichen Wiesen. Bericht 37 Nationales Forschungsprogramm Boden, Liebefeld-Bern.

Thomet, P. & E. Thomet. 1991. Propositions en faveur d'un aménagement et d'une exploitation écologiques du paysage agraire. Rapport du Programme National Sol, Liebefeld-Berne.

Tränkner, A. & R. Liesenfeld. 1990. Unterdrückung von *Pythium ultimum*, einer Keimlingskrankheit der Erbse (*Pisum sativum*) durch Saatgutbehandlung mit Kompostextrakten. Mitteilungen aus der Biologischen Bundesanstalt für Land- und Forstwirtschaft, Berlin-Dahlem. 266:64.

Troxler, C. & J. Zettel. 1987. Der Einfluss verschiedener Bewirtschaftungsweisen auf die Mikro-Arthropodenfauna in Rebbergen bei Twann. Mitteilungen Naturforschende Gesellschaft in Bern. N.F. 44:187-202.

Ulber, B, G. Stippich & W. Wahmhoff. 1990. Möglichkeiten, Grenzen und Auswirkungen des gezielten Pflanzenschutzes im Ackerbau: III. Auswirkungen unterschiedlicher Intensität des chemischen Pflanzenschutzes auf epigäische Raubarthropoden in Winterweizen, Zuckerrüben und Winterraps. Zeitschrift für Pflanzenkrankheiten und Pflanzenschutz 97:263-283.

USDA. 1990. The Low-Input/Sustainable Agriculture (LISA) Program. *In* Alternative opportunities for U.S. farmers. USDA-Cooperative State Research Service, Washington D.C.

USDA. 1991. The basic principles of Sustainable Agriculture (also Called "Alternative Agriculture" and "LISA"). USDA-Cooperative State Research Service, Washington, D.C.

van Emden, H. F. 1981. Wild plants in the ecology of insect pests, pp. 251-261. *In* J. M. Thresh (ed.). Pests, pathogens and vegetation: the role of weeds and wild plants in the ecology of crop pests and diseases. Pitman Publishers, Massaschusetts.

Van Lenteren, J. C. 1987a. Environmental manipulation advantageous to natural enemies of pests, pp. 123-163. *In* V. Delucchi (ed.). Integrated pest management—quo vadis? Parasitis, CH-Geneva.

Van Lenteren, J. C. 1987b. Biologischer Pflanzenschutz in Gwächshäusern: Wohin? Schriftenreihe "Biologischer Pflanzenschutz" des Bundesministers für Ernährung, Landwirtschaft und Forsten (Germany). 344:226-244.

Van Lenteren, J.C. 1988. Biological and integrated pest control in greenhouses. Annu. Rev. Entomol. 33:239-269.

Van Wingerden, W.K.R.E., A.R. Van Kreveld & W. Bongers. 1992. Analysis of species composition and abundance of grasshoppers (Orth., Acrididae) in natural and fertilised grassland. J. Appl. Entomol. 113:138-152.

Vereijken, P. 1990a. Research on integrated arable farming and organic mixed farming in the Netherlands. Schweizerische Landwirtschaftliche Forschung 29:249-256.

Vereijken, P. 1990b. Methodology of (arable) farming systems research. Schweizerische Landwirtschaftliche Forschung 29:349-358.

Vereijken, P. (ed.). 1995. Designing and testing prototypes. Progress Report 2. Research Institute of Agrobiology and Soil Fertility, Wageningen, The Netherlands.

Vereijken, P., C. Edwards, A. El Titi, A. Fougeroux, M. Way. 1986. The management of arable farming systems for integrated crop protection. Bulletin of the International Organization for Biological Control of Noxious Animals and Plants / West Palaearctic Regional Section 9(2).

Vereijken, P. & D. J. Royle (eds.). 1989. Current status of integrated farming systems research in Western Europe. Bulletin of the International Organization for Biological Control of Noxious Animals and Plants / West Palaearctic Regional Section 12(5).

Völkl, W. 1988. Food relations in *Cirsium*-aphid-system: The role of the thistles as a base for aphidophagous insects, pp. 145-148. *In* E. Niemczyk and A.F.G. Dixon [eds.]. Ecology and effectiveness of aphidophaga. SPB Academic Publishing, The Hague.

Völkl, W. & S. Keller 1991. Insekten als Bioindikatoren in der Landwirtschaft: Eignen sich spezialisierte Phytophage und Parasitoide? Landwirtschaft Schweiz 4:493-498.

Welling, M. 1990. Förderung von Nutzinsekten, insbesondere Carabidae, durch Feldraine und herbizidfreie Ackerränder und Auswirkungen auf den Blattlausbefall im Winterweizen. Dissertation. Universität Mainz.

Welling, M., C. Kokta, H. Bathon, F. Klingauf & G.A. Langenbruch. 1987. Die Rolle der Feldraine für Naturschutz und Landwirtschaft. Nachrichtenblatt für den Deutschen Pflanzenschutzdienst (Braunschweig) 39:90-93.

Weiss, E. & C. Stettmer. 1991. Unkräuter in der Agrarlandschaft locken blütensuchende Nutzinsekten an. Volume 1, 104 pp. *In* W. Nentwig & H.M. Poehling [eds.]. Agrarökologie. Haupt, Bern.

Wermelinger, B., J.J. Oertli & J. Baumgärtner. 1991. Environmental factors affecting the life-tables of *Tetranychus urticae* (Acari: Tetranychidae). III. Host-plant nutrition. Exp. Appl. Acarol. 12:259-274.

Wichtrup, L. G., H. Steiner & T. Wipperfürth. 1985. Der Einfluss von Klee als Untersaat auf die Populationsdynamik von Blattläusen (Homoptera, Aphididae) und epigäischen Arthropoden bei Winterweizen im Lautenbach-Projekt. Mitteilungen Deutsche Gesellschaft für allgemeine Entomologie 4, 430-432.

Wiedmer, U. & E. Boller. 1990. Blühende Rebberge in der Ostschweiz: 2. Zum Pollenangebot auf den Rebenblättern. Schweizerische Zeitschrift Obst-Weinbau 126:426-431.

Wilding, N. & P. Brobyn. 1980. Effects of fungicides on the development of *Entomophthora aphidis*. Trans Br. Mycol. Soc. 75:297-302.

Yarham, D.J. 1979. The effect on soil-borne diseases of changes in crop and soil management, pp. 371-383. *In* Schippers B. & Gams W. [eds.]. Soil-borne plant pathogens. Academic Press, London, New York.

Yarris, L. 1982. Hessian fly attacks spring wheat. Rev. Appl. Entomol. 70 (6973).

Zadoks, J. C. 1989. Does IPM give adequate attention to disease control, in particular at the farmer level? FAO-Plant Protection Bull. 37 (4):144-150.

Zadoks, J. C. (Ed.), et al. 1989. Development of farming systems. Pudoc, Wageningen, 90 pp.

Zimmermann, G. & T. Basedow. 1980. Freilanduntersuchungen zum Einfluss von Fungiziden auf die durch Entomophthoraceen (Zygomycetes) verursachte Mortalität bei Getreideblattläusen. Zeitschift für Pflanzenkrankeiten und Pflanzenschutz. 87:65-72.

The Role of Spiders and Their Conservation in the Agroecosystem

SUSAN E. RIECHERT—Department of Zoology, University of Tennessee
Knoxville TN 37996-0810

Aldo Leopold (1949) stated, "A system of conservation based solely on economic self-interest...tends to ignore, and thus eventually eliminate many elements in the land community that lack commercial value, but that are...essential to healthy functioning." One element of the agroecosystem that has been severely limited by our technological wizardry is the invertebrate generalist predator (Kajak 1978a; Luczak 1979). These include insects, mites, and spiders that feed on a wide variety of arthropods, many of which are important agricultural pests. Generalist predators are polyphagous in that they eat anything that they may encounter that can be captured, including other predators.

Generalist predators including the spiders, emphasized here, control prey largely through an assemblage effect: species of varied sizes and habits need to be present throughout the growing season of the pest species to limit the growth of associated pest populations (Riechert 1992, Riechert and Lawrence 1997). In part, species assemblages have a greater effect because spider species composing a community have varied phenologies: different species will have different peaks in population numbers during the course of a growing season (e.g., Howell & Pienkowski 1971). Spiders are also known to partition foraging sites within a habitat such that different species encounter different prey types (e.g., Olive 1980, Pasquet 1984). Both the application of chemical insecticides (Pfrimmer 1964, Chu & Okuma 1970, Hirano & Kiritani 1976, Dondale et al. 1979, Kajak 1981, Brown et al. 1983, Bostanian et al. 1984, Naton 1993, Olzak et al. 1994, White & Hassall 1994) and mechanical disturbance (Kajak 1978a, LeSar & Unzicker 1978, Robertson et al. 1994, Schaber & Entz 1994, White & Hassall 1994) represent major sources of mortality to spider populations in crop systems. Both practices suppress spider densities and spider species diversity in these systems (Luczak 1979, Nyffeler & Benz 1981).

There is evidence from natural communities that the arthropod generalist predator serves as a buffer to exponential increases in the population sizes of their prey. Clarke and Grant (1968), for instance, demonstrated that when spiders were removed from an enclosure erected in a forest floor litter

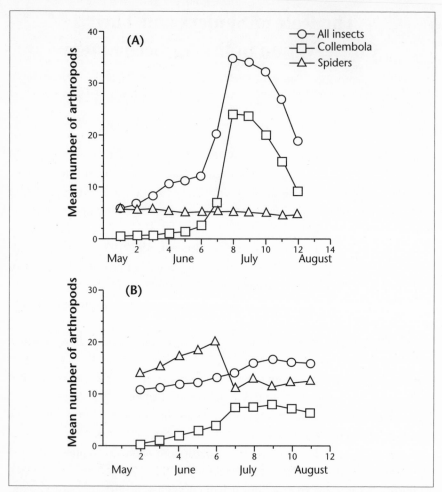

Figure 1. Arthopod abundances over time in a forest litter community: (A) in absence of spiders (removal enclosure). (B) in presence of spiders (control enclosure). (From Clarke & Grant 1968).

community, phytophagous insect populations showed exponential rises that were not observed in the presence of spiders (Figure 1). Unfortunately, this early study was not properly replicated. Other evidence comes from the ecosystem literature: Van Hook (1971) found that 25% of the arthropod biomass in an old-field system traveled through (was consumed by members of) one genus of spider *Lycosa* (Araneae: Lycosidae). Moulder and Reichle (1972) report similar results in their study of the energy dynamics of forest floor arthropod communities.

Yet the literature is replete with references to the view that generalist predators such as spiders have no role in the agroecosystem. This view is based on the following points: (1) generalist predators feed on beneficial insects (i.e., pollinators, specialist parasitoids and predators, and each other) (e.g., Spiller 1986, Moran & Hurd 1994); (2) their interactions with other members of the food web are so complex as to limit theoretical and empirical testing of their potential effects (Andow 1991); and (3) generalist predators that are largely self-damped through territorial and even cannibalistic behavior are not expected to reach high enough densities to influence pest numbers in agroecosystems.

Our failure to recognize the potential of generalist predators as agents of pest control is ultimately related to two problems: (1) evaluation of their biological control potential has centered on the performance of a single (prominent) species in each agroecosystem, thereby equating this predator to the prey specialists (e.g., parasitoids); and (2) few studies have considered their conservation in agroecosystems. This chapter considers the role of spiders and their conservation in agroecosystems.

Generalist vs. Specialist Predators

I have reviewed the first of these problems at various times (Riechert & Lockley 1984; Riechert & Harp 1987; Riechert 1992). Thus, I will provide only a brief review of the important points here.

Emphasis has been placed in natural enemies research on prey specialists rather than generalists to a large extent, because it is easier to mathematically deal with single prey–single predator or host-parasite systems than with prey and predator assemblages. Theorists have difficulty dealing with real-world complexity, so most predator-prey modeling efforts have been limited to the interaction between a single predator and a single prey species (Hassell 1978). Stability in the predator-prey interaction (limitation of prey numbers without mutual extinction) is achieved mathematically only when the predator or parasitoid population tracks the density of its prey population—its population density increases and decreases along with that of its prey as in the limit cycle (Figure 2A). Such density-dependent tracking is observed in predators and parasitoids that have short life cycles and feed exclusively on a given prey type. Limit cycle control is not exhibited by spiders because, with annual and biennial life cycles, they are much longer-lived than are the insect pests that they feed on (Schaefer 1987). Further, spiders rarely limit feeding to a particular prey type (reviewed in Riechert & Luczak 1982).

Theoretical developments available from the food web literature

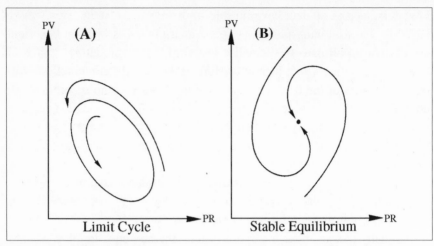

Figure 2. Graphical representations of two kinds of stable predator-prey interactions: (A) limit cycle in which prey and predator population numbers cycle out of phase with one another. (B) Stable equilibrium point control in which the corresponding population densities approach an equilibrium point. Arrows refer to respective population trajectories (PR = Prey and PV = Predator), approaching limit cycle (—) in (A) and equilibrium point (•) in (B).

(DeAngelis et al. 1975; Post & Travis 1979) suggest that generalist predators may limit prey population growth without density-dependent tracking of these prey. If there is an assemblage of predator species that feed on an assemblage of prey species, the predators can exhibit what is referred to as equilibrium point control of these prey (Figure 2B). The necessary characteristics of the predators comprising these assemblages include: (1) that they be self-damped—population densities cannot be proximally determined by local prey availabilities; and (2) that they specialize temporally on certain prey types. Territorial behavior is a form of self-damping behavior that is exhibited by spiders throughout their lives (see Riechert 1992 for a review). Territory size is adjusted to the lows in prey levels experienced in a given habitat over evolutionary time. Where tested, for instance, we find that there is a genetic basis to territory size (*Metepeira:* Uetz et al. 1982; *Agelenopsis aperta:* Riechert 1987; Riechert & Smith 1989). Thus, spider population growth is limited to the number of territories of specific size that are accommodated by a given habitat. At the upper limit, then, fewer spiders may be present than might be accommodated by available prey levels.

To achieve equilibrium point control of prey, generalist predators must also adjust their foraging choices according to local patterns of prey availability (Murdoch et al. 1985). Work completed in Tennessee indicates that

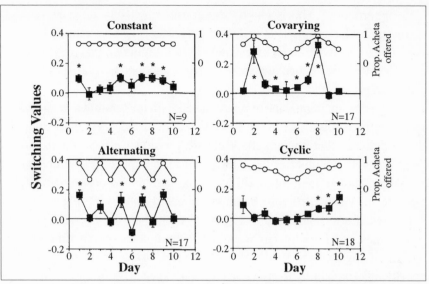

Figure 3. Switching values between 2 prey types exhibited by the wolf spider, *Rabidosa rabida,* when offered different prey variability patterns over time (o). Prey variability pattern expressed in figure as proportion of prey that are of preferred prey type in a 2 prey species system (*Acheta domesticus* & *Gryllus pennsylvanicus*). (Means and Standard error bars). *'s denote switching values that are significantly (P < 0.05) different from 0. Positive values indicate density dependent switching (prey control), and negative values inverse density dependent switching (destabilizing).

spiders seek foraging sites that offer "safe" prey availability treatments: wolf spiders (*Hogna rabida* Walckenauer, Lycosidae) avoided foraging locations at which prey availabilty showed a pattern of positive covariance as opposed to constant numbers or negative covariance (Riechert et al., in press). By positive covariance, we mean that on some days individual spiders will receive no prey and on other days a superabundance. Under a prey availability pattern of negative covariance, one prey type will be abundant on some days while another might be abundant on days when the first type was absent. Provencher (1990) has also tested for potential switching among prey types in *H. rabida*. This spider showed different preferences for two orthopteran prey types when offered at equal densities. To achieve equilibrium point control, these generalist feeders should shift their attacks to the prey type that is numerically prominent (positive density-dependent switching) as relative abundances change over time. We found that *Rabidosa rabida* (Walckenauer), (Lycosidae) generally exhibited this switching behavior across a variety of different prey variability patterns (Figure 3). Significant density-dependent switching was observed in 35% of the possible opportunities. This is a high value considering the fact that our test was a conservative one: the less preferred prey was generally offered at lower densities in the experiment and switching to foraging on this prey occurred significantly often when we increased its numbers beyond those of the preferred prey.

Equilibrium point control is associated with the top-down or cascading trophic effects discussed in the food-web literature. Recent work supports the hypothesis that spiders show classic top-down effects on the lower trophic levels. Schmitz (1994), for instance, completed experimental removals of spiders in an old-field food web system and observed a trophic cascade in which herbivore biomass increased and edible plant biomass decreased. Carter and Rypstra (1995) also verified the operation of top-down control and trophic cascades by spiders in another experimental system involving soybean food webs. Although Clark et al. (in press) did not specifically explain their predator removal experiment in corn in terms of trophic theory, they did find: (1) that 13 of the 15 taxa of generalist predators tested fed on the targeted pest (the armyworm, *Pseudaletia unipuncta* Haworth, Lepidoptera: Noctuidae) in feeding trials; and (2) that armyworm damage to corn plants was significantly greater in plots where pitfall trapping had significantly reduced carabid and staphylinid beetles, ants, and spiders than in control plots.

The Agroecosystem—a Spider-Poor Habitat

Given that spider assemblages, theoretically, should have an important limiting influence on associated prey population sizes, the question remains as to why they fail to demonstrate this influence in agroecosystems (See review of biological control literature on spiders in Riechert and Lockley [1984]). It is apparent from the literature that spider species diversity is extremely low in these systems with only a few prominent species present (e.g., Luczak's 1979 review of spider community compositions in agricultural fields in Europe and North America). Two possible explanations for this finding include: (1) spiders are poor colonizers of new habitats (are poor dispersers), and or (2) spiders readily disperse into agroecosystems but die off or emigrate from them as the result of unfavorable conditions existing in them.

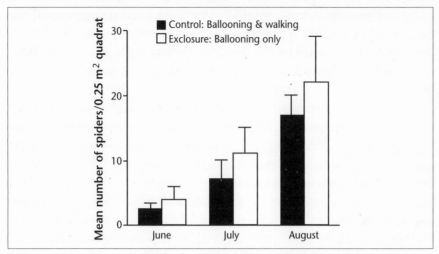

Figure 4. Spider densities in plots permitting the entrance of only aerial ballooning spiders (exclosures) versus plots permitting emigration by both ballooning and cursorial spiders (Controls). Combined means + SEs for 2 replicate plots. Six, 0.25m² samples/month/replicate. No significant differences were observed between exclosure and control plots for any month. Note, however, that the exclosure plots had consistently higher numbers of spiders present. We attribute this result to the fact that spider densities could not equilibrate following dispersal in the exclosures. (From Bishop and Riechert [1990]).

Spider Dispersal

Despite the fact that spiders lack wings, they have excellent dispersal capabilities. Most species disperse as juveniles through aerial ballooning. Silk threads are released from spinners at the rear of the abdomen and when these are caught by air currents, the animal releases its hold on the substrate. This is passive dispersal in that once the individual has committed itself to the ballooning act, it has no choice in the direction of its travel or in the distance it goes. Upon landing, however, the individual spider can reinitiate ballooning if it finds the new environment unfavorable (e.g., Tolbert 1977). Spiders colonizing agroecosystems may then enter such systems in large numbers and from great distances (Plagens 1986; Decae 1987).

Most workers, however, have assumed that natural areas must be located adjacent or in close proximity to crop fields to ensure spider colonization of these fields (e.g., Agnew & Smith 1989). Bishop and Riechert (1990) found this not to be the case. We tested the extent to which spiders moving into disturbed systems come from local versus distant sources. We set up sheet metal flashing exclosures around mulched garden plots that excluded the wandering spider fauna. We found no differences between the numbers of spiders present after dispersal in plots limited to ballooners only (the exclosures) versus those in which both dispersal modes were allowed (Figure 4). We also sampled nearby woodlots and old fields for comparison of spider species compositions between these habitats and the vegetable plots. We found that, minimally, 65% of the spider species collected in the vegetable plots had come from long distances away: these species were not represented in adjacent and nearby natural areas. Thus, both the experimental and habitat censuses suggest that nearby habitat reservoirs are not required for spider colonization of cultivated systems.

Our conclusions agree with those of Luczak (1980) who makes 2 points concerning spider colonization of croplands in Poland: (1) there is little overwintering spider fauna in these agroecosystems; and (2) adjacent natural habitats and agroecosystems do not necessarily have high species similarities. Note that neither Luczak (1980) nor Bishop and Riechert (1990) feel that there is a spider fauna unique to particular agroecosystems or agroecosystems in general. Rather, the species composition of the spider assemblage present in any given year is largely the result of stochastic colonization processes.

The Agroecosystem Environment

Since dispersal does not seem to limit spider colonization of agroecosystems, it is necessary to consider the extent to which spiders moving into these systems die or depart them. Tillage destroys the beneficial fauna each season. More importantly, in this case, it may lead to inhospitable thermal and feeding environments in the early stage of crop growth, which coincides with the period of maximum spider dispersal, i.e., spring–early summer (Bishop 1990). Spiders balloon into cultivated systems in tremendous numbers. They apparently leave these systems or die just as rapidly.

Spider Habitat Requirements

Because spiders lead feast-and-famine existences (Riechert 1992), they have retained a non-sclerotized abdomen for the consumption of large meals. In fact, they can incorporate a body mass exceeding their own within a few hours (Collatz 1987). The trade-off is that spiders are, of necessity, more closely coupled to their thermal environments than are insects, which have heavily sclerotized integuments (Levi 1967). A major problem of having a non-sclerotized abdomen is that spiders are subject to high levels of cuticular water loss: Low (1983) estimates that 88% of the water lost by spiders is through the cuticle.

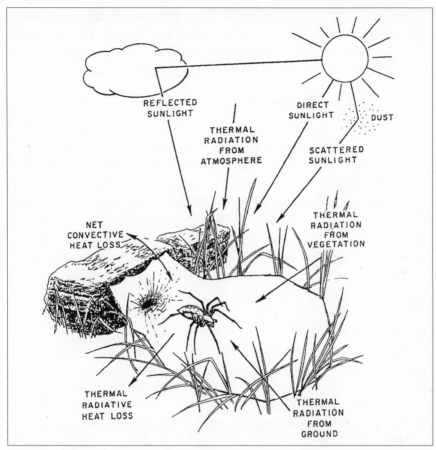

Figure 5. Energy exchange between a spider and its thermal environment: heat sources and sinks.

Spiders lose water as a function of their body temperature. In such a small poikilotherm, body temperature will be closely coupled to the following features of the environment: convection, conduction, solar radiation (UV and visual wavelength heat) and thermal radiation (long-wave or radiant heat) (Figure 5, Riechert & Tracy 1975). Chemical reactions in biological processes proceed best between 25° and 35° C. Spiders, generally, forage within this temperature range, though there is variation among species associated with the conditions under which they exist (see review by Pulz 1987). Thus, spiders are sluggish at low temperatures and prey capture becomes difficult. Temperatures above 35° C cause excessive evaporative water loss and thermal stress. Prolonged exposure to high temperatures causes loss of motor control and subsequent death.

Spiders are, therefore, quite responsive to their thermal environments. They seek shade and favorable temperatures within shade when searching for web-sites (Riechert 1985). Their movement increases where temperatures are unfavorable, increasing the probability of escaping these conditions.

Spider Feeding Environments

Spiders, thus, need and seek favorable thermal environments. They seek favorable feeding environments as well. Though exhibiting many adaptations to starvation (e.g., low metabolic rates, a sit-and-wait foraging strategy, energy-based territoriality), spiders respond to cues that place them in habitat patches affording high prey levels. In locating these foraging sites, they respond to both vibratory (near sound) and olfactory cues emitted by insects (see review in Riechert & Gillespie 1986). Some spiders further test prey levels by laying a few strands of silk on the substrate (Gillespie 1981). Finally, most spider species abandon foraging sites in search of new ones if prey become scarce (Turnbull 1966; Kronk & Riechert 1979).

Habitat Manipulations

Riechert and Bishop (1990) used a mixed vegetable system to test the effects of favorable feeding and physical environments in agroecosystems on spider prey control performance. The vegetable types tested were: spinach (*Spinacea oleracea*), radishes (*Raphanus sativus*), cabbage (*Brassica oleracea* var. *capitata*), brussels sprouts (*Brassica oleracea* var. *gemmifera*), potatoes (*Solanum tuberosum*), beans (*Phaseolus vulgaris*), tomatoes (*Lycopersicon esculentum*) and corn (*Zea mays*). Two habitat manipulations were completed to test this hypothesis: (1) grass hay mulch was laid between rows at the time of planting; and (2) rows of vegetables were alternated with rows of flowering buckwheat (*Fagopyrum esculentum*). We expected both habitat manipulations would limit the emigration of spiders from the treatment plots once they had arrived by natural dispersal, but for different reasons. We expected the mulch would provide the high humidities and protection from high temperatures that spiders seek during the heat of the day (Riechert & Gillespie 1986), while we expected that the flowering plants would attract pollinating insects that could serve as alternate prey to the spiders during periods of low pest density (Hagen & Hale 1974; Gonzáléz et al. 1982).

We applied grass hay mulch rather than some other mulch type (e.g., straw, leaf litter, bark) in this initial study. Mixed grasses have the least binding effect on soil nitrogen during decomposition of mulches available in southeast United States (Waksman 1938; Steward et al. 1966). Note, however, that such effects are very small, regardless of the type of mulch used. We applied the mulch to a depth of 10 cm at the time of planting and renewed the mulch as needed during the course of the study. We used buckwheat to attract pollinating insects, because it could be periodically mowed to keep flowers present for a maximum period of time. To equalize plot sizes between the controls (plots lacking mulch and buckwheat rows) and treatments, we doubled the number of vegetable rows in the controls.

We applied no chemicals to the system during the course of the study and we sampled for plant damage, grazing insect pest numbers, and spider densities at maturity for each vegetable type. We sampled only those vegetable types that exhibited measurable levels of plant damage in at least one plot.

During the first season of study, the treatments were offered in combination with tilled plots serving as controls. We found that spider densities were clearly higher in the treatment plots than in the controls (Figure 6A, Riechert & Bishop 1990). We further found that for all vegetable crops sampled, less damage and fewer grazing insect pests were present in treatment plots than in the control plots (Figure 6B and C respectively).

In the second growing season, we tested among 5 treatments: (1) tilled (control), (2) flowers (rows of vegetables alternated with rows of buckwheat), (3) mulch, (4) mulch and flowers, and (5) mulch and flowers from which we physically removed the spiders. Despite the fact that the different vegetable types exhibited different levels of infestation, the treatment effect was the same. Tilled plots, plots containing flowering buckwheat, and plots containing mulch + flowers but from which spiders were removed clustered together with respect to spider densities, plant damage and grazing insect numbers (Figure 7A, B and C; ANOVA contrasts, Table 1). These plots had consistently higher levels of plant damage and insect pest numbers and lower spider densities than did the mulch and mulch + flower treatments. The differences between the two groups of treatments were highly significant (Riechert & Bishop 1990).

Differences in pest insect numbers between the treatments were only marginally significant because plots lacking high spider numbers exhibited highly variable insect numbers (Table 2). We attribute the high variance in grazing insect numbers to the fact that the plots lacking high spider densities had plants that were totally stripped of vegetation (with few insect pests remaining), but also plants that were currently being fed on by large numbers of grazing insects.

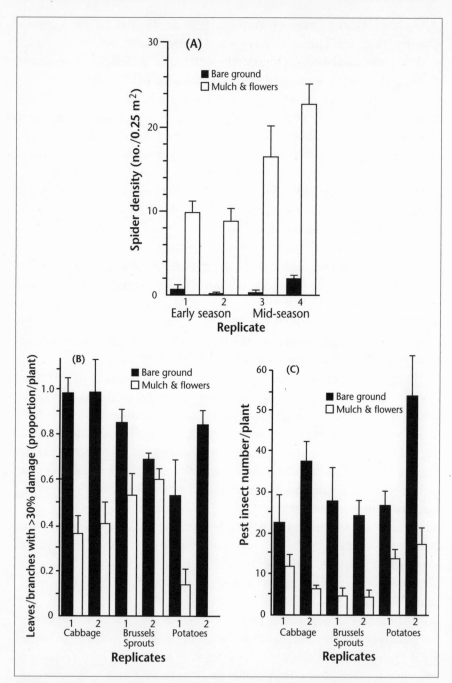

Figure 6. Means ± S.E. for spider densities (A), crop damage (B) and grazing pest numbers (C) in vegetable crops grown under various habitat manipulations: BG = tilled, F = Flowering buckwheat, M = grass hay mulch, M + F = mulch + flowers, M + F - S = spider removal plots (from Riechert & Bishop 1990).

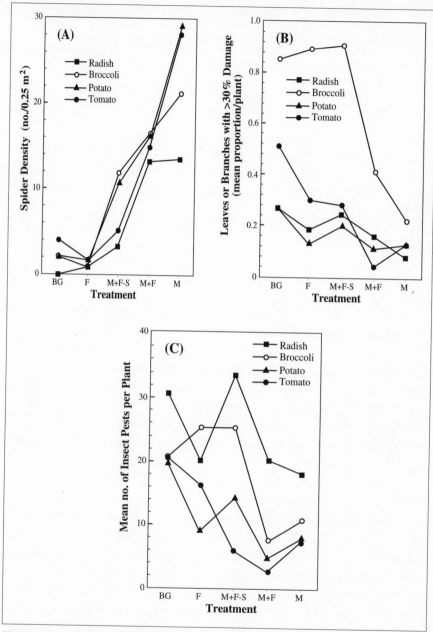

Figure 7. Results of habitat manipulations by vegetable type: combined means for two vegetable garden sites in east Tennessee. Treatments: BG = tilled control plots, F = alternating rows of flowering buckwheat and crop; M = grass hay mulch between rows and around spaced plants (e.g., cabbages); M + F = combined mulch and flower treatment; and M + F – S = combined mulch and flower treatment from which spiders are continuously removed. (A) spider densities, (B) insect numbers, and (C) plant damage (from Riechert & Bishop 1990).

Table 1. Planned contrasts among treatments for spider densities (Nos./0.25m²), plant damage (proportion of leaves or branches with > 30% damage) and grazing insect numbers (Nos./plant) in Year 2 of Riechert & Bishop 1990 study. Overall F-tests: Spider Density- F = 9.89, P = 0.020; Plant Damage- F = 7.72, P = 0.036; Insect Nos.- F = 3.6, P = 0.121. M refers to mean value, F to F value for contrast between tilled control and respective treatments, and P to significance level of the contrast.

Contrast	Spider Density			Plant Damage			Insect Nos.		
	M	F	P	M	F	P	M	F	P
Control	2.0			0.48			23.0		
Flowers	1.3	0.03	0.877	0.38	0.92	0.391	17.6	2.18	.213
Mulch	21.8	23.99	0.008	0.14	18.89	0.012	11.0	8.46	.044
Mulch + Flowers	16.4	12.75	0.023	0.18	15.36	0.017	8.83	11.82	.027
Mulch + Flowers - Spiders	7.7	2.29	0.205	0.41	0.84	0.411	19.9	3.01	.157

Tables 2 and 3 demonstrate that spider biocontrol is an assemblage effect. From the plot analyses (Table 2), we observe that spider foraging had an effect on a broad spectrum of insect pests. Table 3 summarizes predation censuses conducted during 15-minute watches at random locations in the garden plots. Eighty-four percent of the predation events involved spiders preying on pest insects; again, a broad range of insect types was taken. Fifteen families of spiders were observed foraging in the plots during the censuses. Ten of these families were cursorial and 5, web builders.

Riechert and Lawrence (1997) used a partial enclosure experiment to further investigate the hypothesis that assemblages of spider species limit prey to a greater extent than would any single prominent spider species. We found that none of the four species selected for inclusion in the experiment because of numerical or biomass prominence in an old-field habitat performed better than did the assemblage of spider species in limiting total herbivorous insect numbers, per capita numbers, or the diversity of prey taxa. This was despite the fact that all of the single-species effects had lower densities of associated herbivorous insects than did a set of spider removal controls.

The Benefits of Mulching

This study indicates that the early application of mulch to an annual crop system increases spider numbers and species diversity which, in turn, re-

Table 2. Breakdown of total pest insect numbers grazing on sampled plants by taxon to show that an assemblage of prey were controlled by spiders in the Riechert & Bishop (1990) system. Pest types showing low frequencies of occurrence for each vegetable type summarized under category "Others." (Combined means for six samples from each of two study areas for each treatment). Vegetable types sampled were all those showing some plant damage in at least one plot: A) Radish, B) Broccoli, C) Potato, and D) Tomato.

	Treatment									
	Mulch + Bare Ground		Flowers		(Mulch + Flowers) – Spiders		Mulch		Mulch + Flowers	
	M	SE	M	SE	M	SE	M	SE	M	SE
A) RADISH										
Flea Beetles	19.21	5.30	17.16	5.70	25.66	4.98	15.44	4.40	16.53	4.7
Aphids	10.53	1.66	1.88	0.79	6.36	3.97	2.54	0.72	2.35	0.71
Others	1.04	0.48	0.71	0.31	1.66	0.78	0.31	0.13	1.33	0.78
B) BROCCOLI										
Imported Cabbageworm	5.40	1.35	14.33	3.12	17.44	4.24	4.34	0.66	3.02	0.88
Aphids	7.08	2.25	9.35	2.36	0.09	0.09	0.78	0.36	1.76	0.71
Flea Beetles	7.90	1.78	1.46	0.66	7.20	3.33	5.28	1.46	2.05	1.05
Others	0.24	0.10	0.23	0.16	0.63	0.37	0.34	0.15	0.68	0.42
C) POTATO										
Flea Beetles	11.55	6.26	3.80	0.94	11.08	2.74	5.15	1.29	2.66	0.75
Colorado Potato Beetles	5.67	1.31	3.16	0.80	1.74	0.66	0.49	0.28	1.12	0.54
Aphids	1.42	0.72	1.35	0.37	1.03	0.47	0.89	0.42	0.19	0.09
Leafhoppers	1.13	0.25	0.50	0.14	0.38	0.17	0.81	0.26	0.37	0.12
Others	0.11	0.11	0.45	0.21	0.08	0.08	0.55	0.16	0.25	0.07
D) TOMATO										
Aphids	19.90	6.36	14.84	4.66	5.27	1.32	3.89	1.47	2.57	0.22
Flea Beetles	0.35	0.10	0.99	0.33	0.67	0.29	2.49	0.43	0.17	0.05
Striped Blister Beetles	0.22	0.11	0.18	0.07	0.07	0.03	0.30	0.22	0.02	0.02
Tomato Hornworm	0.15	0.11	0.13	0.06	0.20	0.11	0.39	0.13	0.004	0.004
Others	0.11	0.04	0.07	0.05	0.11	0.03	0.22	0.14	0.02	0.01

Table 3. Frequencies of predator activity in plots from 15-min. watches and individual plant censusing. Under predation events: IGW = imported cabbageworm, AP = aphid, LF = leafhopper, HCB = harlequin stink beetle, LY = lygus bug, MI = mirid bug, DP = Diptera, BL = blister beetle, CP = Colorado potato beetle, FB = flea beetle, SCB = striped cucumber beetle, MC = miscellaneous chrysomelid beetles, EL = elaterid beetle. (From Riechert & Bishop 1990).

	Foraging Activity Relative	Predation Events												
		ICW	AP	LF	HCB	LY	MI	DP	BL	CP	FB	SCB	MC	EL
SPIDER FAMILIES														
Aegelenidae[a]	2.2		2							1	2			
Anyphaenidae	0.4										2			
Araneidae[a]	13.4				8		1	1	25		4	4	5	5
Clubionidae	7.4		4										1	
Gnaphosidae	0.7		1											1
Hahniidae	5.1			1										
Linyphiidae[a]	20.2		4	1		1		25						
Lycosidae	5.4		3		3									
Oxyopidae	0.4													
Pisauridae	0.3													
Salticidae	13.2	4	4			1		4	1	1	2	1		
Tetragnathidae[a]	0.6													
Theridiidae[a]	10.7			6	4		1		8	1	1	2		1
Thomisidae	3.4		2	3			2		1	1		1	1	
Uloboridae[a]	0.1													
Unknowns	1.2		2											

Table 3. *continued*

OTHER ARTHOPOD PREDATORS	Foraging Activity Relative	Predation Events												
		ICW	AP	LF	HCB	LY	MI	DP	BL	CP	FB	SCB	MC	EL
Carabidae	0.1													
Chilopoda	0.1													
Chrysopidae	0.1													
Coccinellidae	6.8	1		1	1									
Formicidae[b]														
Mantidae	0.1													
Nabidae	0.1													
Opiliones	0.2													
Parasitoids	4.6	1												
Reduviidae	0.7													
Staphylinidae	0.4		2											
Wasps	2.1													

[a]Family that uses web to trap prey.
[b]Excluded since usually seen in association with aphids.

duces phytophagous insect numbers and plant damage. Spiders require high humidities and moderate temperatures. These conditions are not typically met at planting of the annual crop system where there is no initial vegetative cover. But it is precisely this time (spring and early summer) in east Tennessee when spiders exhibit peaks in aerial dispersal (Bishop 1989) and are colonizing croplands. The grass hay mulch used in the Riechert and Bishop system provided more favorable midday temperatures and humidities than did tilled ground (Figure 8A, B).

Chinese scientists accomplished a similar increase in spider densities in rice paddies by providing bamboo retreats (teepees) that the predators sheltered in during the heat of the day (R. Jones, personal communication). Workers also moved the teepees, as needed, to areas in the paddies exhibiting high pest numbers. The teepees were transported at midday when the inactive spiders were sheltered within them.

Nentwig (1988) also augmented spider numbers and diversity through ground cover manipulation. He based his experiment on the observation that the ecotone between two habitat patches contains a greater abundance and diversity of arthropods than do the centers of the respective habitat patches (Mader et al. 1986). In an agroecosystem context, such edges might serve as refuges for natural enemies during periods of habitat disturbance, as well as centers of dispersal (Heydemann 1981; Muhlenberg & Werres 1983). In a meadow system, Nentwig (1988) produced a treatment that alternated broad mown strips with small unmown strips. He found that, as predicted for edges, this strip management treatment contained increasingly greater species diversity over time when compared with the mowed control that lacked stripping. He also found that there was a higher ratio of predators to polyphagous insects in the strip-managed area than in the mowed control area. He concluded that the augmentation of predators (especially spiders) reduces phytophagous insects.

Observational reports also have implicated mulch as important to pest control. Sugar beet farmers in Oregon that applied wheat straw mulch at levels 5–6 times those recommended for water retention (1lb./100 feet of row), reported markedly reduced pest levels (Hobson, personal communication). In East Tennessee, Louis Provencher (personal communication) noted that aphid populations on tomato were decimated within a few days following the application of grass hay mulch and subsequent colonization of the crop by spiders. Kobel-Lamparski et al., (1993) compared the spider community occupying tilled ground versus mulched vineyards in southwest Germany. As in the other studies already described, these workers found

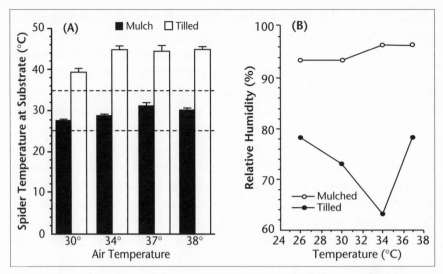

Figure 8. Comparison of thermal environments between mulch and bareground plots of mixed vegetable system from Riechert and Bishop (1990) study: a) estimated spider body temperature. b) relative humidity at the ground surface.

that the spider communities in the mulched vineyards were significantly richer both in terms of number of species and numbers of individuals than in the vineyards lacking mulch.

Not all substrate covers equally conserve the local spider fauna in crop systems, however. Using 5 m^2 vegetable garden plots, I tested three mulch types against a tilled control for their potential to conserve the spider fauna: grass hay, oak-hickory leaf litter and pine bark. Six weeks after establishment of the plots, quadrat sampling was completed within the various substrate types and spider density determinations made. Analysis of variance completed on spider densities indicates a highly significant treatment effect ($P < 0.00001$) and a nonsignificant replicate effect ($P = 0.50$). The multiple range test for homogeneous groups (Sokal & Rohlf 1969) indicates that grass hay had significantly more spiders than did other substrate types ($\alpha=0.05$), that leaf litter and bark substrates were homogeneous and that leaf litter and tilled substrates were similar.

On Polycultures and Intercropping

While the application of a mulch ground cover clearly has a beneficial effect on spider communities and consequent deleterious effect on their prey, the effects of other methods of agroecosystem diversification/enrichment on spider population densities are less clear. This is true both of the addition of weed strips to agricultural systems and of polyculture techniques. Riechert and Bishop (1990) did not observe an increase in spider densities with the addition of rows of flowering plants to the system. In fact, we found that the presence of flowering plants detracted from the beneficial effects of mulch on spider densities (Figure 7A) and the consequent effects these spiders had on prey numbers and plant damage (Figures 7B and C). This result is consistent with those of some other studies involving the addition of flowering weeds to systems in order to enhance natural enemies (e.g., Bugg et al. 1987), though Wyss (1995) reported that apple orchard plots with strips of weeds added between tree rows had higher spider densities and lower apple aphid densities than equivalent plots lacking the treatment. One plausible explanation for the fact that the addition of weeds to a field crop does not necessarily enhance prey control by natural enemies is that weeds harbor high densities of prey types that are favored by some predators, which results in predator aggregation in the weeds rather than in the crop (Ables et al. 1978). These latter workers found that the presence in a system of alternative prey to a targeted pest species detracts from a predator's effect on the targeted species.

If this explanation is upheld, polyculture, in general, might not be the best management technique for enhancing biological control by generalist predators. Although it is well known that the density of beneficial arthropods is higher in mixed cultures than in monocultures (Andow 1991), their absolute densities will be of less importance in the agroecosystem than the effects they might have on associated polyphagous insects. See Corbett (this volume) for a discussion of relevant sink-source dynamics.

Spiders vs. Other Arthropod Generalist Predators

In the Riechert and Bishop (1990) study, plant damage and insect pest numbers were significantly reduced in plots containing grass hay mulch. The results of spider removals indicate that they were responsible for the reductions in crop damage in the mulched plots. This does not mean that ground beetles and other generalist predators do not contribute to prey reduction in agroecosystems. It only means that effects by spiders overshadowed those of other predators in this system in which we greatly enhanced spider densities (i.e., to 10–20 times those of other generalist predators). The habitat manipulation already described for augmenting spider densities in rice paddies in the Hunan Province of China has resulted in a reported reduction in pesticide use of 60% for this province. Affording shelter through ground cover, thus, can be a very effective tool to increasing natural control of pest insect numbers and their effects in agroecosystems.

No-Till Agriculture

Spiders as agents of biocontrol can be used in combination with no-till agriculture for the kind of whole ecosystem approach to habitat management that Leopold encouraged in his treatise "Sand County Almanac." Laub and Luna (1991), for instance, found that by mowing a rye winter cover crop rather than spraying it at corn planting time, they achieved a net benefit of $91–$113/hectare. They cite significant suppression of armyworm populations through conservation of natural enemies in the mowed treatment as the underlying cause of this economic benefit. The common practice in no-till has been to apply herbicide to the cover crop, a practice that obviously has led to concerns about insect problems (Musick 1970). The mowing of the cover crop has two advantages to spiders over herbicide application: (1) the spider fauna is not decimated by the herbicide; and (2) the mulch left behind offers a favorable microhabitat to spiders.

Although the phenomenon has not been well studied, herbicide applications have been found to have a limiting influence on soil arthropods. In their review of the subject, Eijsackers and Bund (1980) concluded that the more mobile predaceous fauna (beetles, spiders and ants) has the highest incidence of contact with spray drops and vapors from herbicides. They also noted that the food chain accumulation phenomenon increases herbicide impact on these predators. In addition to the lethal effects of the herbicide itself, Eijsackers and Bund (1980) reported that these animals avoid treated areas and discard prey items that have herbicide residues on them. They list numerous population studies showing that predator numbers are greatly reduced following spraying. Only one study pertained specifically to spiders but a significant limiting effect was observed in this study (Davis 1965). Luczak (1986) and Riechert (personal observations) have also observed the negative effects of herbicide applications on spider numbers, and Kajak (1978b) reported similar effects for fertilizers.

Given the common practice of herbicide removal of the cover crop in no-till agriculture, it is not surprising that spiders have failed to be identified as important natural enemies in the "classic" no-till systems. Srikanth et al. (1997) found that both conventional furrow irrigation and post-harvest trash-burning reduced spider abundance in sugarcane (*Saccharum officinarum*) fields. The burning of sugarcane refuse would kill spiders directly and reduce cover and the benefits it provides. If the sugarcane refuse were mowed and left as ground cover in the fields, the effect might be equivalent to the application of mulch. In this way the sugarcane system is similar to the no-till system described above in which a winter cover crop is mowed instead of being chemically removed. Bugg et. al. (1991) observed only one predation event by spiders on egg masses of fall armyworm (*Spodoptera frugiperda* [J. E. Smith], Lepidoptera: Noctuidae) in an experiment involving cool-season cover crops in cantaloupe fields. The cover crops received limited applications of conventional chemical fertilizer, fungicides, and herbicides, which may have impacted spider densities. However, as Bugg has pointed out (personal communication), more intensive nocturnal observations probably would have detected more spider predation of lepidopterous eggs than that observed. This is because the nocturnal sac spiders (Anyphaenidae, Clubionidae, and Gnaphosidae) have been identified as the major spider predators on eggs. Recent work on relay intercropping vegetables and cotton in southern Georgia (S. C. Phatak personal communication) has suggested that surface management of crop residues has led to the buildup of high densities of spiders that are commonly found on sentinel lepidopterous egg masses. This pattern has been observed both in growers' fields and in experimental research plots.

Conclusion

Legume and grass hay mulches appear to have beneficial effects on natural enemies of pests, most notably spiders. The use of cover crops in no-till agriculture can provide the mulch favorable to these natural enemies if they are mowed rather than treated with herbicides at seeding of the crop. Mulch provides additional benefits as well. It affords greater moisture retention by the soil and contributes organic matter to the soil through its decomposition (Laub & Luna 1991), and it conserves phosphorus and nitrogen (Joe Hobson, personal communication). While it may be difficult to suggest that a grower apply mulch for its benefits to natural enemies alone, with the added advantages of water retention, organic matter accumulation and a favorable cost/benefit ratio, this whole ecosystem approach may well be practical.

References

Ables, J. R., S. L. Jones & D. W. McCommas Jr. 1978. Responses of selected predator species to different densities of *Aphis gossypii* and *Heliothis virescens* eggs. Environ. Entomol. 7:402-404.

Agnew, C. W. & J. W. Smith, Jr. 1989. Ecology of spiders (Araneae) in a peanut agroecosystem. Environ. Entomol. 18:30-42.

Andow, D. A. 1991. Vegetational diversity and arthropod population responses. Annu. Rev. Entomol. 36:561-586.

Bishop, L. 1989. The mechanisms and effects of spider dispersal on community structure. Ph.D. Dissertation, University of Tennessee, Knoxville.

1990. Meteorological aspects of spider ballooning. Environ. Entomol. 19:1381-1387.

Bishop, L. & S. E. Riechert. 1990. Spider colonization of agroecosystems: source and mode. Environ. Entomol. 19:1738-1745.

Bostanian, N. J., Dondale, C. D., Binns, M.R. & D. Pitre. 1984. Effects of pesticide use on spiders (Araneae) in Quebec apple orchards. Can. Entomol. 116:663-675.

Brown, K. C., Lawton, J. H. & S. W. Shires. 1983. Effects of insecticides on invertebrate predators and their cereal aphid (Hemiptera: Aphididae) prey: Laboratory experiments. Environ. Entomol. 12:1747-1750.

Bugg, R. L., L. E. Ehler & L. T. Wilson. 1987. Effect of common knotweed (*Polygonum aviculare*) on abundance and efficiency of insect predators of crop pests. Hilgardia 55(7):1-53.

Bugg, R. L., F. L. Wackers, K. E. Brunson, J. D. Dutcher & S. C. Phatak. 1991. Cool-season cover crops relay intercropped with cantaloupe: influence on a generalist predator, *Geocoris punctipes* (Hemiptera: Lygaeidae). J. Econ. Entomol. 84:408-416.

Carter, P. E. & A. L. Rypstra. 1995. Top-down effects in soybean agroecosystems: Spider density affects herbivore damage. Oikos 72:433-439

Chu,Y. & C. Okuma. 1970. Preliminary survey on the spider fauna of the paddy fields in Taiwan. Mushi 44:65-88.

Clark, M. S., J. M. Luna, N. D. Stone & R. R. Youngman. In press. Generalist predator consumption of armyworm (Lepidoptera: Noctuidae) and effect of predator removal on damage in no-till corn. Environ. Entomol. 23:617-622.

Clarke, R. D. & P. R. Grant. 1968. An experimental study of the role of spiders as predators in a forest litter community. Part 1. Ecology 49:1152-1154.

Collatz, K. G. 1987. Structure and function of the digestive tract, pp. 229-238. *In* W. Nentwig [ed.]: Ecophysiology of spiders. Springer Verlag, New York.

Davis, B. N. K. 1965. The immediate and long-term effects of the herbicide MCPA on soil arthropods. Bull. Entomol. Res. 56:357-366.

DeAngelis, D. L., R. A. Goldstein & R. V. O'Neill. 1975. A model for trophic interaction. Ecology 56:881-892.

Decae, A. 1987. Dispersal: ballooning and other mechanisms, pp. 348-356. *In* W. Nentwig [ed.], Ecophysiology of spiders. Springer, Berlin.

Dondale, C. D., B. Parent & D. Pitre. 1979. A 6-year study of spiders (Araneae) in a Quebec apple orchard. Can. Entomol. 111:377-380.

Eijsackers, H. & C. F. Van de Bund. 1980. Effects on soil fauna, pp. 255-306. *In* R. J. Hance [ed.], Interactions between herbicides and the soil. Acad. Press, London.

Gillespie, R. G. 1981. The quest for prey by the web building spider *Amaurobius similis* (Blackwall). Anim. Behav. 29:953-954.

González, D., B. R. Patterson, T. F. Leigh & L. T. Wilson. 1982. Mites: a primary food source for two predators in San Joaquin Valley cotton. Calif. Agric. 36: 18-20.

Hagen, K. S. & Hale, R. 1974. Increasing natural enemies through use of supplementary feeding and non-target prey, pp. 170-181. *In* F. G. Maxwell & F. A. Harris [eds.], Proceedings of the Summer Institute on Biological Control of Plant Insects and Diseases. University Press of Mississippi, Jackson, Miss., USA.

Hassell, M. P. 1978. The dynamics of arthropod predator-prey systems. Princeton University Press, Princeton, New Jersey, USA.

Heydemann, B. 1981. Zur Frage der Flächengröße von Biotopbeständen für den Arten-und Ökosystemschutz. Jahrbuch für Naturschutz und Landschaftspflege (Germany) 31:21-51.

Hirano, C. & K. Kiritani. 1996. Paddy ecosystem affected by nitrogenous fertilizer and insecticides, pp. 197-206. *In* Science For Better Environment (Proceedings of the International Congress on the Human Environment, Kyoto 1975). Asahi Evening News, Tokyo.

Howell, J. O. & R. L. Pienkowski. 1971. Spider populations in alfalfa with notes on spider prey and effect on harvest. J. Econ. Entomol. 64:163-168.

Jones, R. L. 1981. Report of the USDA Biological Control of Stem Borers Study Team's visit to the People's Republic of China. Univ. Minnesota, Minneapolis, Minnesota, USA.

Kajak, A. 1978a. Invertebrate predator subsystem, pp. 539-89. *In* A. J. Breymeyer, G. M. Van Dyne [eds.], Grassland systems and man, Cambridge Univ. Press, London.

1978b. Th effect of fertilizers on number and biomass of spiders in a meadow. Symp. Zool. Soc. London 42:125-129.

1981. Analysis of the effect of mineral fertilizations on the meadow spider community. Ekologia Polska 29:313-326.

Kobel-Lamparski, A., C. Gack, & F. Lamparski. 1993. Influence of mulching treatment on epigeic spiders in vineyards of the Kaiserstuhl area (SW Germany) . Arachnol. MITT 5:15-32.

Kronk, A. E. & S. E. Riechert. 1979. Parameters affecting the habitat choice of a desert wolf spider, *Lycosa santrita* Chamberlin & Ivie. J. Arachnol. 7:155-166.

Laub, C. A. & J. M. Luna. 1991. Influence of winter cover crop suppression practices on seasonal abundance of armyworm (Lepidoptera: Noctuidae), cover crop regrowth, and yield in no-till corn. Environ. Entomol 20:749-754.

Leopold, A.1949. A Sand County almanac and sketches here and there. New York, Oxford University Press.

LeSar, C. D. & J. D. Unzicker. 1978. Soybean spiders: species composition, population densities and vertical distribution. Ill. Nat. Hist. Surv. Biol. Notes 107:3-14.

Levi, H. W. 1967. Adaptations of respiratory systems of spiders. Evolution 21:571-583.

Low, A. M. 1983. Untersuchungen zur cuticularen und pulmonaren Transpiration der Volgelspinne *Eurypelma californicum*. Dipl. Thesis Univ. München. (As cited in Pulz 1987).

Luczak, J. 1979. Spiders in agroecoenoses. Pol. Ecol.Studies 5:151-200.

1980. Spider communities in crop fields and forests of different landscapes of Poland. Pol. Ecol. Stud. 6:735-762.

1986. The distribution of spiders and the structure of their communities under the pressure of agriculture and industry. Poznan 9-14 Septembre 1985. Ed. INRA Paris. Les Colloques de l'INRA 36:85-96.

Mader, H. J., R. Klüppel & H. Overmeyer. 1986. Experimente zum Biotopverbundsystem-tierökologische Untersuchungen einer Anpflanzung. Scrihftenreihe für Landschaftspflege und Naturschutz. 27:1-136.

Moulder, B. C. & D. E. Reichle. 1972. Significance of spider predation in the energy dynamics of forest floor arthopod communities. Ecol. Monogr. 42:473-478.

Moran, M. D. & L. E. Hurd. 1994. Short-term responses to elevated predator densities: noncompetitive intraguild interaction and behaviour. Oecologia 98:269-273.

Mühlenberg, M. & W. Werres. 1983. Lebensraumverkleinerung und ihre Folgen für einzelne Tiergemeinschaften. Naturschutz und Landschaftsplanung 58:43-50.

Murdoch, W. W., J. Chesson & P. L. Chesson. 1985. Biological control in theory and practice. Am. Nat. 125:344-366.

Musick, G. J. 1970. Insect problems with no-till crops. Problems with no-till crops, will it work everywhere? Soils Mag. 23:18-19.

Naton, E. 1993. Investigations during 10 years on the dominant spider fauna of a field. 10 dominant species (Arachnida, Araneae). Spixiana 16:247-282.

Nentwig, W. 1988. Augmentation of beneficial arthopods by strip management. Oecologia. 76:597-606.

Nyffeler, M. & G. Benz. 1981. Okologische bedeutung der spinnen als insektenpradatoren in wiesen und getreidefeldern. Mitteilungen der Deutschen Gesellschaft für allegemeine entomologie. 9:33-35.

Olive, C.W. 1980. Foraging specializations in orb-weaving spiders. Ecology 61:1133-1144.

Olszak, R. W., Luczak J., Niemczyk, E. & R. Z. Zajac. 1994. The spider community associated with apple trees under different pressure of pesticides. Ekologia Polska 40:265-286.

Pasquet, A. 1984. Predatory-site selection and adaptation of the trap in four species of orb-weaving spiders. Biol. Behav. 9:3-19

Pfrimmer, T. R. 1964. Populations of certain insects and spiders on cotton plants following insecticide application. J. Econ. Entomol. 57:640-44.

Plagens, M. J. 1986. Aerial dispersal of spiders (Araneae) in Florida cornfield ecosystem. Environ. Entomol. 15:1225-1233.

Post, W. M. & C. C. Travis. 1979. Quantitative stability in models of ecological communities. J. Theor. Biol. 79:547-553.

Provencher, L. 1990. An individual-based approach to understanding prey limitation by spiders. Ph.D. dissertation. University of Tennessee, Knoxville.

Pulz, R. 1987. Thermal and water relations, pp. 26-55. In W. Nentwig [ed.]. Ecophysiology of spiders. Springer Verlag, New York.

Riechert, S. E. 1985. Decision problems in multiple goal contexts: Spider habitat selection. Zeitschrift für Tierpsychologie 70:53-69.

Riechert, S. E. 1987. Between population variation in spider territorial behavior: hybrid-pure population line comparisons, pp. 33-41. *In* M. Huettel [ed.], Arthropod behavioral genetics. Plenum Press, New York.

Riechert, S. E. 1992. Spiders as representative sit-and-wait predators, pp. 313-328. *In* M. J. Crawley [ed.], Biology of natural enemies. Blackwell, London.

Riechert, S. E. & L. Bishop. 1990. Prey control by an assemblage of generalist predators in a garden test system. Ecology 71:1441-1450.

Riechert, S. E. & R. Gillespie. 1986. Habitat choice and utilization in web building spiders, pp. 23-48. *In* W.B. Shear [ed.], Spiders: webs, behavior and evolution. Stanford Univ. Press, Stanford, California.

Riechert, S. E. & J. Harp. 1987. Nutritional ecology of spiders, in arthropod nutrition, pp. 318-328. *In* F. Slansky & J. G. Rodriguez [eds.], Academic Press, New York.

Riechert, S. E. & K. Lawrence. 1997. Test for predation effects of single versus multiple species of generalist predators: Spiders and their insect prey. Entomol. Exp. Appl. 84:147-155.

Riechert, S. E. & T. Lockley. 1984. Spiders as biological control agents. Annu. Rev. Entomol. 29:299-320.

Riechert, S. E. & J. Luczak. 1982. Spider foraging: behavioral responses to prey, pp. 353-384. *In* P.N. Witt & J. Rovner [eds.], Spider communication: mechanisms and ecological significance. Princeton University Press.

Riechert, S. E. & J. Maynard Smith. 1989. Genetic analyses of two behavioural traits linked to individual fitness in the desert spider, *Agelenopsis aperta.* Anim. Behav. 37:624-637.

Riechert, S. E. , L. Provencher & K. Lawrence (in press) The significance of generalist predators in agroecosystems: potential equilibrium point control of prey by spiders. Ecol. Applic.

Riechert, S. E. & C. R. Tracy. 1975. Thermal balance and prey availability: bases for a model relating web-site characteristics to spider reproductive success. Ecology 56:265-284.

Risch, S. J., D. Andrew & M. A. Altieri. 1983. Agroecosystem diversity and pest control: data, tentative conclusions, and new research directions. Environ. Entomol. 12:625- 629.

Robertson, L. N. Kettle, B. A. & G. B. Simpson. 1994. The influence of tillage practices on soil macrofauna in a semi-arid agroecosystem in northeastern Australia. Agric. Ecosystems Environ. 48:149-156.

Schaber, B. D. & T. Entz. 1994. Effect of annual and biennial burning of seed alfalfa (Lucerne) stubble on populations of lygus (*Lygus* sp) and alfalfa plant bug (*Adelphocoris lineolatus* [Goeze]) and their predators. Ann. Appl. Biol. 124:1-9.

Schaefer, M. 1987. Life cycles and diapause, pp. 331-347. *In* W. Nentwig [ed.], Ecophysiology of spiders. Springer Verlag, New York

Schmitz, O. J. 1994. Resource availability and trophic exploitation in an old-field food web. Proc. Nat. Acad. Sci. 91:5364-5367.

Sokal, R. R. & F. J. Rohlf. 1969. Biometry. W.H. Freeman, San Francisco.

Spiller, D. A. 1986. Interspecific competition between spiders and its relevance to biological control by general predators. Environ. Entomol. 15:177-181.

Srikanth, J., S. Easwaramoothy, N.K. Kurup & G. Santhalakshmi. 1997. Spider abundance in sugarcane: impact of cultural practices, irrigation and post-harvest trash burning. Biol. Agric. Hortic. 14:343-356

Steward, B. A., L. K. Porter & F. G. Viets. 1966. Sulphur requirements for decomposition of cellulose and glucose in soil. Soil Sci. Amer. Proc. 30:453.

Tolbert, W. W. 1977. Aerial dispersal behavior of two orb weaving spiders. Psyche 84:13-27.

Turnbull, A. L. 1966. A population of spiders and their potential prey in an overgrazed pasture in eastern Ontario. Can. Zool. 44:557-583.

Uetz, G. W., T. C. Kane & G. E. Stratton. 1982. Variation in the social grouping tendency of a communal web-building spider. Science 217:547-549.

Van Hook, R. I. 1971. Energy and nutrient dynamics of spider and orthopteran populations in a grassland ecosystem. Ecol. Monogr. 41:1-26.

Waksman, S. A. 1938. Humus. Bailliere, Tindall and Cox, London.

White, P. C. L. & M. Hassall. 1994. Effects of management on spider communities of headlands in cereal fields. Pedobiologia 38:169-184.

Wyss, E. 1995. The effects of weed strips on aphids and aphidophagous predators in an apple orchard. Entomol. Exp. Appl.75:43-49.

Natural and Artificial Shelter to Enhance Arthropod Biological Control Agents

KERRY A. BEANE—2199 Monterey Drive, South Lake Tahoe, CA 96150

ROBERT L. BUGG—University of California Sustainable Agriculture Research and Education Program, University of California, Davis, CA 95616-8716

Introduction

Predators and parasites that attack arthropod pests require shelter from the elements. Appropriate habitat may promote foraging, resting, seasonal dormancy (e.g., overwintering), or nesting. Here, we present a selective review of literature documenting the importance of such habitat, including discussion of both natural and artificial shelter.

Foraging

Physical environmental conditions profoundly affect natural enemy activity during the growing season, as excessive wind is thought to limit foraging by adult hover flies (Diptera: Syrphidae). Hedgerows, windbreaks, or shelterbelts can protect croplands in windy areas, and provide some protection to windward as well as to leeward. Shelter can reduce soil erosion, and improve photosynthetic and water-use efficiency by crop plants, and can lead to locally elevated temperatures in the sheltered areas (Van Eimern 1964). These practical considerations raise the possibility of using wind shelter to enhance biological control by aphidophagous syrphids.

Using segregating traps and painted pan traps containing water and detergent, Lewis (1965a) showed that syrphids occurred in areas sheltered by artificial windbreaks (0.915 m in height, made with horizontal slats, 45 percent open area), and that of all 13 insect taxa assessed (diurnal and nocturnal), syrphids showed by far the greatest tendency to concentrate in the sheltered area. Aphids (Homoptera: Aphididae) also settle selectively near shelter (Lewis 1965b), so the net effect of windbreaks on aphid control is in question.

Because hedgerows and windbreaks often contain flowering plants used by syrphids, effects of shelter and of flowers may be confounded. Pollard (1971) believed that shelter influenced syrphid oviposition, but that flowers did not. In Pollard's experiment, potted brussels sprouts (*Brassica oleracea* var. *gemmifera*) plants were placed in various habitats, then retrieved and inspected for syrphid eggs. Adult syrphids were more abundant in areas with flowers, but oviposition was depressed in unsheltered areas, regardless of whether flowers occurred nearby. However, Pollard's was not a true factorial experiment, and the two factors of interest were neither manipulated nor controlled. The experiment also lacked systematic interspersion of treatments and rigorous statistical analysis.

Bowden and Dean (1977) used suction traps to assess the distribution of adult syrphids on both sides of and at two distances from a high (7 m) hedgerow. Prevailing wind did not seem to influence the distribution: syrphids were consistently more abundant on the western side, which was more diverse floristically.

In a windy section of the San Joaquin Valley of California, Pickett et al. (1990) used windbreaks of organdy to shelter French prune (*Prunus domestica*) trees. This shelter led to increased densities of prune leafhopper (*Edwardsiana prunicola* [Edwards], Homoptera: Cicadellidae) and the egg-parasitic *Anagrus epos* Girault (Hymenoptera: Mymaridae). Sheltered trees showed a highly significant, 22-fold, increase in the densities of parasitized leafhopper eggs. Rows of French prune trees may themselves provide wind shelter to adjoining grapevines (*Vitis vinifera*), and this is believed to be important in promoting activity by *A. epos* against grape leafhopper (*Erythroneura elegantula* Osborn, Homoptera: Cicadellidae) (Corbett & Rosenheim 1996),

Resting

Overnight resting ("sleeping") aggregations of various entomophagous aculeate Hymenoptera (e.g., Sphecidae, Eumenidae) are often observed in the field (Linsley 1962). Typical sites for such aggregations vary among species, and include herbaceous and woody plants as well as human-made structures (e.g., eaves of buildings). These are believed to provide some shelter from wind and rain, as well as favorable illumination during the morning or afternoon. Sleeping aggregations of some Sphecidae may recur for many successive years in the same precise locations (R.L. Bugg & J.A. Rosenheim pers. obs. 1997), suggesting that some sort of persistent pheromonal marking may be involved, although this has not been demonstrated. To our knowledge "sleeping sites" have never been explored as a tool for enhancing biological control, perhaps because, with many species, the aggregations contain exclusively males, which have no direct function in pest control.

The earwig shelter provides year-round habitat for predacious earwigs (Dermaptera: Forficulidae) in gardens, orchards and commercial vegetable fields. The commercially available dome-shaped plastic unit was preceded in design by homemade shelters composed of inverted clay pots stuffed with various natural packing materials and fitted with a saucer-like bottom secured to the top with wire. The pre-stuffed plastic unit developed by Biotechnik, Inc. (Kevelaer, Germany) can be purchased individually or in bulk units of 100 (Beane 1992). When hung in an orchard or staked near field crops, the shelters attract and concentrate earwigs. The European earwig (*Forficula auricularia* L., Dermaptera: Forficulidae), although a pest in some crops, attacks aphids, psyllids and worm pests, and the shelters afford this earwig protection from their natural enemies, e.g., birds and spiders. The shelter's attractancy is based on the principle that earwigs seek out tightly confined, protected spaces.

Seasonal Dormancy

We define seasonal dormancy as any of several states of torpor entered into by organisms or subpopulations thereof during certain times of the year. This term is intended to comprise reproductive and other forms of diapause, including both aestival and hibernal diapause. Seasonal dormancy may be prompted by changes in food availability, photoperiod, or weather. In many cases, arthropods seek shelter prior to entering seasonal dormancy.

Overwintering success by parasites or predators may determine whether biological control is attained, as indicated by Dreistadt & Dahlsten (1991) in the case of *Tetrastichus gallerucae* (Fonscolmbe) (Hymenoptera: Eulophidae). In northern California, this introduced egg parasite of elm leaf beetle (*Xanthogaleruca luteola* [Müller], Coleoptera: Chrsyomelidae) survives the winter in extremely low numbers, insufficient to provide early-season biological control of the pest.

Overwintering sites of entomophagous taxa may at times coincide with those of the target pest. In such cases, cultural and biological controls may appear to be in conflict. As mentioned by Bugg & Pickett (this volume), in California, in almond (*Prunus dulcis* var. *dulcis*) orchards sanitation procedures dislodge and destroy "mummy nuts," thus reducing overwintering both by the pest navel orangeworm (*Amyelois transitella* [Walker], Lepidoptera: Pyralidae) (Barnett et al. 1989), and by its parasite *Goniozus legneri* Gordh (Hymenoptera: Bethylidae) (Legner & Warkentin 1988; Legner & Gordh 1992; W. Bentley pers. comm. 1997).

In other instances, overwintering sites of pests and beneficial arthropods may differ. For example, in the Yakima Valley of Washington state, Tamaki (1972) found that during peach (*Prunus persica*) leaf abscission in autumn, many green peach aphid (*Myzus persicae* [Sulzer], Homoptera: Aphididae) are deposited on the ground, but must find their way back up the tree to oviposit, because the species overwinters in the egg stage, attached to peach wood (branches, twigs). While on the ground, green peach aphid is subject to intense predation by the bigeyed bug *Geocoris bullatus* (Say) (Hemiptera: Lygaeidae). Tamaki and Weeks (1972) found that *G. bullatus* overwinters in the egg stage on the ground, amid understory stands of the perennial bunch grass orchardgrass (*Dactylis glomerata*) and decaying peach leaves. Late fall tillage of the orchard understory apparently greatly disrupts overwintering by the predator.

Karban et al. (1997) reported that Willamette mite (*Eotetranychus willamettei* [Ewing], Acari: Tetranychidae), a pest, overwinters under bud scales of grape, as does its predator *Metaseiulus occidentalis* (Nesbitt) (Acari: Phytoseiidae). However, two-spotted mite (*Tetranychus urticae* Koch, Acari: Tetranychidae), another pest, does not overwinter on grapevines, but rather on associated cover crops.

In another example from grapes, *Anagrus epos* is unable to overwinter in California vineyards. Provision of French prune trees harboring overwintering eggs of prune leafhopper (*Edwardsiana prunicola* [Edwards], Homoptera: Cicadellidae) as overwintering habitat may lead to improved colonization by the parasite of nearby vineyards (Corbett & Rosenheim 1996; Murphy et al. 1996). The evaluation by Corbett and Rosenheim (1996) was made possible through the use of rubidium labelling, permitting the identification of parasites that had overwintered on French prune (Corbett et al. 1996). In this instance, the biotic environment provided by the overwintering site was all-important; in other cases, the physical environment is critical.

Although lady beetles (Coleoptera: Coccinellidae) have long been recognized as predators of agricultural pests, generalizations on their usefulness have been difficult, because of their highly variable feeding habits, seasonal movement, and hibernation patterns (Hemptinne 1988; Honek 1989; LaMana & Miller 1996). Despite these complexities, distance from overwintering sites has been seen as determining abundance of coccinellids in fields (Hodek 1967).

The planting of trees or perennial bunch grasses near agricultural sites and the use of burlap or cloth tree wraps have been used to aid overwintering by various coccinellids (Hodek 1967). In the Sacramento Valley of California, one of us (R.L. Bugg personal observation) has found mixed overwintering aggregations of lady beetles amid stands of native bunch grasses (e.g., deer grass (*Muhlenbergia rigens*), purple needle grass (*Nassella pulchra*), and blue wildrye (*Elymus glaucus*) planted in rural rights-of-way (see Bugg et al. this volume). The introduced *Coccinella septempunctata* (L.) was observed along with the two native species *Coccinella novemnotata* Herbst and *Hippodamia convergens* Guerin-Meneville.

In a replicated study, Wratten et al. (this volume) found that tussock-forming grasses supported greater densities of overwintering predaceous beetles than did mat-forming grasses. Hourly thermal measurements at ground level showed that the two grass forms led to similar mean temperatures, but that tussocks had less variability. Thus, tussocks provided refuge from extreme temperatures.

In England, a three-year study in a hawthorn (*Crataegus monogyna*) hedgerow bordering open fields revealed that the total number of fauna was reduced when the vegetation beneath the hedge was removed with a paraquat-diquat herbicide (Pollard 1968). The insects most consistently affected by the treatment were predators. Evaluations of predacious Hemiptera concerned three species in particular, *Anthocoris nemorum* L. (Anthocoridae), *Phytocoris ulmi* (L.) (Miridae) and a hemipteran termed by Pollard (1968) *Heterotoma planicornis* (Pallis) (Miridae), which was probably actually *Heterotoma merioptera* (Scopoli) (John H. Lattin, personal communication). These predators were reduced due to removal of the understory flora. Further study of effects of hawthorn hedgerow border on sugar beet and wheat crops indicated that *P. ulmi* and *Heterotoma* sp. prob. *merioptera* remained within the immediate area of the hedgerow, but first-generation *A. nemorum* moved into the adjoining wheat crop as far as 15 and 30 yards from the hedgerow. *Anthocoris nemorum* was also found in the nearby beet field, its numbers declining as sampling distance from the hedge increased. The pattern of *A. nemorum* occurrence in cereal crops indicated that the predator seeks shelter from the sun and prevailing winds. For this reason, the species could not be expected to disperse readily throughout cereal crops. In contrast, rape, providing shade and wind shelter, contained abundant, well distributed *A. nemorum*. The highest numbers of *A. nemorum* still occurred close to the edges of the field. Although overall *A. nemorum* movement from the hedgerow during the season was not significant in this study, the results support the conclusion that large-scale removal of hedges and understory flora can reduce the numbers of predators found in crops (Moore et al. 1967).

Anderson (1962a) reported that the pirate bugs *Anthocoris nemorum*,

Anthocoris confusus Reut., and *Anthocoris gallarum-ulmi* (DeG.) commonly overwinter beneath the bark scales of sycamore (*Platanus* sp.) and other trees. Anderson (1962b) described a simple laboratory method for sustaining overwintering *Anthocoris* spp. that enables up to 75% female survival. Bugs were kept alive in small tubes for up to six months at 34° F. The small corked tubes were embedded in saturated plaster of paris and provided bugs with resting sites protected from excess moisture.

"Trap bands" and "trunk traps" have long been known as artificial over-wintering sites for lady beetles, lacewings (Neuroptera: Chrysopidae), and predatory bugs (Hemiptera). Some references indicate the usefulness of trap bands for the control of codling moth (*Cydia pomonella* [L.] Lepidoptera: Tortricidae) and *Anthonomus pictus* Blatchley (Coleoptera: Curculionidae) in apples (*Malus pumila*) and other orchard systems (New 1967). Tests in peach, pear (*Pyrus communis*) and pecan (*Carya illinoinensis*) orchards have shown their effectiveness for harboring both insect pests (caterpillars [Lepidoptera], aphids, psyllids [Homoptera: Psyllidae], stink bugs [Hemiptera: Pentatomidae]) and their natural enemies (Tamaki & Halfhill 1968; Fye 1985; Mizell & Schiffhauer 1987). In an unsprayed English walnut (*Juglans regia)* orchard in Esparto, Yolo County, California (James and Claire Haag, owners), fluted cardboard trunk bands were affixed to trees in October, and sampled in late November (1991). Several predators were found in the bands (Robert L. Bugg and Lawrence A. Allen personal observation). European earwig (*Forficula auricularia* L., Dermaptera: Forficulidae) was abundant. Spiders (Araneae) included Clubionidae (*Cheiracanthium* sp. and *Strotarchus* sp.), Gnaphosidae (*Drassyllus* spp., and *Trachyzelotes* sp.), and a house spider *Oecobius* sp (Oecobiidae). In addition, the predatory mite *Bechsteinia terminalis* (Banks) (Acari: Anystidae) occurred by the hundreds in each band (Robert L. Bugg and Jurgen Otto personal observation). The precise roles of these predators in pest control of walnut is still unclear.

In northern Florida, Mizell and Schiffhauer (1987) tested trap designs for the collection of overwintering predators of stinkbugs in pecan orchards. The researchers used roofing nails to attach both cardboard and filter traps to the tree trunks. The traps each covered about 1,500 cm of tree trunk space and were placed about one meter above the ground facing north. Three trap designs in all were tested, including a burlap trap, a cardboard trap, and a filter trap, and all were occupied by overwintering beneficial insects and spiders as well as stinkbugs. To build the burlap trap, one must cut a 45x100 cm burlap strip, then fold it over three equal times and staple it to the tree at the end closure points (Tedders 1974). The cardboard trap consists of 3,

15x100cm strips of three different materials glued together. Plastic Screen Glass (Warp Bros., Chicago, IL) composes the outermost strip, covering a center of heavy, corrugated cardboard, and an inner layer of burlap. The trap was modified to the strip design for use in pecan orchards (Fye 1985). A third type, the filter trap, is composed of a UV-proof plastic bag (15x5x100 cm) lined with a 15x100 cm strip of greenhouse Coolpad (CASSCO, Inc., Montgomery, Alabama). Twenty-five to fifty holes of about one cm in diameter are punched through the plastic bag to allow insects to enter.

Tamaki and Halfhill (1968) evaluated trap bands that were painted black on outer surfaces and secured on primary scaffold branches of peach trees. The bands were composed of 0.0165 cm thick x 8 cm wide sheet-aluminum bands and lined with tarred burlap and heavy paper. When placed in peach orchards, the bands were occupied by earwigs, lacewings, and spiders. Overwintering mortality of predators was reduced, enabling them to reach higher populations earlier in the season and provide better control of green peach aphid.

The corrugated fiberboard traps originally developed by Fye (1985) for use in pears were composed of more layers of corrugated cardboard than the band used by Mizell and Schiffhauer (1987), forming instead a thick rectangular unit easily anchored in place by wedging in the crotch of the pear trees. Trapping results led Fye (1985) to assert that suitable overwintering sites for natural enemies are often inadequate in orchards. He concluded that the enhancement of fall orchard populations with corrugated fiberboard shelters can be achieved, but only if sufficient numbers of predators are already present in the area.

Burlap and corrugated cardboard trunk wraps have been tested informally by California pecan growers and University of California Cooperative Extension Farm Advisor Steve Sibbett. The aim has been to enhance overwintering success by *Olla v-nigrum* (Mulsant) (Coleoptera: Coccinellidae), a principally arboreal aphidophagous lady beetle. The wraps were used heavily by this species during aestival-hibernal dormancy, often with hundreds of lady beetles occupying a single wrap. However, the small area used in this study has made for uncertain results in terms of enhancement of biological control (Steve Sibbett personal communication 1992).

Under unprotected winter conditions in Germany, mortality of common green lacewing (*Chrysoperla carnea* [Stephens], Neuroptera: Chrysopidae) can reach 60–90%. Sengonça & Frings (1987) developed a wooden shelter ("hibernation box") for lacewing adults that can increase lacewing survivorship in winter with the aim to consequently boost spring population levels.

The "house for lacewings" is a rectangular red or brown painted wooden unit that is placed at a height of 150–180 cm in open fields. With its horizontal wood slat entranceway facing counter to the prevailing wind direction, it provides adult lacewings a single chamber tightly stuffed with wheat straw (barley straw and leaves are not recommended). Lacewing hibernation boxes should be located away from houses, trees and other structures to have the greatest effect. Once lacewings have entered the hibernation boxes and the cold, wet weather begins, the boxes should be brought into a rain-protected room then replaced in open field locations in spring. The shelters have been offered as a tool for augmenting biological pest management in orchards and gardens. Sengonça & Henze (1992) found that, depending on the associated crop in the fall, mean density of common green lacewing per box might range from 117 to 224, with both sugar beet and fallow fields harboring significantly higher densities than winter barley (*Hordeum vulgare*) fields. Few spiders or flies overwintered in the boxes, and mean winter mortality of adult common green lacewing ranged from 0.7–4.0%. In light of the long preovipository flights demonstrated for common green lacewing by Duelli (1980a,b) ("ranging" in the terminology of Dingle [1996]), it appears questionable whether the use of hibernation boxes can lead to local enhancement of common green lacewing density or effectiveness at pest control.

Nesting and Other Ovipositional Sites

Predatory wasps (Hymenoptera), whether solitary (Eumenidae, Sphecidae) or social (Vespidae), have diverse nesting requirements. These wasps nest variously in burrows of their own excavation, pre-existing burrows or soil crevices, pre-existing cavities in wood or twigs, hollow straws, stone or wood crevices, or in mud or paper structures that they themselves build. Some Vespidae construct paper nests underground; others build them aloft, under eaves or other overhangs, or in crevices or treeholes. Many predatory wasps transport paralyzed or dead prey to their nest sites for larval use. Straightforward logistical considerations suggest that the local provision of appropriate structures or materials may enhance the activity of wasps that attack agricultural pests. Several papers are available that explore the complexities in foraging patterns of social Hymenoptera in relation to nest site (e.g. McIver 1991; Acosta et al. 1995)

Some sphecid wasps (Hymenoptera: Sphecidae) are predators of plant pests in agricultural systems. Kolesnikov (1978) relocated cocoons of the sphecid digger wasp *Cercerus arenaria* (L.), to a young pine (*Pinus* sp.) plantation in Bryansk Province, Russia, thereby establishing a new colony.

Kolesnikov (1978) exhumed cocoons containing the young larvae of the wasps and placed them 20–25cm deep and approximately 15cm apart under lightly packed soil in a newly dug trench at the edge of a pine forest study area (near Novozybkov). During an eight-year period, the introduced colony increased by a factor of eight in one area and nine in another. The colony apparently reduced local weevil densities in non-replicated studies. Density effect was most pronounced within 50m of the colony, where only 16 weevils per 50 sweep sample were detected. At a distance between 50 and 200m, 29 weevils were detected per sweep, and at 200m, 30 to 40 weevils were caught per sweep. Such data should be viewed critically, however, in that initial observations to provide baseline information on weevil density were lacking, as is replication (Schoenig et al. this volume).

Kolesnikov (1978) observed that solitary aphid wasps (Hymenoptera: Sphecidae: Pemphredoninae) seek nesting sites in the straw thatch of buildings. He established 10x10cm wire frames of straw, fitting them over the projecting straws at the edge of buildings in Bryansk province. One third of all straw houses occupied by insects contained Pemphredoninae. An estimated 90,000 garden and orchard insect pests were stored in just 615 straws. The replacement of straw with modern roofing materials that are inhospitable to solitary wasps was cited as a partial explanation for increasing aphid damage to orchards in the area. This led Koesnikov (1978) to create artificial sphecid nesting sites in orchards and vegetable gardens. Each artificial nesting site consisted of undamaged straw in lengths of 35–40cm, bundled and tied between two pairs of stakes and placed in orchards. These easy-to-assemble units, tested at the Novozybkov Pedagogical Institute, suggested that the Pemphredoninae readily use and benefit from artificial shelters.

Social wasps (Vespidae) also benefit from appropriate nest sites. Work in the southwestern United States has focused on managing *Polistes* spp, (Hymenoptera: Vespidae) to control bollweevil, caterpillars, and other pests of cotton. Gillaspy (1979) demonstrated that *Polistes* spp. colonies can be transported to new sites and will thrive in artificial shelters such as modified ice cream cartons or downward opening wooden boxes. *Polistes apachus* (Sausssure) seemed to adapt especially well to human manipulation in comparison to other species. Failures of *Polistes* wasps to successfully capture prey often result in the prey falling to the ground where it is eaten by carabid beetles, earwigs, or crickets (Nakasuji et al. 1976). Gillaspy's success in feeding budworms to wasps at the nest site suggests that the wasps may possibly be "trained" to hunt for certain prey. Problems develop when a wasp colony becomes "fixed" to a prey type that is not the critical pest in the system. More work remains to be done on "training" colonies to target specific prey.

Polistes exclamans arizonensis Snelling colonies have overwintered successfully on a diet of tobacco budworm (*Heliothis viriscens* [F.], Lepidoptera: Noctuidae) in a specially designed outdoor cage in Tucson, Arizona (Fye 1972). Naturally occurring nests could be transferred successfully into the cage as long as some of the cells were capped. Natural nests placed in the cage in late October were found to be much larger than those constructed within the cage. It was estimated that the larvae from the larger, naturally occurring nests (means ranging from 103 to 503 cells) consumed about 3,400 larvae while larvae of the smallest nest consumed about 626 larvae. *Polistes exclamans arizonensis* seems to be a good candidate for introduction into cotton fields for controlling tobacco budworm. Both at the overwintering cage site and in the field where *P.e. arizonensis* nests were placed, wasps of many other species were notably attracted and induced to hibernate in the area. The tendency for different wasp species to aggregate seems to indicate the existence of "assembly scents" (= aggregation pheromones), which if artificially synthesized could facilitate wasp attraction and nest building in sites where their presence for biological control purposes is desirable (Fye 1972).

Various other predators that lack elaborate nesting behaviors may yet have ovipositional site preferences. Dunbar (1971) reported that the three *Geocoris* spp. (Hemiptera: Lygaeidae) that he studied did not prefer to oviposit on green foliage. His observations indicated that *Geocoris atricolor* Montandon and *Geocoris pallens* Stål preferred to lay eggs on dried leaves and stems and on the ground, whereas *Geocoris punctipes* (Say) preferred to oviposit in hollow stems.

A domatium is a plant-produced chamber that houses animals, and its formation is not induced by its residents (as galls are) (O'Dowd & Willson 1991). Agrawal & Karban (1997) found that when simulated domatia were adhered to cotton (*Gossypium hirsutum*) foliage, these served as ovipositional sites for bigeyed bugs (*Geocoris* spp.) and minute pirate bug (*Orius tristicolor* White, Hemiptera: Anthocoridae), leading to enhanced densities of these predators, and improved pest control. Domatia occur on close relatives of cotton, and the genes for domatia could potentially be bred into cotton. This possibility is being explored (R. Karban personal communication).

Integrating Shelter with Other Habitat Components and with Practical Farm Management

The apparent interdependency of shelter and nectar effects (Pollard 1971; Bowden and Dean 1977) suggests that both may be needed to attain the highest level of hover fly activity. Therefore, appropriate flowering tree or understory plants should be included in windbreaks that border farming systems where syrphids are likely to be important biological control agents and where strong winds are likely at critical times of the year.

The different densities of common green lacewing encountered in the overwintering boxes placed amid various crops suggest that overwintering sites should be concentrated amid plants that show substantial late-season buildups of the target insects. For example, in northern California, numerous species of agriculturally important entomophagous insects occur abundantly on flowering coyote brush (*Baccharis pilularis* ssp. *consanquinea*) from mid-September through mid-October (Steffan 1997). Overwintering boxes could be placed amid stands of such plants, then removed in late autumn or early winter for storage under controlled conditions. As mentioned earlier, there is some concern as to whether overwintering shelter will reliably lead to local enhancement of beneficial arthropods. This may depend on the vagility of the species in question and on whether foraging (short-distance movement) or ranging (long-distance movement) ensues upon emergence (Dingle 1996). This represents an important area of scientific inquiry. In the event that foraging occurs immediately after emergence, local provision of early-season foods may be important in preventing further dispersal, although cool-season cover cropping to provide these has not reliably led to enhancement of lady beetles in Georgia pecan orchards (Bugg et al. 1991). Perhaps the use of rubidium, ^{15}N, or some other marker could be used to label insects in local overwintering sites and determine their relative contributions to the initial colonists of fields. Hagler (1997) showed that sprayed-on rabbit immunoglobulin could be used to label overwintering convergent lady beetle (*Hippodamia convergens* Guerin-Menéville, Coleoptera: Coccinellidae). Retention of this marker by the predator appears superior to that of fluorescent dust or elemental markers. Immunological evaluation of insects could reveal patterns of dispersal and the importance of local overwintering sites. This work could be analogous to that of Corbett and Rosenheim (1996).

Selective destruction of pest overwintering habitat may be important in cultural control, and ideally should be complemented by an awareness of and respect for the overwintering habitat of beneficial arthropods. In cases where the overwintering site is the same for pest and beneficial arthropod, there may be ways of selectively destroying the former while preserving and releasing the latter. Decision plans based on natural enemy–pest ratios could be developed that trigger the removal of the overwintering site. In the case of the trunk traps mentioned for various crops and mummy nuts mentioned in the case of almonds, screened enclosures could be used to containing the mummies, culls, or trunk traps. Mesh size should be appropriate to permit emergence and escape of the beneficial arthropods but not of the pest.

Enhancement of stinging social wasps on the farm may have undesirable side effects, and may put farmworkers at risk (Gillaspy 1979). The latter problem appears most likely if nesting boxes are placed near areas where farm workers congregate.

As with all proposed enhancement programs, reasonable cost-benefit analyses should be obtained before initiating major projects to provide shelter. At this time, the needed quantitative data are lacking.

References

Acosta F. J., F. Lopez & J. M. Serrano. 1995. Dispersed versus central-place foraging — intra- and intercolonial competition in the strategy of trunk trail arrangement of a harvester ant. Amer. Naturalist 145:389-411.

Agrawal, A. A. & R. Karban. 1997. Domatia mediate plant-arthropod mutualism. Nature 387:562-563.

Anderson, N. H. 1962a. Bionomics of six species of *Anthocoris* (Heteroptera: Anthocoridae). Trans. Royal Entomol. Soc. London 114:67-95.

Anderson, N. H. 1962b. Studies on overwintering of *Anthocoris* (Hem., Anthocoridae). Entomol. Monthly Mag. 98:1-13.

Barnett, W. W., L. C. Hendricks, W. K. Asai, R.B. Elkins, D. Boquist & C. L. Elmore. 1989. Management of navel orangeworm and ants. California Agric. 43(4):21-22.

Beane, K. 1992. Shelter for predacious earwigs available in the U.S. IPM Practitioner 14:13.

Bowden, J. & G. J.W. Dean. 1977. The distribution of flying insects in and near a tall hedgerow. J. Appl. Ecol. 14:343-354.

Bugg, R. L., J. D. Dutcher & P. J. McNeill. 1991. Cool-season cover crops in the pecan orchard understory: effects on Coccinellidae (Coleoptera) and pecan aphids (Homoptera: Aphididae). Biol. Control 1:8-15.

Corbett, A., B. C. Murphy, J.A. Rosenheim & P. Bruins. 1996. Labeling an egg parasitoid, *Anagrus epos* (Hymenoptera: Mymaridae), with rubidium within an overwintering refuge. Environ. Entomol. 25:29-38.

Corbett, A. & J. A. Rosenheim. 1996. Impact of a natural enemy overwintering refuge and its interaction with the surrounding landscape. Ecol. Entomol. 21:155-164.

Dingle, H. 1996. Migration: the biology of life on the move, Oxford University Press, New York, 474 pp.

Dreistadt, S. H. & D. L. Dahlsten. 1991. Establishment and overwintering of *Tetrastichus gallerucae* (Hymenoptera: Eulophidae), an egg parasitoid of the elm leaf beetle (Coleoptera: Chrysomelidae) in northern California. Environ. Entomol. 20:1711-1719.

Duelli, P. 1980a. Preovipository migration flights in the green lacewing, *Chrysopa carnea* (Planipennia, Chrysopidae). Behav. Ecol. Sociol. 7:239-246.

Duelli, P. 1980b. Adaptive dispersal and appetitive flight in the green lacewing, *Chrysoperla carnea*, Ecol. Entomol. 5:213-220.

Dunbar, D. M. 1971. The biology and ecology of *Geocoris atricolor* Montandon, *G. pallens* Stal, and *G. punctipes* (Say). Ph.D. Dissertation. University of California, Davis.

Fye, R. E. 1972. Manipulation of *Polistes exclamans arizonensis*. Environ Entomol, 1:55-57.

Fye, R. E. 1985. Corrugated fiberboard traps for predators overwintering in pear orchards. J. Econ. Entomol. 78:1511-1514.

Gillaspy, J. E. 1979. Management of *Polistes* wasps for caterpillar predation. Southw. Entomol. 4:334-348.

Hagler, J. R. 1997. Field retention of a novel mark-release-recapture method. Environ. Entomol. 26:1079-1086.

Hemptinne, J. L. 1988. Ecological requirements for hibernating *Propylea quatuordecimpunctata* (L.) and *Coccinella septempunctata* L. (Col.: Coccinellidae). Entomophaga 33:505-515.

Hodek, I. 1967. Bionomics and ecology of predaceous Coccinellidae. Annu. Rev. Entomol. 12:79-105.

Honek, A. 1989. Overwintering and annual changes of abundance of *Coccinella septempunctata* in Czechoslovakia (Coleoptera: Coccinellidae). Acta Entomologica Bohemoslovaca. 86:179-192.

Karban, R., G. English-Loeb & D. Hougen-Eitzman. 1997. Mite vaccinations for sustainable management of spider mites in vineyards. Ecol. Applications 7:183-193.

Kolesnikov, V. A. 1978. The sphecid wasps (Hymenoptera, Sphecidae) of Bryansk province as entomophages. Entomol. Rev. 56:57-65.

LaMana, M. L. & J. C. Miller. 1996. Field observations on *Harmonia axyridis* Pallas (Coleoptera: Coccinellidae) in Oregon. Biol. Control 6:232-237.

Legner, E. F. & G. Gordh. 1992. Lower navel orangeworm (Lepidoptera: Phycitidae) population densities following establishment of *Goniozus legneri* (Hymenoptera: Bethylidae) in California. J. Econ. Entomol. 85:2153-2160.

Legner, E. F. & R. W. Warkentin. 1988. Parasitizations of *Goniozus legneri* (Hymenoptera: Bethylidae) at increasing parasite and host, *Amyelois transitella* (Lepidoptera: Phycitidae), densities. Ann. Entomol. Soc. Am. 81:774-776.

Lewis, T. 1965a. The effects of an artificial windbreak on the aerial distribution of flying insects. Ann. Appl. Biol. 55:503-512.

Lewis, T. 1965b. The effect of an artificial windbreak on the distribution of aphids in a lettuce crop. Ann. Appl. Biol. 55:513-518.

Linsley, E. G. 1962. Sleeping aggregations of aculeate Hymenoptera-II. Ann. Entomol. Soc. Amer. 55:149-164.

McIver, J. D. 1991. Dispersed central place foraging in Australian meat ants. Insectes Sociaux 38:129-137.

Mizell, R. F. & D. E. Schiffhauer. 1987. Trunk traps and overwintering predators in pecan orchards: survey of species and emergence times. Florida Entomol. 70:238-244.

Moore, N. W., M. D. Hooper & B. N. K. Davis. 1967. Hedges. I. Introduction and reconnaissance studies. J. Appl. Ecol. 4:201-220.

Murphy, B. C., J. A. Rosenheim & J. Granett. 1996. Habitat diversification for improving biological control: abundance of *Anagrus epos* (Hymenoptera: Mymaridae) in grape vineyards. Environ. Entomol. 25:495-504.

Nakasuji, F., H. Yamanaka & K. Kiritani. 1976. Predation of larvae of the tobacco cutworm *Spodoptera litura* (Lepidoptera, Noctuidae) by Polistes wasps. Kontyû, Tokyo 44:205-213.

New, T. R. 1967. Trap-banding as a collecting method for Neuroptera and their parasites, and some results obtained. Entomol. Gazette 18:37-44.

O'Dowd, D.J. & M. F. Willson. 1991. Associations between mites and domatia. Tree 6:179-182.

Pickett, C. H., L. T. Wilson & D. L. Flaherty. 1990. The role of refuges in crop protection, with reference to plantings of French prune trees in a grape agroecosystem, pp. 151-165. *In* N.J. Bostanian, L. T. Wilson & T. J. Dennehy (eds.). Monitoring and integrated management of arthropod pests of small fruit crops. Intercept Ltd. Andover, Hampshire, United Kingdom.

Pollard, E. 1968. Hedges. II. The effect of removal of the bottom flora of a hawthorn hedgerow on the fauna of the hawthorn. J. Appl. Ecol. 5:109-123.

Pollard, E. 1971. Hedges. VI. Habitat diversity and crop pests: a study of *Brevicoryne brassicae* and its syrphid predators. J. Appl. Ecol. 8:751-780.

Sengonça, Ç. & B. Frings. 1987. Ein künstliches Überwinterungsquartier für die räuberische Florfliege. Die konstruktion eines Florfliegenhäuschens. (An artificial hibernation-accommodation for the predacious lacewing. The construction of a house for lacewings.) DLG-Mitteilungen Sonderdruck aus den DLG-Mitteilungen Heft 12.

Sengonça, Ç. & M. Henze. 1992. Conservation and enhancement of *Chrysoperla carnea* (Stephens) (Neuroptera, Chrysopidae) in the field by providing hibernation shelters. J. Appl. Entomol. 114:497-501.

Steffan, S. A. 1997. Flower-visitors of *Baccharis pilularis* De Candolle subsp. *consanguinea* (De Candolle) C. B. Wolf (Asteraceae) in Berkeley, California. Pan-Pacific Entomol. 52-54.

Tamaki, G. 1972. The biology of *Geocoris bullatus* inhabiting orchard floors and its impact on *Myzus persicae* on peaches. Environ. Entomol. 1:559-565.

Tamaki, G. & J. E. Halfhill. 1968. Bands on peach trees as shelters for predators of the green peach aphid. J. Econ. Entomol. 61:707-711.

Tamaki, G., and R. E. Weeks. 1972. Biology and ecology of two predators, *Geocoris pallens* Stål and *G. bullatus* (Say). Agricultural Research Service, United States Department of Agriculture, Technical Bulletin No. 1446. 46 pp.

Tedders, W. L. 1974. Bands detect weevils. Pecan Quarterly 8:24-25.

Van Eimern, J. (Chairman). 1964. Windbreaks and shelterbelts. World Meteorological Association Technical Note No. 59. 188 pp.

Habitat Management for Biological Control, Examples from China

WILLIAM OLKOWSKI & ANGHE ZHANG—Bio Integral Resource Center (BIRC),
P.O. Box 7414, Berkeley, CA 94706

Habitat manipulation for pest control has been an integral feature of Chinese agriculture, the oldest continuous agricultural system in the world. This resilient traditional agriculture, now impacted in part by western technology, can be characterized as highly productive, labor and land intensive, diverse in crops per unit area, and largely based on locally obtained renewable resources, particularly human and animal manures. Today, the Chinese feed five times as many people as the U.S. on one seventh the production area. Thus it seems worthwhile to study this system to learn why it has been so successful (Metcalf & Kelman 1980; Plucknett & Beemer, Jr. 1981; Wittwer, et al. 1987).

Part of the reason Chinese agriculture has been so successful is its ability to develop and disseminate innovations many of which are habitat manipulations for enhancing biological control. Many Chinese innovations have been adopted by European agriculture without the Europeans knowing the original source. Documentation in English of the original Chinese inventions and the technology transfer from China to the West has been the life work of Joseph Needham. Together with a large group of associates he has been publishing this information in a series of volumes for many years (see for example the summary volume prepared by Temple, 1986). This paper gathers together examples of habitat manipulations for pest control from current Chinese literature in the same spirit. Another objective is to encourage re-examination of U.S. pest control research objectives particularly with regard to generalist predators.

Habitat manipulation for pest control offers an approach which can reduce off-farm inputs, particularly purchase of pesticides. Although such techniques are frequently well-known they have been abandoned due to the dominance of pesticide technology. Habitat management can focus on the immediate environment of the pest or that of the pest's natural controls, for ways to interfere with pest biology and enhance natural enemy populations.

The following discussion summarizes information published in the IPM Practitioner (produced by BIRC, address above) that is related to habitat manipulation in China. Information has been gathered over the last four years and has been grouped into the following categories: On-farm Cropping Practices, Field Augmentation of Natural Enemies, and Field Transport of Natural Enemies and Conservation of natural enemies.

On-farm Cropping Practices

This category includes methods for increasing natural enemies of pest insects by introduction of forest trees and shrubs, intercropping, cover cropping, raising of pesticide crops, and use of seaweed for pest control purposes.

Agroforestry

From a pest management perspective it is useful to organize agroforestry-pest interactions into three major categories, called primary, secondary, and tertiary interactions. These terms originated when considering how the control strategy interacts with the pest population.

Primary Interactions

An example of a primary interaction is a system of intercropping where at least one plant species provides a botanical pesticide. For example, a tree species producing secondary plant metabolites with insecticidal activity (botanicals) is grown with a crop to directly kill or repel pests of that crop, or for processing as a botanical insecticide. This type of interaction is the simplest to understand, the easiest to document, and is closest to conventional pest control thinking.

The Chinese grow over 50 botanicals that are involved in primary interactions. Among these are: oiltea trees, e.g., *Camellia oleifera*; rotenone sources such as *Derris* spp.; chinaberry trees (*Melia azedarach*) false indigo (*Amorpha fruticosa*) and walnut trees (*Juglans* spp.). For citations on most of these species as well as many others see Zhang and Olkowski, (1992). Only the false indigo example will be discussed here. Table 1 lists some promising botanicals based on work by Chiu (1985).

False Indigo. Amorpha fruticosa, originated in North America, and was introduced to China in the early twentieth century. False indigo is widely planted throughout regions in the northern and southern parts of China and has become one of the shrub species that generates a high economic profit. It suffers little if any pest damage. This resistance is due to the chemical content of its stems and leaves (Zheng et al. 1978). Extracts from its leaves and seeds can be used to control aphids (Homoptera: Aphididae), grain moth *Sitotrogra cerealella* (Oliv.) (Lepidoptera: Gelechiidae), cotton boll-

Table 1. Other promising Chinese botanical insecticides[a].

Plant Species	Activity[b]	Pests Affected	Type of Material
Celastraceae			
Celastrus angulata	R, SP	cabbage butterfly (Pieridae, *Pieris rapae*) Noctuidae, *Spodoptera litura* Chrysomelidae, *Raphidopalpa femoralis* cucumber beetle (Curculionidae, *Diabrotica*)	seed oil spray, root bark
	R, SP, Sys	rice weevils (Curculionidae, *Sitophilus*)	root powder
Tripterygium wilfordii	R, SP, Sys	cucumber beetle	root powder
Tripterygium foresstii		rice yellow stem borer, (Pyralidae, *Chilo* spp.) gypsy moth (Lymantriidae, *Lymantria dispar*)	root powder + soap
Ericaceae			
Rhododendron molle	CT, SP, Sys	cabbage butterfly bean plastaspid (Plataspidae,*Copotosoma cribaria*) root weevil (Curculionidae, *Sciopithes obscurus*)	
Lauraceae			
Litsea cubeba	F, A	broad bean weevil (Bruchidae, *Bruchus rufimanus*)	seed oil

[a] Source: Chiu (1985).

[b] A = adulticide; CT = contact toxicant; F = fumigant; R = repellent; SP = stomach poison; Sys = systemic.

worms *Heliothis virescens* (Fabricius) (Lepidoptera: Noctuidae) and other pests on many crops. Ke and Lu (1989) reported that farmers crush one kg of false indigo leaves in 25kg water, filter, and spray the material on crops to control aphids. Qui (1989) extracted four ingredients with pesticidal properties from leaves which, when mixed, control several pest species including chafers, *Melolontha* spp. (Coleoptera: Scarabaeidae) found on desert plants. False indigo is also planted as an intercrop or mixed cropped with other desert plants where it has been shown to repel chafers and other species. It is used as an intercrop in strips with wheat, around lakes, along roads, ditches and rivers to serve as windbreaks or shelterbelts. In mountain regions it is mixed with poplar (*Populus* spp.), pine (*Pinus* spp.) and other trees.

Secondary Interactions

Secondary interactions refer to methods to enhance natural enemy populations which reduce or prevent pest problems. Since this is focused on the second trophic level the category name "Secondary" is used. The following examples are listed under the specific natural enemy groups enhanced.

Yellow Citrus Ants. The oldest example of biological control is manipulation of yellow citrus ants (*Oecophylla smaragdina*) (Fabricus), Hymenoptera: Formicidae. Known for more than 1,700 years, nests of this ant are still sold in marketplaces for establishment in orchards in Southeast China. The ant efficiently eliminates many herbivore pests found in citrus as well as in other tropical and subtropical orchards where citrus (*Citrus* spp.), coconut (*Cocos nucifera*), cocoa, coffee (*Coffea arabica*), lychee (*Litchi chinensis*), mango (*Mangifera indica*), rubber, and tea are cultivated. Apparently the key to the success of these ants is that they attack and kill larger pests without interference to the existing natural controls for scales, aphids and mites (Huang & Yang 1987).

The nests are woven from tree leaves by the ants in mountain forests. Citrus farmers move the nests to their orchards in early spring. Water moats constructed at the base of tree trunks prevent ants from reaching the ground and escaping (Figure 1). Bamboo strips serve as bridges for ants moving from tree to tree. The ants are violent, skilled attackers, that challenge even large pests. Orchards with the ants have about a 62% reduction in fruit damage compared to those in which chemical pesticides are used (Yang 1990). However, yellow citrus ants cannot survive winters in many areas. They become immobile when the temperature drops to 7° C (44.6° F), and die when the temperature reaches 0-2° C (32 to 35.6° F) for three days.

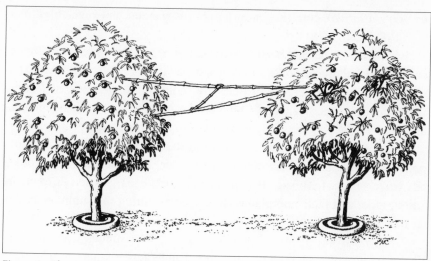

Figure 1. The moats are constructed from the clay soils and are kept filled with water. Bamboo poles or hemp ropes connect trees providing ant trails.

If no artificial shelters or food are provided in winter in many areas they die or migrate. They normally overwinter in mountain valleys, foothills, or halfway up the mountain in trees such as *Cleistocalyx operculatus*; white canary tree (*Canarium album*), smallfruit fig (*Ficus retusa*), longan (*Dimocarpus longan*), lychee, and other densely foliated species (Pu 1984). In Fugian Provence, Huang and Yang (1987) described a refinement of the above technique where ant nests are moved from orange to pomelo trees within the same orchard before harvesting oranges. The ants overwinter on the pomelo trees which are more densely foliated through the winter. In the spring the ants recolonized the orange trees. This procedure eliminated the need to purchase ant nests in the spring thereby making the technique more feasible, but requires mixed orchards with oranges and pomelo.

Lady Beetles. Lady beetles (Coleoptera: Coccinellidae) prey on numerous pest groups found in forests, orchards, and field crops. The migrating habits of many species, flying long distances to favored overwintering sites in mountain forests, is well known. It is less well appreciated that a number of species depend upon local trees for overwintering or summer feeding sites. In China, the planting of poplar (*Populus* spp.), elm (*Ulmus* sp.), pine (*Pinus* spp.), apple (*Malus pumila*), citrus (*Citrus* spp.), pear (*Pyrus* spp.), and *Paulownia* spp. along with field crops, offers summer habitat and overwintering sites to these species. This practice thereby assists in the control of pests found on cotton (*Gossypium hirsutum*), wheat (*Triticum aestivum*), rice (*Oryza sativa*), corn (*Zea mays*), millet, sugarcane and vegetables (Yan et al. 1989; Zhu 1978).

For example, the lady beetle *Propylea japonica* (Thunberg) is an important natural enemy of cotton aphids and paddy rice (*Oryza*) pests. This beetle overwinters in cracks of tree bark, holes of living trees, leaf litter, soil, and plant roots. In Central China beetles become active when temperatures reach 10° C (50° F). They leave intercropped peach and pear orchards, hibiscus, pagoda, willow, and bamboo forests, and move into adjacent cotton fields during the hottest months of summer, June and July, when other lady beetles move to forests. It is the main natural enemy of aphids found in cotton, rape, and their interplanted tree species during the summer season (Yan et al. 1989).

Lacewings. Green lacewings (Neuroptera: Chrysopidae) provide another example of a secondary agroforestry interaction where a beneficial species is provided food or refuge near a forest. As with migrating lady beetles, the preservation or restoration of forests within certain distances of crop areas is essential to the natural maintenance of some beneficial species on the farm. This is particularly true of the lacewing *Chrysopa sinica* (Tjeder) in North China, which overwinters in forests on southeastern mountain slopes. By contrast, in South China, the same species can overwinter on broad bean (*Vicia faba*) and wheat, later moving to cotton, sorghum (*Sorghum halepense* ssp. *bicolor*), corn, and soybean (*Glycine max*).

Five generations of the lacewing overwinter in the forests of Zhejiang (central coast, Figure 2) in one season. In late April, *C. sinica* colonizes aphid-infested trees such as willow (*Salix* spp.), Chinese wingnut (*Pterocarya stenoptera*), white poplar (*Populus albus*), and black locust (*Robinia pseudoacacia*) (Yan et al. 1989). Later, the lacewing moves to orchard trees and vegetable crops. Population densities peak between late June and late August, and the lacewing appears important in controlling cotton aphid (*Aphis gossypii* Glover, Homoptera: Aphididae). Lacewing densities in cotton appear approximately proportional to the numbers of trees planted around the field (Zhao 1989).

Trichogramma. Egg parasitoids in the genus *Trichogramma* (Hymenoptera: Trichogrammatidae) also benefit from the presence of trees near crops. In China, there are about 20 known species of *Trichogramma*. At least five species are mass reared and released on many crops (Bao & Chen 1988; Olkowski & Zhang 1990).

Trichogramma ostriniae (Pang & Chen), which attacks oriental corn borer (*Ostrinia furnacalis* [Guenée], Lepidoptera: Pyralidae), is well adapted

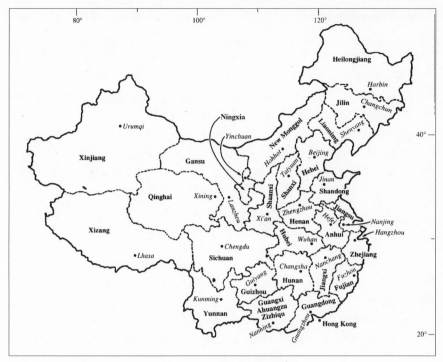

Figure 2. The provinces and capital cities of mainland China

to a discontinuous agricultural system, cycling among caterpillar pests of trees, rice, cotton, and vegetables. In northern China, overwintering occurs in the forest. In Hebei Province (Central China), during the third generation of oriental corn borer, *T. ostriniae* disperses from corn fields to the forest, where it attacks the eggs of poplar caterpillar (*Clostera anachoreta* [Fabricius], Lepidoptera: Notodontidae). Next, elm tussock moth (*Ivela ochropoda* [Eversmann], Lepidoptera: Lymantriidae) serves as host. In October, the wasp larva forms a cocoon inside host eggs, in preparation for overwintering. Emergence typically occurs in April, after six months of dormancy. The first spring host is poplar caterpillar; thereafter, the parasite attacks elm tussock moth eggs for a short period in June. In July, population density of *T. ostriniae* peaks, and movement into corn begins. By September, densities have declined, and dispersal to the forest occurs again. *Trichogramma* spp. are small and have limited flight capacities. Therefore, the development of tree/crop or shelterbelt systems and the protection of natural forest edges is needed to ensure survival in sufficient numbers to aid pest control in field crops.

Tertiary Interactions

Tertiary agroforestry-pest interactions might be called higher order interactions. Such dynamics are difficult to document, because of the many variables involved. An example was outlined by Chen (1990), who reported that mixed cropping of pine and smoke tree (*Cotinus coggygria*) increased the soil nitrogen level by 65%, leading to increased pest resistance of pines. Caterpillar pests that fed on pine needles from a mixed forest of pine and smoke tree had growth rates 5 to 41% lower than those fed on needles from a pure pine forest. Reproduction rates of the pests were also reduced. Other examples are detailed by Zhang & Olkowski (1992).

Intercropping

Intercropping safflower (*Carthamus tinctorius*) with cotton has proved an effective way to control aphid in cotton in Liaoning Province (North Coastal China) (Li & He 1987). During the three seasons of monitoring (1983-1985), the safflower increased lacewing numbers in cotton by an average of 13 times compared with the no-safflower cotton, which had to be chemically treated. Overall profit from the two regimes, was similar, since, although there is less cotton produced per area in the intercropped fields, the sale of the safflower and the savings from pesticide purchase (application costs were not included in the comparison) made up the difference.

Safflower was interplanted, two rows to eight of cotton starting on April 5th. Rows of cotton were 50 cm apart, plants within rows were 18-20 cm apart, giving ca. 105,000 to 120,000 plants per hectare. The safflower was more closely planted, 56 cm apart, giving a density of 225,000/ha. By May 25th green peach aphid (*Myzus persicae* [Sulzer], Homoptera: Aphididae), and three lacewing species were present. Lacewing larval peaks took place between June 14th and 29th, reaching as high as 615,000 per ha. When the lacewings moved into the cotton their numbers averaged 43,000 to 50,000/ ha, with many other natural enemies present. In contrast, in areas without safflower, the lacewing numbers only rose to ca. 3,375/ha, and other natural enemies were much lower.

This example suggests that intercropping can enhance natural enemy numbers in a field cropping situation. The following example demonstrates how intercropping can operate in an orchard setting.

Cover Cropping

Tropic ageratum (*Ageratum conyzoides*) is a key component in the biological and integrated control systems for pest mites in citrus orchards in south and south central China. This species is distributed over lower mountains, hills, and flatlands in areas south of the Yangtze River. Planted among citrus trees, this composite assists in the control of citrus red mites (*Panonychus citri* [McGregor], Acari: Tetranychidae), and other plant-feeding mites by enhancing numbers and effectiveness of several predatory mites. Tropic ageratum provides a pollen source, alternate prey (Psocoptera: Psocidae), and habitat for predatory mite species which are then available to suppress pest mite numbers on the fruit trees. The alternate food sources are valuable for preventing starvation of predatory mites after they have eaten all prey mites on the citrus.

Also, the cover plant interplanted amongst citrus trees provides an environment-modifying benefit. Jiang et al. (1988) reported that *Amblyseius ehariai* (Amitai and Swirski) (Acari: Phytoseiidae), a voracious, fast-growing common predatory mite species in Hunan Province (Central China) where citrus is a major crop, does not like high temperatures or dryness. However, in this area the summers are hot and dry. The cover crop mediates the temperature and moisture which benefits the predatory mites.

Deng et al. (1983) reported that *Amblyseius nicholsi* Ehara (Acari: Phytoseiidae) grows 12 times faster than the prey red mites in the temperature range of 10 to 28° C (68 to 82.4° F). Chen et al. (1982) reported the preferred temperature and humidity ranges for the ovipositing female of *Amblyseius delioni* (Muma and Denmark) are 23 to 28° C (73.4 to 82.4° F) and RH 75-90%. In laboratory tests when temperatures are higher than 30° C (86° F), the adults moved out of cultures. Above 33° C (91.4° F), the adults died after one day of exposure, and below 18° C (64.4° F), oviposition was reduced and ceased entirely when the temperature was below 11° C (51.8° F).

When moved from tropic ageratum to citrus trees, the predatory mite *Amblyseius okinawanus* (Ehara) has been reported producing 100% control after 21 days (Table 2; Huang & Yang 1987). After failing to find adequate food sources in the citrus, the predatory mites move down to the ageratum. Some of the Chinese common names for this plant include character components meaning white or white-colored down, which covers the whole plant, i.e., stem, leaves, and flowers. This down provides a preferred oviposition site for predatory mites, apparently since the down is similar to the silk of the prey mites. As cover-cropping becomes more conventional in farming, the selection of plant materials to produce pollen and alternate food sources for mite predators in citrus and other crops should be considered. The technique of using ageratum is further documented by Zhang and Olkowski (1989).

Table 2. Comparing releases of field-reared and laboratory-reared predatory mites with chemical controls for suppression of red mites in citrus orchards.[a]

	Day 0		7 days		14 days		21 days	
Treatments	pest[b]	pred[c]	pest	pred.	pest	pred.	pest	pred.
Number released (lab-reared)	500	4	60	6	6	7	0	11
Number released[d] (field-reared)	150	3	75	4	4	7	0	10
Chemical	590	2	0	0	10	3	50	1
No treatment	300	1	400	1.3	630	1.5	150	2

[a] From Huang and Yang 1987; Anyuan citrus orchard, Jianxi province on 5 yr. old trees, numbers indicate mites per 100 leaves; observations started on September 4, 1985.
[b] Pest = *Panonychus citri*, citrus red mite
[c] Predator: *A. okinawanus* released to achieve 1:38 predator:pest mite ratios
[d] Predatory mites are transferred to citrus with ageratum and perilla plant cuttings

Field Enhancement of Natural Enemies

Predatory Wasps

The idea of providing human-made structures for predatory paper wasps in the genus *Polistes* (Hymenoptera: Vespidae) is not novel. However, getting an advanced emergence and an additional generation by moving over-wintering queens in their nests within controlled temperature and humidity greenhouses is an innovation. This warmer, protected overwintering habitat facilitates earlier spring egg-laying and emergence of the first generation of the predatory wasp *Polistes hebraeus* Fabricius (Walker) (Lepidoptera: Noctuidae). This species is moved in the spring within its nests into cotton fields for control of the old world bollworm (*Heliothis armigera* [Hübner], Lepidoptera: Noctuidae), and the small cotton measuring worm (*Anomis flava* F., Lepidoptera: Noctuidae).

Pest control has been increased 61 and 57%, respectively, using this method. At the same time, suppression of the wheat pests, *Mythimna separata* (Walker) (Noctuidae) and wheat sawfly (*Dolerus tritici* [Chu], Hymenoptera: Tenthredinidae) was obtained in nearby fields (Li & He 1987.

Comparable work in the U.S., with colonizing and field enhancement by providing nest sites, was conducted for a short time by three investigators, but has been abandoned. The abandonment was due to the lack of total control of the target pests by this method alone, based on personal communication with two of the three U.S. researchers. In China, enhancing field populations of *Polistes* species is seen as one of a group of techniques that

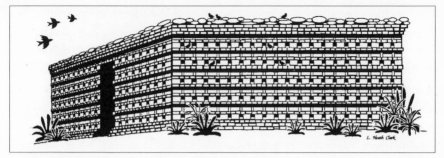

Figure 3. This square bird house is about 3 meters high built with nests built into the walls. Water sources and shrubs for pearching are nearby.

are used together for control. For example, *Polistes* enhancement is used along with *Trichogramma* releases, intercropping, and other techniques. Also, the Chinese work has developed the technique further by using greenhouses for overwintering queens, gaining an additional field generation rather than merely augmenting field populations by providing nest sites. *Polistes* enhancement could be considered as part of an IPM program which stresses biological control for many lepidopteran pests.

Enhancing Bird Numbers for Grasshopper Control

Native populations of rosy starling (*Sturnus roseus*) have been increased by building artificial bird houses and habitat on 3,320 ha. This has let to suppression of grasshoppers in grasslands of Northwestern China. Increase rosy starling densities were achieved by building rock piles that simulate the favorite nesting sites, planting shrubs, and digging water troughs (Figure 3). In 1981, 7,823 experimental sites were found in the area and grasshopper numbers were reduced from an average of 39 per m^2 to 1.3 per m^2 before the birds left to continue their migration in August. In May 1982, the same pattern was observed: grasshopper numbers were reduced to 0.2 per m^2. The birds normally stay in the Tianshan area about 80 days. Subsequent to the data collection as indicated above, most of the sites were destroyed because of the construction of a canal. Using the previous results as a model, some permanent artificial nest sites were again constructed in an area that had no natural shelter. These shelters were constructed like small houses without roofs (see Figure 3). Grasshopper numbers around these shelters in 1986, were greatly reduced by June 20, from 42 to 2.3 per m^2 in a 500 m radius around the nest area.

This work is significant for U.S. agriculture because it demonstrates that population enhancement of native predators could have a great impact on grasshopper numbers, something which has not been demonstrated previously. Although we cannot import the rosy starling, we can evaluate similar native or naturalized birds or other grasshopper predators that can be augmented to help prevent outbreaks. Grasshopper numbers are kept under control regularly by spraying large areas of western U.S. by airplane. Recent work with the protozoan *Nosema locustae* has shown great promise as becoming an important component of such programs, helping to reduce broad spectrum pesticide use (Daar 1987a, b). However, other work enhancing existing predators, particularly birds, certainly seems possible based on this Chinese work.

Field Transport of Natural Enemies

Use of Host Eggs to Enhance Pine Caterpillar Parasites

Experiments on supplying sterile oak silkworm (*Antheraea pernyi* [Cuérin-Méneville], Lepidoptera: Saturniidae) eggs to forests where the pine caterpillar (*Dendrolimus punctatus* [Walker], Lepidoptera: Lasiocampidae) causes damage indicate an overall increase in the population density of parasites by 68 to 140%. The oak silkworm, which is an alternative host for many parasites of the pine caterpillar, is collected from the wild. Eggs are supplied to the forest on cards three times per month throughout the year. In host-supplemented forests, a 5-29%, 4-25%, and 8-18% increase in egg parasitization was observed on the first, second and third generation, respectively (Tong et al. 1988). Nine of the 14 native parasites can develop on the artificially supplied eggs.

In the U.S., enhancement using host eggs is known from field experiments on cabbage for increasing numbers of *Trichogramma* species attacking cabbage lepidopterans, and in stored product pest control for augmentation of *Trichogramma* species. No operational programs for this technique are currently known in the U.S. There is no known comparable work in the U.S. which uses a naturally occurring herbivore for augmenting field populations of an egg parasite that attacks a native pest species. Although this Chinese work was carried out against a forest pest, the approach may be useful against a wide variety of lepidopteran pests if commercial sources of appropriate host eggs (sterilized to prevent pest exaccerbation) become available.

Conservation of Natural Enemies

Generalist Predators

Researchers in China have shown that wolf spiders (Araneae: Lycosidae) are of major importance in suppressing planthoppers in rice (Ge & Chen 1989), and they have managed other species to control certain sawflies in pine (Xie 1988). The pest control role of spiders has been noted in small-scale plots in the U.S. (Reichert & Bishop 1990), in Israel on cotton (Mansour 1987a), apple (*Malus pumila*) and citrus (Mansour 1987b), and in other countries, as well. The literature documenting spider importance is rapidly expanding world wide (see the bibliography listed under Thompson 1990). However, the Chinese spider research has advanced beyond the mere observational, to that of practical management. If spider populations can be conserved, and even augmented by relatively small changes on the farm, this has the potential of being useful and cost-effective for growers in many cropping systems.

Other generalist predators that the Chinese have experience in managing include ducks, frogs, mantids (Orthoptera: Mantidae), various bird species, and others.

Restricted Predators

The term restricted refers to predators that are somewhat more limited in their prey ranges than generalists. Lady beetles (Coleoptera: Coccinellidae), for example, are relatively restricted compared to lycosid spiders. Some lady beetle species are even highly specific, feeding upon a single prey genus. The seven-spotted lady beetle (*Coccinella septempunctata* L.), which has been widely distributed in the U.S. in the last few years, is native to Asia, including China, where it has been extensively studied. Chinese work with this and other species is summarized by Olkowski et al. (1990). One unusual augmentative system developed in China uses small plastic greenhouses with spring aphid populations growing on radish (*Raphanus sativus*), rape (*Brassica napus*) or cabbage (*Brassica oleracea* var. *capitata*). These are constructed in field sites and inoculated with overwintered *C. septempunctata* adults. After a suitable period the beetles are collected and released elsewhere or the covering of the greenhouse is removed and the beetles allowed to spread throughout the field. Using this technique a 15 m^2 greenhouse can produce 2.25 million larvae from 250 adults with minimal costs.

268 ENHANCING BIOLOGICAL CONTROL

Summary

The Chinese pest management literature describes the routine use of a number of pest management strategies that could well be implemented on many U.S. farms, including intercropping to feed the beneficial insects and on-farm biological control enhancement techniques. While some Chinese examples could be of specific application here (e.g., greenhouse conservation of *Polistes*), and others by analogy (e.g., pollen cover crops for mites in orchards), the principles behind these approaches already have been accepted by many in the U.S. However, in one area the Chinese approach has truly changed our thinking and may have a similar influence on others. That is in the deliberate incorporation of generalist predators into their pest management programs (see Helenius, this volume). One of the advantages to this strategy is that the farmer can manage implementation with minimal imports of energy, fertilizer, and labor. The techniques will require adaptation to U.S. conditions, but the potential for similarly successful pest control methods should be explored.

Acknowledgment

We thank Linda Heath-Clark for providing drawings of the citrus trees with ant nests (Figure1) and bird house (Figure 3).

References

Bao, J. & X. Chen. 1988. Research and applications of *Trichogramma* in China. Science and Technology Literature Pub. House, Beijing (in Chinese).

Chen, C. 1990. Integrated managmenet of pine caterpillars in China (in Chinese). Chinese Forestry Publication House, Beijing.

Chen, Shou-Jian, Fen-wei Chou, Sheng-gai Zhuang, & Li-ya Liao. 1982. Bionomics and utilization of *Amblyseius deleoni* Muma et Demark (Acarina, Phytoseiidae). Acta Entomol. Sinica 25:49-55.

Chiu, S-F. 1985. Recent research findings on Meliaceae and other promising botanical insecticides in China. Zeitschrift fur Pflanzenkrankheiten und Planzenschutz 92:310-319.

Daar, S. 1987a. New federal IPM program for grasshoppers. IPM Practitioner 9:1-3.

Daar, S. 1987b. Promising data on biological control for grasshoppers. IPM Practitioner 9:5.

Deng, Zhen-hua, & Ying-min Li. 1983. Raising and releasing experiments with *Amblyseius nicholsi*. Acta Entomologica. Sinica 26:208.

Ge, F., & C. Chen. 1989. Laboratory and field studies on the predation of *Nilaparvata lugens* (Hom.: Delphacidae) by *Theridion octomaculatum* [Araneae: Theridiiae]. Chinese J. Biol. Control 5:84-88.

Huang, H. T. & P. Yang. 1987. The ancient cultured citrus ant, a tropical ant is used to control insect pests in southern China. BioScience 37:665-671.

Jiang, H., X. Li, Z. Yao, Y. Tan & G. Yan. 1988. Biology of *Amblyseius eharai* (Acari: Phytoseiidae) and functional response to citrus red mite. Natural Enemies of Insects 10:165-169.

Ke, Z. & Lu. 1989. Development and application of botanical pesticides. J. Farmer's Consultant 1989:19-22 (in Chinese).

Li, T.S. & J.G. He. 1987. Zoological Research Institute, Chinese Academy of Sciences, Beijing, China. Kunchong Zhishi 28:168-189.

Mansour, F. 1987a. Spiders in sprayed and unsprayed cotton fields in Israel, their interaction with cotton pests and their importance as predators of the Egyptian cotton leafworm, *Spodoptera littoralis*. Phytoparasitica 15:31-41.

1987b. Effects of pesticides on spiders occurring on apple and citrus. Phytoparasitica 15:43-50.

Metcalf, R. L. & A. Kelman. 1980. Plant protection, pp. 313-344. *In* L.A. Orleans [ed.], Science in contemporary China. Stanford University Press, Palo Alto, CA.

1990. *Trichogramma*—a modern day frontier in Biological Control. IPM Practitioner 12:1-15.

Olkowski, W., A. Zhang & P. Thiers. 1990. Improved biocontrol techniques with lady beetles. IPM Practitioner 12:1-12.

Plucknett, D. L. & H. L. Beemer, Jr. 1981. Vegetable farming systems in China. Westview Press, Boulder Colorado.

Pu, Z. 1984. Principles and techniques of biological control of insect pests. Science Pub. House, Beijing (in Chinese).

Qui, Q. 1989. Study on pest control effect and extract from *Amorpha fruticosa*. J. For. Sci. Technol. Information 3:6-7 (in Chinese).

Reichert, S.E. & L. Bishop. 1990. Prey control by an assemblage of generalist predators: spiders in garden test systems. Ecology 71:1444-1450.

Temple, R. 1986. The genius of China, 3,000 years of science, discovery and invention. Simon and Schuster, New York.

Thompson, M. 1990. Spiders in agriculture bibliography. Distributed by author. East Bay Regional Park District, PO Box 930, Pleasanton, CA 94566.

Tong, Xinwang, Lexiang Ni & Xianming Lao. 1988. Enhancement of egg parasitization of *Dendrolimus punctatus* (Lep.: Lasiocampidae) by supplementing host eggs in the forest. Chinese J. Biol. Control 4:118.

Wittwer, S. Y. Youtai, S. Han & W. Lian Zheng. 1987. Feeding a billion. Michigan State University Press, East Lansing, MI.

Xie, Q. 1988. A preliminary report on moving spiders to control *Neodiprion sertifer Geoffroy*. Zhejiang For. Sci. Technol. 8:40-41.

Yan, J. C. Xu, G. Li, P. Zhang, W. Gao, D. Yao & Y. Li. 1989. Insect natural enemies of forest pests. Chinese Forestry Pub. House, Beijing (in Chinese).

Yang, P. 1990. Pest control with ants, p. 456. *In* Z. Zhang & J. Cao [eds.], Pest management: strategy and method. Science Publishing House, Beijing (in Chinese).

Zhang, A. & W. Olkowski. 1989. Ageratum cover crop aids citrus biocontrol in China. IPM Practitioner 11:8-10.
  1992. Agroforestry and IPM in China. IPM Practitioner 14:1-12.
Zhejiang U. Entomology Faculty, [ed.]. 1982. *In* Agricultural entomology, pp 32-34. Department of Plant Protection, Zhejiang Agricultural University, Shanghai Science and Technology Publishing House.
Zheng, W. 1978. Silviculture of major species in China. Agricultural Pub. House, Beijing. (in Chinese).
Zhao, J. 1989. Protection and application of lacewings. Wuhan University Press. (in Chinese).
Zhu, R. 1978. Illustrations of insect natural enemies. Science Pub. House, Beijing. (in Chinese).

The Role of Experimentation in the Development of Enhancement Strategies

STEVE E. SCHOENIG—Biological Control Program, California Department of
Food & Agriculture, 3288 Meadowview Road, Sacramento, CA 95832

ROBERT L. BUGG—University of California, Sustainable Agriculture, Research
& Education Program, University of California, Davis, CA 95616-8716

JESSICA UTTS—Division of Statistics, University of California
Davis, CA 95616

Introduction

Experimentation has a role in both the theory and implementation of enhancement techniques at the farm level. Ideally, farming practices thought to increase the abundance of natural enemies should be adopted only when there is sound experimental evidence that these practices justify their costs. Planting cover crop A one year and cover crop B the following year in the same field and then trying to compare the two cover crops is only one example of poor experimentation. Here we discuss principles and potential pitfalls for experimentation in general, and biological control enhancement schemes in particular. We will also compare a group of different statistical methods by using them to analyze the same data set illustrating that there are often multiple ways to approach an analysis. Sometimes the assumptions necessary for an analysis will limit its use with particular data sets. However, when more than one method is appropriate it is useful to compare and contrast the results.

Before the development of modern agricultural experimentation and statistics, most field experiments were conducted haphazardly with no attempt to deal with interference from random variability or effects from unmeasured variables, and no inclusion of proper experimental controls. The twentieth century has seen the gradual adoption of many techniques for removing ambiguity from the experimental conclusions. We have also seen the adoption of an agricultural production paradigm that emphasizes external inputs into farming systems, i.e., non-sustainable and energy-intensive management practices (Edens & Haynes 1982). Most agricultural research has been directed towards maximizing yields and profits in the short term. Furthermore, the components of these farming systems were often studied in isolation from other components within the system, reflecting an underlying reductionist philosophy. Modern experimental design and analysis was integral to this so-called revolution, but did not cause it, nor are the two inextricably linked. There is a call for a more holistic approach to the investigation of agricultural alternatives, and for scientists and farmers to expand their horizons and awareness (MacRae et al. 1989). This does not mean that reductionist or statistical approaches to research will not be useful in the search for a new agriculture. Under any paradigm there will be a need to measure efficacy, choose between alternatives, understand fundamentals, and refine new systems.

Evaluating enhancement techniques with respect to efficacy and optimal application is analogous to the operation of our legal system. Generally, one starts off with the assumption that an enhancement technique is ineffective (innocent) until proven effective (guilty). Then there is the collection and review of the data (the evidence). Finally, a conclusion is reached (the verdict). Sometimes the evidence is compelling and incontrovertible. At other times, the evidence is ambiguous or circumstantial. Guilt can be determined only through inferentially eliminating all (or most) reasonable doubt that the evidence could have arisen under another set of circumstances— i.e., innocence.

In modern experimentation, observational and controlled studies supply the data for examination. Descriptive statistical methods are employed to summarize the data in the most useful way to the experimenter; inferential statistics are used to separate and measure systematic and random patterns within the data. Statistical tests and their associated p-values are not the juries that deliver a verdict: this is the role of the experimenter and the research/farming community as a whole. Rather, statistical tests are the first stage in measuring an effect in which the question is asked: Could this observed pattern be explained simply as random noise alone? If the statistical methods reject the null hypothesis of a random pattern alone, then the experimenter must interpret the practical significance of the observed pattern. One should not blindly substitute high-powered statistics for common sense, yet well-thought-out experimental design and analysis are really a refinement and extension of common sense itself.

Experimental Design

The term "experimental design" has come to be almost synonymous with complex analysis of variance designs. In a broader sense, however, experimental design refers to the planning of studies so that agricultural and biological questions can be answered in an unambiguous way, free of the interference of confounding variables, hidden biases, and random noise. Furthermore, the behaviors and the biologies of the plants and animals must be considered in setting up experiments on enhancement of biological control. Rarely can an elegant statistical analysis compensate for a flawed experimental design. The following are several issues that should be addressed in most such studies.

Issues of Spatial Scale

Spatial scale plays an important role in the dynamics of predator-prey systems (Heads & Lawton 1983; Kareiva 1990) and should be taken into account when designing studies of enhancement of biological control (Corbett this volume; Rzewnicki et al. 1989). Many enhancement strategies involve movement of natural enemies into the target crop from adjacent vegetation (e.g., weeds, nectar sources, nursery plants). The distances traversed will vary among natural enemies. Experimental plots need to be the right size to evaluate the effects of non-crop plants. Highly mobile species such as green lacewings (Neuroptera: Chrysopidae), hoverflies (Diptera: Syrphidae), ichneumonid wasps (Hymenoptera: Ichneumonidae) and lady beetles (Coleoptera: Coccinellidae) may disperse between closely-spaced plots in landscape-scale experiments; therefore widely-spaced farms as replicates may be needed to show the effects of cultural practices. By contrast, surface-dwelling predators such as ground beetles (Coleoptera: Carabidae) and bigeyed bugs (Hemiptera: Lygaeidae) may prove tractable by smaller-scale experiments (Wratten this volume). Corbett (this volume) demonstrated, with the use of simulation models, that spatial scale is important in measuring enhancement. A modeling approach can benefit experimental design if the population dynamics and movement are well understood. Unfortunately, this will not always be the case.

Issues of Temporal Scale

Phenological aspects of crop and non-crop plants and behavioral and phenological aspects of the herbivore and natural enemies are important in designing enhancement schemes and the experiments to evaluate them. Seasonal and diel patterns of movement must be considered in setting up a sampling plan. If different treatments of an experiment involve different non-crop plants or different types of natural enemies, the timing issue can be complex and care must be taken when making comparisons.

The duration of an experiment is also important in the evaluation of enhancement strategies. It may not be possible to see results from habitat management or changes in cultural practices until several seasons have elapsed. Furthermore, year-to-year differences may be large in comparison to enhancement effects, and on a practical basis, the pattern of variability in enhancement may be as important as the overall mean. For example, in years with late frosts an insectary plant may do poorly and not lead to much enhancement, whereas in good years complete control of the pests may be achieved. Even though the overall enhancement averaged over many years is higher compared to no insectary plants, the damage in bad years could prove intolerable.

Issues of Replication

Replication is the inclusion of more than one valid experimental unit of each treatment type in an experimental design. Experimental designs need replication for two major reasons (Hurlbert 1984). Firstly, most experiments are set up to discover treatment effects that apply to all fields farmed similarly, not just the plots in which the experiment was conducted. To do this, it is necessary to minimize interference from unmeasured variables that may differ among the different experimental units. With replication and interspersion of treatments, background heterogeneity of the experimental units can be distributed evenly among treatments either by random assignment or by blocking spatially with or without another variable. Completely random assignment of treatments to experimental units is most appropriate when there is no knowledge of background heterogeneity. However, in situations where a small number of experimental units are laid out contiguously, such as plots within a field, replicates are better situated systematically, or with restricted randomization. This avoids the not-unlikely possibility that random assignment might group all like treatments together. Failure to replicate and intersperse treatments makes the general applicability of results doubtful. Results may reflect effect of plot location and impingement of chance events rather than actual enhancement treatments. Andow and Hidaka (1989), in comparing two syndromes of rice production, essentially compared two different farms, which must have differed in other respects besides production syndrome. Such a comparison should be done on a wider scale as, for example, in the prune-refuge work discussed by Murphy et al. (this volume). The second reason for replicating treatments is to allow the use of inferential statistics in evaluating and understanding biological control enhancement techniques. Most statistical techniques require replication of treatments to estimate random background variability.

This "error term" is often necessary for performing hypothesis tests and creating confidence intervals. Increasing the number of replicates increases the statistical power or accuracy of the investigation.

Multiple observations do not always qualify as replications. A replicate is a single or composite observation from one experimental unit. There must be at least one other experimental unit with the same treatment applied. The experimental units should be discrete physical objects to which any of the treatments could have been independently assigned. Pseudoreplication (Hurlbert 1984) is the use of multiple observations from an experimental unit as true replicates. The problems with pseudoreplication are that interference from unmeasured variables is more likely, background variability will be underestimated, statistical tests will be too liberal, and hypothesis tests may not be possible. Another definition of pseudoreplication, in the context of analysis of variance, is the testing of treatment effects with an incorrect error term (Hurlbert 1984). In such tests, multiple observations for a given experimental unit should be averaged and treated as a single observation from an experimental unit. If there is no true replication, then inferential statistics should not be used. In a review of the ecological literature, Hurlbert (1984) found that pseudoreplication occurred in 48% of the papers that used inferential statistics.

Confounded Variables

Sometimes the primary aim of experimentation is the investigation of cause-and-effect relationships. In other situations, cause-and-effect relationships are assumptions that are implicit in hypotheses being tested. There are inherent difficulties in identifying the true causes of an effect. Factors that are highly correlated with true causes, but are not themselves direct causes, can be mistakenly identified as such. These mistakes are the result of the confounding of variables in an experiment. Confounding is the co-occurrence of factors (either intentionally or unintentionally) in an experiment such that their effects cannot be distinguished. Treatments can be unknowingly contaminated, thereby giving misleading results. On a more subtle level, treatment effects may result from mechanisms other than those that the experimenter has assumed, leading to erroneous conclusions.

Confounding between variables can be identified and eliminated through the construction of treatments and controls, exploratory data analysis, and an understanding of the mechanisms by which factors are interacting in the system under study. An experimental control cannot always be merely a treatment in which nothing is done. If a complex enhancement technique is being tested, then one should try to decompose the technique into more elemental components that can each serve as a series of treatment/controls. For example, if a researcher was interested in the effects of adding posts with holes drilled in them for wasp nesting sites, it might be necessary to include an intermediate control treatment consisting of posts without nesting holes, because the post might be used as a perching site for other insect or vertebrate predators, thereby confounding the nesthole effect. This would be in addition to control plots with no posts.

Exploratory data analysis (EDA) (Tukey 1977) is a suite of techniques for viewing patterns in data that can give insight into the interrelationships between variables. Correlation analysis, regression analysis, scatter plots, and principal components analysis, along with other techniques, are used in EDA to identify potential confounding of variables. Above all, an experimenter must realize that one or even a few experiments may not establish an hypothesized cause-and-effect relationship beyond reasonable doubt.

Execution

Overview of Sampling Issues

The objective of most enhancement experiments is to make comparisons of the densities of pests or natural enemies in plots affected by different treatments. It is neither desirable nor feasible to count every individual in the population to make these comparisons. Rather, samples from the plots will be examined, and based on these, total (or mean) population densities can be estimated and compared. Thorough reviews of sampling design are given by Green (1979), Southwood (1978), and Kogan and Herzog (1980).

There are two broad issues in sampling: (1) How will each sampling unit be physically examined? and (2) How many samples should be taken?

Sampling Techniques

The sampling technique used should be appropriate to the structure of the habitat involved and to the biologies and behaviors of the target organisms. It is more important to choose a technique with low variability than one that captures the highest proportion of organisms on average. Observations can be made directly in the field, or material can be collected and moved to the lab for future examination. Specific sampling techniques, which are numerous, are reviewed in several books on the subject such as those by Southwood (1978) and Kogan and Herzog (1980). It is important to draw a distinction between random sampling and haphazard sampling. Random sampling is achieved when each sampling unit within a treatment has an equal chance of being selected, the choice of which unit to sample made by a random number generator or random event. Haphazard sampling, often mistaken for random sampling, occurs when the experimenter chooses which unit to sample seemingly without "paying attention" or without overt preference. The problem with haphazard sampling is that biases occur unconsciously. Random sampling is preferable in most studies.

Sample Size Determination, Population Estimation,
Hypothesis Testing and Statistical Power

For estimating population parameters, such as the mean or the difference between two treatment means, a sample size may be chosen to achieve a specific level of precision (width of the confidence interval). Information on (1) the amount of variance among samples units, (2) confidence level, and (3) length of confidence interval desired must be known ahead of time. Methods for carrying out sample size calculations for estimation are discussed by Karandinos (1976), Southwood (1978), Ruesink (1980), Wilson et al. (1989), and Odeh and Fox (1991).

Determining how many samples units are needed to test an hypothesis requires knowledge of (1) the amount of variability among samples, (2) the desired rate of falsely rejecting the null hypothesis (type I error), and (3) the desired probability of detecting the alternative hypothesis if true or the minimum size of the effect to be detected. The probability of detecting the alternative hypothesis, when true, is called the statistical power or simply the power of the test. Often, individual experiments are conducted with sample sizes that are not large enough to detect important effects that are actually present (Toft & Shea 1983). The power is determined by the amount of variability, the sample size, and the fixed probability that the null hypothesis is falsely rejected (the type I error or α level). Experimenters aim to maximize the amount of power in an experiment by altering one of these three factors. Once experimental techniques are refined and blocking and additional variables (covariates) are used, the variability in the system cannot be lowered further by the experimenter. The alpha error level is determined by the experimenter (the lower the type I error rate, the lower the power); however, scientific convention has dictated that type I error rate be no higher than 0.05 for most tests. This practice has often been dogmatically followed when not appropriate (Salsburg 1985; Carmer & Walker 1988). In final evaluations of agricultural and ecological theories, a conservative approach (i.e., Type I error < 0.05) is probably best. For preliminary or exploratory research, however, and for applied studies, an error rate greater than

0.05 may be justified (Salsburg 1985). With the amount of variability minimized and type I error rate fixed, increasing sample size is the only way to increase statistical power.

Determining the sample size needed to achieve a given power varies from somewhat technical to essentially impossible. For ANOVA designs Neter et al. (1990) and Odeh and Fox (1991) detailed the methodology for hand calculations using supplied power tables. Computer programs (e.g., O'Brien 1986a,1986b) can calculate sample sizes for specified power in certain designs of simple and intermediate complexity. More complex designs will require the assistance of a statistician or the creation of a simulation program. While there is no hard convention, it would seem that experiments done with power less than 0.9 for the relevant minimum biological effect size are compromising the ability to conclusively dismiss the treatment hypothesis.

A series of low power tests conducted to test essentially the same hypothesis can be analyzed with a new technique called meta-analysis detailed later in this chapter.

Analysis

The analysis should begin with a graphical exploration of the data (i.e., graphs, charts, etc.) and the calculation of simple descriptive statistics (i.e., means, variances, etc.). Statistical tests should be chosen for their appropriateness to the hypothesis being tested, and the appropriateness of underlying assumptions, not solely because others have used them. Often appropriateness of the statistical model cannot be tested until the model is fitted and examined for problems. In most statistical models this is carried out by examining the residuals or estimated error term for each observation. All of these activities are now carried out most efficiently on computers.

Computers have freed researchers from the tedium of hand calculations, and have put mathematically complicated techniques at most researchers' fingertips. However, this presents the danger of allowing the use of complex techniques without a sufficient awareness of what is actually being done. Statistical manuals, which cover most common methods, such as Steel and Torrie (1980), Sokal and Rohlf (1981), and Zar (1984) are a starting point to intelligent and conservative use of statistics.

Appropriateness of Statistical Methods

Most statistical techniques have a set of underlying assumptions, which, if not fulfilled, can jeopardize the validity of an analysis. For the standard

ANOVA and regression models, the basic assumptions are that the error variances are independently and identically normally distributed with a mean of zero and a constant variance (Eisenhardt 1947; Neter et al. 1990). These models are fairly robust to violations of normality and moderately robust to non-equality of variance; however, they are not very robust to non-independence. Remedial measures include; (1) transformations or non-parametric methods for violations of non-normality; (2) transformations and weighted least squares for non-equality of variance; and (3) techniques that deal explicitly with correlated data, e.g., repeated measures, time series, geostatistics. Milliken and Johnson (1984) discuss many solutions for problems with "messy data."

Most other statistical methods have underlying assumptions that will dictate the validity of these methods. We will not catalog these assumptions further herein; however, we stress the importance of learning and evaluating the assumptions behind any statistical test with which one is unfamiliar.

Case Study #1

Congruence of Results Using Different Analytical Methods—Cover Crops and a Bigeyed Bug (Hemiptera: Lygaeidae) in Cantaloupe

Often one experiment can yield different types of data, and one type of data can often be analyzed many different ways. In this example we analyze data from an experiment that measured the effects of eight cover crop treatments and of cantaloupe vigor on the abundance of the bigeyed bug, *Geocoris punctipes* (Say) (Hemiptera: Lygaeidae). We present this example to show the diversity of techniques that can be employed and their complementary use in interpreting data.

Experimental Design

Bugg et al. (1991) relay intercropped cantaloupe (*cucumis melo*) with eight cover crop treatments in a randomized complete block design (Table 1) . There were three blocks in the study. Detailed methods and cultivars are given in the paper cited. Cantaloupe stand quality (vigor) was rated on a 0 to 1 scale twice during the season and then averaged into a single value for each plot. Prior studies suggest that crop stand quality may influence predators (e.g., Mack et al. 1988). On six dates, densities of *G. punctipes* were measured (1) on cantaloupe leaves, (2) amid cover-crops or their residues, and (3) on laboratory-reared fall armyworm (*Spodoptera frugiperda* [J. E. Smith] Lepidoptera: Noctuidae) sentinel egg masses. Egg mass baits, affixed to the upper foliage of randomly-chosen cantaloupe plants, were used

Table 1. Cover crops used in study of bigeyed bug (*Geocoris punctipes* [Say]) in Cantaloupe (from Bugg et al. 1991).

Common Name	Scientific Name
Crimson clover	*Trifolium incarnatum*
Lentil	*Lens culinaris*
Rye	*Secale cereale*
Subterranean clover	*Trifolium subterraneum*
'Vantage' vetch	*Vicia sativa* x *Vicia cordata*
White mustard	*Brassica hirta*
Polyculture	Mixture of all the above species
Weedy fallow	Resident winter-annual weeds

to evaluate predation. On each sampling date, four subsamples were taken per plot. Means over the six sampling dates are presented in Table 2.

Additionally, on eight dates, baits were placed on cantaloupe leaves and observed for the presence or absence of *Geocoris*.

Tests for Cover Crop Effects Using Season Means— Univariate and Multivariate

Overall effects due to cover-crop regime were assessed through analysis of variance (ANOVA), using season-long means from each plot (Table 2). Results are displayed in Table 3. Univariate ANOVAs were conducted using *Geocoris* per egg mass as the response variable. Because analyses showed no significant effects due to block, the analyses were repeated with that term deleted from the models, and the error terms were suitably modified. Cover crop effect was statistically significant ($P = 0.0481$). The means for each cover crop treatment, averaged across blocks, are displayed in Table 4. Means were compared using Fisher's protected least significant difference test (Steel & Torrie 1980). The least significant difference was calculated as 0.139 *Geocoris* per egg mass. The largest mean was observed for the subterranean clover treatment; however, the means for the next three largest treatments are not significantly different from that of subterranean clover. We can conclude that cover crops do affect the abundance of *Geocoris*, but we cannot recommend one treatment as being the best enhancer.

A multivariate analysis of variance (MANOVA) was performed to test for a collective or multivariate response of all four *Geocoris* response vari-

Table 2. Per-plot season-long means for response variables in relay-intercropping study (based on Bugg et al. 1991).

Plot	Crop	Block	Geocoris per Sentinel Egg Mass	Geocoris amid Cantaloupe	Geocoris amid Cover Crops	Cantaloupe Stand Quality
101	Fallow	1	0.321	1.200	1.429	0.771
102	Rye	1	0.106	0.300	0.143	0.757
103	Crimson	1	0.090	0.600	0.857	0.786
104	Lentil	1	0.174	2.000	2.571	0.757
105	Subclover	1	0.358	3.100	7.000	0.543
106	Vantage	1	0.096	1.700	2.143	0.686
107	Mustard	1	0.232	1.100	0.571	0.614
108	Polycult.	1	0.088	0.600	0.429	0.814
201	Polycult.	2	0.062	0.800	0.143	0.700
202	Fallow	2	0.204	1.100	2.714	0.671
203	Subclover	2	0.247	4.200	8.000	0.586
204	Lentil	2	0.109	0.700	1.286	0.771
205	Crimson	2	0.184	0.900	1.571	0.914
206	Mustard	2	0.208	1.300	0.429	0.500
207	Rye	2	0.016	0.300	0.143	0.700
208	Vantage	2	0.194	1.700	6.429	0.814
301	Crimson	3	0.033	1.00	0.714	0.957
302	Lentil	3	0.100	1.00	0.857	0.786
303	Vantage	3	0.334	1.200	4.429	0.729
304	Polycult.	3	0.069	0.600	1.571	0.886
305	Rye	3	0.044	0.300	0.000	0.829
306	Fallow	3	0.118	1.600	2.000	0.657
307	Mustard	3	0.038	1.000	0.143	0.471
308	Subclover	3	0.213	4.200	5.714	0.586

ables to cover crop treatment. The four response variables are *Geocoris* amid cover crop, *Geocoris* amid cantaloupe, *Geocoris* per egg mass, and proportion of egg masses discovered (observed damaged, or occupied by one or more predators). Cover-crop type was found to have highly statistically significant ($P<0.003$) relationship with the aggregate *Geocoris* abundance variables. It is harder to display the multivariate means, and also there is no readily available method for doing mean separations as a sequel to the test for main effects.

Table 3. Summary of statistical tests for relationships among cover-cropping regime, cantaloupe stand quality, densities of *Geocoris punctipes,* and discovery of sentinel egg masses of fall armyworm, *Spodoptera frugiperda.*

Analysis	Dependent Variable = Effects (P)
One-Way ANOVA —Analysis of Variance	*G. punctipes* per egg mass of fall armyworm = m + cover crop (P=0.0481).
Two-Way ANOVA	*G. punctipes* per egg mass of fall armyworm = m + cover crop (P=0.0475) + Block (P=0.2897).
ANCOVA—Analysis of Covariance	*G. punctipes* per egg mass of fall armyworm = m + cover crop (P=0.0487) + Block (P=0.1842) + cantaloupe stand quality (P=0.1528).
	G. punctipes per egg mass of fall armyworm = m + cover crop (P=0.0766) + cantaloupe stand quality (P=0.2812).
Simple Regression	*G. punctipes* per sentinel egg mass = m + cantaloupe stand quality (P=0.1197).
	G. punctipes amid cantaloupe = m + *G. punctipes* amid cover crops (P=0.0001)
	G. punctipes per egg mass of fall armyworm = m + *G. punctipes* amid cover crops (P=0.0005).
	G. punctipes per egg mass of fall armyworm = m + *G. punctipes* amid cantaloupe plants (P=0.0052).
	G. punctipes amid cover crops = m + cantaloupe stand quality (P=0.2066).
	G. punctipes amid cantaloupe = m + cantaloupe stand quality (P=0.0161).
Stepwise regression	*G. punctipes* per egg mass of fall armyworm = m + *G. punctipes* amid cover crops (P=0.0004) (*G. punctipes* amid cantaloupe plants and cantaloupe stand quality showed non-significant effects after entry of *G. punctipes* amid cover crops).
MANOVA—Multivariate Analysis of Variance	*G. punctipes* per egg mass of fall armyworm, *G. punctipes* amid cover crops, *G. punctipes* amid cantaloupe plants = m + cover crop (P=0.0001)
Logistic Regression	Proportion of egg masses damaged by predators = m + cover crop (P=0.004) + date (P=0.0014).

Repeated Measures Analysis Tests for Cover Crop Effects Using Individual Dates as Replicates

Each plot in the study was measured on six dates over the season. When multiple observations are taken on the same experimental units this is referred to as a repeated measures design. The use of these individual dates as replicates and also treating date as a factor or split-plot is usually not permissible because of the correlated nature of observations taken on the same experimental units. There are a number of strategies for correctly analyzing

Table 4. Cover crop treatment means of *Geocoris punctipes* on sentinel egg masses of fall armyworm, *Spodoptera frugiperda*, averaged across blocks and dates.

Crop	Mean	Std. Dev.
Rye	0.055	0.046
Fallow	0.214	0.102
Lentil	0.128	0.040
Subterranean clover	0.273	0.076
'Vantage' vetch	0.208	0.120
Crimson clover	0.102	0.076
Polyculture	0.073	0.008
White mustard	0.159	0.106

repeated measures data. Most are discussed by Milliken and Johnson (1984). We chose a univariate method that employs a series of linear contrasts over the dates and is detailed by Gurevitch and Chester (1986). In this procedure, polynomial contrasts are created to reduce the multiple observations over date into single numbers, first using the zero order power, then the first, second, etc. This method allows for the testing of normal treatment effects without inflating degrees of freedom, the testing of trends in the response over time, and the testing of differences of time trends between treatments. In our example treatments are the cover crop regimes. The test for treatment effects yielded a statistically significant ($P = 0.0485$) result. Mean separations were identical to that for the univariate ANOVA discussed above. The effect of date was evaluated by a series of model fits with the linear, quadratic, and cubic contrast variables used one at a time as the response variable. The only statistically significant response was for the linear contrast main effect. In this experiment, *Geocoris* abundance increased as the season progressed.

It was not possible to reject the hypothesis that no higher order curvature in this trend was present. Furthermore, it was not possible to reject the null hypothesis that there was no interaction between the slope of the linear time trend and cover-crop treatment.

Tests for Influence of Stand Quality Using Season Means

Different kinds of cover crops may harbor differing densities of predators that later disperse to adjoining cantaloupe, possibly affecting predation of pests on cantaloupe. However, predation could also be affected by stand quality of cantaloupe. If cantaloupe stand quality is not uniform across all cover-cropping regimes, various influences on predation may be confounded. We used regression analysis and analysis of covariance to explore relationships among predator abundance, predation of fall armyworm egg masses, and cantaloupe stand quality. Simple regression models were used to determine whether there were statistical relationships among the various measurements of density for *G. punctipes* (i.e., amid cover crops, amid cantaloupe, or on sentinel egg masses) and cantaloupe stand quality. It was found that a statistically significant relationship existed between cantaloupe stand quality and *Geocoris* amid cantaloupe, and between cantaloupe stand quality and *Geocoris* amid the cover crop, but not with *Geocoris* on egg masses. Analysis of covariance (ANCOVA) (Neter et al. 1990), was also used, with mean *G. punctipes* per egg mass bait as the dependent variable, cover-crop regime as the independent variable, and overall mean coefficient of stand quality as the concomitant variable. Cantaloupe stand quality was not a statistically significant covariate (P = 0.2812) and did not change the significance of the cover crop regime variable.

Relationships Between G. punctipes in the Cover Crop, on Cantaloupe and on Sentinel Egg Masses Using Season Mean Data—
Correlations, Principal Components, and Regression Analysis

A correlation matrix of the seasonal means for *Geocoris* per egg mass, *Geocoris* amid cantaloupe, *Geocoris* amid cover crops, and cantaloupe stand quality is displayed in Table 5. The three *Geocoris* variables are all positively correlated with each other. The strongest correlation was between *Geocoris* in the cover crop and *Geocoris* amid the cantaloupe. These relationships are echoed in the simple linear regression results where *Geocoris* on egg masses were related to *Geocoris* amid cantaloupe (*P* = 0.0052), and to *Geocoris* amid the cover-crop (*P* = 0.0005). The values of the regression coefficients are not really interpretable here, because different areas were measured when sampling the different categories. Of the apparent negative correlations between the *Geocoris* variables and stand quality, only one was statistically significant. This was between *Geocoris* amid cantaloupe and stand quality. Apparently there were more *Geocoris* on poorer quality leaves. This was probably due to the fact that the later-maturing cover crops impacted the cantaloupes more negatively than did early-maturing cover crops yet produced *Geocoris* more abundantly.

A principal components analysis (PCA) was performed on the correlation matrix to further investigate the patterns of correlation. The results are displayed in Table 6. This technique breaks the overall variation among the

Table 5. Correlation matrix using means for each plot.

	Geocoris per Sentinel Egg Mass	Geocoris amid Cantaloupe	Geocoris amid Cover Crops	Cantaloupe Stand Quality
Geocoris per Sentinel Egg Mass	1.000	0.551	0.659	–0.328
Geocoris amid Cantaloupe	0.551	1.000	0.847	–0.485
Geocoris amid Cover Crops	0.659	0.847	1.000	–0.267
Cantaloupe Stand Quality	–0.328	–0.485	–0.267	1.000

data into components of variation that are uncorrelated and one hopes represent underlying causes or mechanisms influencing multiple variables in concert. In this example the first principal component explained 65.6% of the variability in the data. It can be deduced from the eigenvector values that this component represents the positively correlated combined abundance of *Geocoris* in each plot and the negatively correlated stand quality. The second component was primarily made up of the stand quality and added another 20.0% to the explained variability. The third component represented a minor pattern of low *Geocoris* densities on egg masses in some plots with higher densities on cantaloupe. Overall, what the PCA tells us is that most of the variability in the data is explained by general *Geocoris* abundance and stand quality. Only 15 % more of the variability is related to more complex associations.

Logistic Regression Modeling of Probability of Egg Mass Discovery by *Geocoris*

Logistic regression is a statistical technique for modeling the proportions or probabilities of discrete outcomes (see section below). We wished to model the probability of discovery of *Spodoptera* egg masses placed on the cantaloupe leaves by *Geocoris* as a function of date, cover crop, stand quality, and number of *Geocoris* on the leaves. We used the *Geocoris* egg bait presence/absence data collected over eight dates. Data were analyzed using the BMDP-LR statistical software program (Dixon 1988). Our dependent variable was coded 0 if an egg mass was undiscovered and 1 if discovered. We conducted the analysis in a forward stepwise manner including variables into the model one at a time only if they led to a statistically significant improvement in the goodness-of-fit of the model. The cover crop treat-

Table 6. Principal Components Analysis using mean for each plot.

	Principal Component 1	Principal Component 2	Principal Component 3	Principal Component 4
EIGEN VALUE	2.6223	0.8001	0.4745	0.1031
% of Variability Explained	65.5572	20.0020	11.8622	2.5787
Cumulative % of Variability Explained	65.5572	85.5591	97.4213	100.0000
EIGEN VECTORS				
Geocoris per Sentinel Egg Mass	0.49220	0.23347	-0.81816	0.18396
Geocoris amid Cantaloupe	0.56570	0.01686	0.49361	0.66034
Geocoris amid Cover Crops	0.55539	0.36043	0.28074	–0.69485
Cantaloupe Stand Quality	–0.35952	0.90294	0.09027	0.21747

ment and date of sampling were the only two variables found to be significant ($P<0.01$ for both). Table 7 displays the observed and fitted (predicted) probabilities for each combination of date and cover crop. Also displayed in Table 7 are the fitted probabilities from the univariate models where each variable was included independently. Unlike traditional linear models, this procedure provides no convenient way to do follow-up testing to see which cover crop treatments and which dates are significantly different. The parameters of the logistic regression model can be interpreted (Hosmer & Lemeshow 1989) as the natural logarithm of the ratio of the odds with reference to either a control treatment or mean treatment response. This interpretation ties into other statistical techniques for categorical data analysis. However, it is much more straightforward to interpret the predicted probabilities. For example, using the univariate estimates for Table 7 we can calculate that egg masses in the subterranean clover treatment were $0.703/0.577 = 1.23$ times more likely to be discovered than those in the polyculture treatment. Subterranean clover appears to be the best cover crop in promoting *Geocoris* discovery of *Spodoptera* egg-masses.

Table 7. Observed and logistic regression fitted proportions of *Spodoptera* egg masses discovered by *Geocoris punctipes* on relay-intercropped cantaloupe. (observed = normal text, fitted = *italic text*)

Cover Crop	June 3	June 10	June 18	June 24	July 1	July 8	July 15	July 22	Row Averages
Fallow	0.800 *0.695*	0.400 *0.558*	0.933 *0.760*	0.733 *0.777*	0.733 *0.567*	0.533 *0.583*	0.667 *0.625*	0.533 *0.698*	*0.655*
Rye	0.500 *0.526*	0.467 *0.381*	0.600 *0.607*	0.667 *0.630*	0.467 *0.389*	0.267 *0.406*	0.333 *0.448*	0.600 *0.530*	*0.486*
Crimson clover	0.667 *0.535*	0.333 *0.390*	0.733 *0.616*	0.733 *0.638*	0.333 *0.400*	0.333 *0.415*	0.400 *0.458*	0.533 *0.539*	*0.496*
Lentil	0.667 *0.706*	0.333 *0.571*	0.733 *0.770*	0.800 *0.786*	0.733 *0.580*	0.733 *0.596*	0.667 *0.638*	0.667 *0.710*	*0.667*
Subclover	0.667 *0.741*	0.600 *0.613*	0.667 *0.800*	0.933 *0.815*	0.600 *0.621*	0.600 *0.637*	0.667 *0.680*	0.867 *0.744*	*0.703*
Vantage vetch	0.833 *0.689*	0.800 *0.551*	0.600 *0.755*	0.600 *0.772*	0.600 *0.560*	0.600 *0.576*	0.667 *0.618*	0.600 *0.692*	*0.649*
Mustard	0.800 *0.728*	0.733 *0.597*	0.857 *0.788*	0.800 *0.804*	0.400 *0.606*	0.600 *0.622*	0.533 *0.662*	0.867 *0.731*	*0.688*
Polyculture	0.333 *0.618*	0.467 *0.473*	0.667 *0.692*	0.667 *0.712*	0.333 *0.481*	0.667 *0.498*	0.733 *0.542*	0.600 *0.622*	*0.577*
Column averages	*0.652*	*0.517*	*0.723*	*0.742*	*0.525*	*0.542*	*0.583*	*0.658*	

Case Study #2

*Reviving a Data Analysis—Cover Cropping and
Biological Control of Arthropod Pests of Apple*

Sometimes statistical methods are performed on data when the assumptions
necessary to use the methods have not been met. One of the common errors
in enhancement studies is to use subsamples as replicates when the study
was actually unreplicated. Graphical techniques and the reporting of means
are the appropriate techniques for these situations. Here we consider the
opposite problem. Data that could have been statistically analyzed were
merely reported as means, minimum and maximum values. We
re-analyze the data to illustrate the procedures and to make available the
revised analysis on an important enhancement study.

Various parasitic wasps that attack orchard pests also feed on nectar. In
Ontario, Canada, Leius (1967) stated that mean parasitism of codling moth
(*Cydia pomonella* [L.], Lepidoptera: Tortricidae) larvae and tent caterpillar
(*Malacosoma americanum* [Fabricius], Lepidoptera: Lasiocampidae) pupae
and eggs was higher in apple (*Malus pumila*) orchards with rich assemblages of flowering understory plants. Leius rated 15 orchards as possessing
poorly, intermediately, or richly flowering understory. The rich orchards
included a number of spring and summer flowers. The hymenopterous parasitoids of tent caterpillar larvae were *Itoplectis conquisitor* (Say), *Glypta
simplicipes* Cresson, and *Scambus hispae* (Harris) (Ichneumonidae); and of
tent caterpillar eggs were *Telenomus* sp. (Scelionidae), *Ooencyrtus clisioca*
(Ashmead) (Encyrtidae), and *Eupelmus spongipartus* Foerster
(Eupelmidae). Parasites of gypsy moth larvae were not identified. Leius
(1967) presented only means, minima, and maxima for each host type under
each flowering regime: data were not labelled as to locality, nor were inferential statistics calculated.

We re-analyzed Leius's (1967) data using a combination of ANOVA
and regression. Analysis of variance (ANOVA) employing only maxima
and minima indicated non-significant differences among flower treatments
for parasitism of codling moth larvae ($P = 0.2223$), tent caterpillar larvae
($P = 0.1932$), and tent caterpillar eggs ($P = 0.6507$). Simple regression
analysis of the same data showed that increasing richness of flowering had a
marginally non-significant influence on untransformed percentage of parasitism of codling moth larvae ($P = 0.0621$) and tent caterpillar larvae
($P = 0.0589$), and a distinctly non-significant effect on parasitism of tent
caterpillar eggs ($P = 0.3316$). In all cases, slopes of the fitted lines were

positive. Nine of the orchards were ignored and only the extremes (minima and maxima) employed in these analyses (i.e., error variance is maximized): from these standpoints, the reassessment is conservative. However, information that can be inferred from the overall means and the extremes is neglected.

Further re-analysis of Leius's (1967) data entailed ANOVA employing maxima, minima and a composite value representing the mean of the remaining three plots in each treatment. Significant differences were found using untransformed percentage parasitism of codling moth larvae ($P = 0.0282$) and tent caterpillar larvae ($P = 0.0119$); there was no significant effect on parasitism of tent caterpillar eggs ($P = 0.2891$). Fisher's protected least significant difference indicated that parasitism of tent caterpillar larvae was significantly greater in orchards with richly flowering understories than in those of poor or intermediate status. Parasitism of codling moth larvae was significantly greater in richly-flowering understories than in those of poor status. Simple regression analyses of the same data showed that increasing richness of flowering had a highly significant influence on parasitism of codling moth larvae ($P = 0.0057$) and tent caterpillar larvae ($P = 0.0037$), and a non-significant effect on parasitism of tent caterpillar eggs ($P = 0.1135$). In all cases, slopes of the fitted lines were positive. These ANOVA and regression analyses effectively decrease the replication by two, because the data from three plots are pooled and regarded as being from a single plot. From this standpoint, the reassessments are conservative; however, such pooling also reduces error variance, so the net effect on the power of the test is unclear.

New and Promising Analytical Techniques

Geostatistics

Geostatistics is a set of statistical techniques for describing and analyzing spatial variability in data. Although initially developed for use in mining and geological sciences, geostatistical techniques are now being adopted by ecologists (Robertson 1987; Rossi et al. 1992) and pest management researchers (Kemp et al. 1989; Schotzko & O'Keeffe 1989, Liebhold et al. 1991). Traditional statistical techniques generally require the assumptions of independent observations and identically distributed errors. Often though, sample independence does not hold in agricultural settings where variables are usually spatially correlated as a function of proximity. Geostatistical methods model spatial dependence explicitly. Typically two steps are involved in a standard analysis (Robertson 1987). The first step characterizes the relationship between variability and distance. The semivariances within or between variables at successive distances are measured and a semivariagram is created that displays this relationship across either space or time. Semivariance is calculated as half of the sum of squared differences between paired data points that are separated by an amount, designated by the letter h, in either space or time. A semivariagram is a two-dimensional graph with semivariances plotted on the y-axis and increasing values of h on the x-axis. The second step is to use the data and the semivariance relationship to interpolate values between actual sampling points. The most common form of interpolation is a technique called kriging. Rossi et al. (1992) argued that a different type (non-ergodic) of covariance and correlogram could provide a more effective description of lag-to-lag spatial dependence because local changes in means and variances (often present in ecological data) can cause the traditional statistics to give misleading results.

Geostatistical results can be the primary focus of an agricultural study, but they will be more useful when used in conjunction with other experimental and statistical techniques. Knowledge of patterns of spatial dependence in a field can be used to optimize the placement of blocks and treatments in an ANOVA design. These techniques can also be used to model the joint spatial dependence between an organism such as a pest or natural enemy and factors in its environment such as a cover crop or companion planting. It is likely that geostatistics will be used increasingly in the agricultural sciences.

Meta-analysis

Meta-analysis is a class of statistical techniques for combining the results of many individual experiments that test the same effect into one analysis yielding one general conclusion (Rosenthal 1991; Hedges & Olkin 1985). Often, individual studies are too small and have too little power to detect biologically important results. By treating several such studies in concert, meta-analysis can raise the overall power of detecting a real effect. This technique would allow many small studies done by individual researchers to answer some of the questions as to the efficacy of certain biological control enhancement strategies. So far the technique has been primarily used for psychological and biomedical studies. There has been some controversy about its use (Mann 1990), concerning the validity of combining results from studies that were not conducted with precisely similar methods.

Categorical Data Analysis

Most data encountered in agricultural situations are quantitative—i.e., measured on a continuous (e.g., crop yields in kilograms) or discretely ordered (e.g., insect enumerations) scale. Many variables, however, are qualitative—i.e., they fall in discrete categories that have no intrinsic order (e.g., types of fertilizer, parasitized vs. non-parasitized hosts). If the outcome (dependent) variable in a study is categorical, then standard ANOVA and regression techniques are not always appropriate and optimal. Techniques have been developed to analyze categorical data of many types. We will discuss two techniques—logistic regression and log-linear modeling.

Logistic regression is a technique that models the probability of categorical outcomes as a function of other discrete and continuous explanatory variables. Standard linear regression does not do a good job with data that fall only between 0 and 1 and are not normally distributed. Proportions and probabilities violate both of these conditions. Logistic regression is applicable under these conditions. As an example, one could use this method to describe the relationship between visitation of leaves by a predator and factors such as crop vigor and surrounding vegetation (see case study #2). Logistic regression is ideally suited for studying correlations between parasitism of hosts and other biological and environmental factors. Because these other factors can be either categorical or not, the method is very flexible. For researchers, the interest will probably not be on modeling probabilities directly, but rather on measuring the strength of the effect each explanatory variable has on determining these probabilities. The theory and practical methodology of logistic regression are detailed by Hosmer and Lemeshow (1989).

Log-linear models are similar to logistic regression models but all variables used must be categorical. Log-linear analyses are used to describe associations or relationships in complex assemblages of categorical variables. Generally, there is no one variable singled out as a dependent or response variable. A log-linear analysis is often used as an initial analysis of complex data sets to understand the level of complexity of interactions among variables. Schoenig and Wilson (1992) used a log-linear model to describe the associations between spider mites, *Tetranychus* sp. (Acari) and their natural enemies on cotton. Fienberg (1985) described the theory and basic methodology of fitting log-linear models.

Path Analysis and Structural Equation Modeling

Path analysis is a form of structured linear regression, most frequently used when there are a number of variables linked to each other in a web of one-way cause-and-effect and mutual correlation relationships. Usually, there is one overall response variable placed at the endpoint of the path diagram. The analyst creates the path diagram in a somewhat subjective manner based upon previous knowledge of the variables in the system. Through performing a series of standardized linear regressions, using different response variables, it is possible to calculate path coefficients that represent the directional influence of a variable with other connected variables. The overall goal is to determine which variables are most influential in affecting the response variable, and through which paths the influence is strongest.

An example of a situation where this approach might be employed is a farm-level study looking at factors affecting predation of a crop pest. Other variables that could be included in the path are other predators, alternative prey, crop variables, non-crop vegetation, other control measures, etc. The standard manual of path analysis is that by Li (1975). There have been some criticism of path analysis (Mitchell 1992) and suggestions that a more comprehensive approach to analyzing interacting variables can be obtained from a methodology called structural equation modeling (SEM). SEM is a complex structure of equations that are solved in such a way that different path structures and coefficients can be evaluated empirically. Mitchell (1992) contrasted the two methodologies and illustrated the issues with an example from pollination ecology. LISREL and SAS statistical package perform SEM.

Interpretation

After the data are collected and have been statistically analyzed, the ultimate questions are at hand. Namely, what do the results mean? What have we learned about enhancement of biological control? Did we succeed in creating an experiment that helped answer questions beyond a reasonable doubt? Generally, researchers focus on the "significance" of the results. In straightforward hypothesis testing there are two types of significance, which are often confused. We shall consider these separately.

Statistical Significance

Statistical significance is obtained when there is little possibility that an experimental result could have arisen due to chance alone. In judging whether a result is significant we observe the p-value associated with that result. The p-value is the probability that you would observe a result as or more discrepant than your result was from the null hypothesis, assuming that the null hypothesis is true. In other words, how likely is the observed result under the null hypothesis? By scientific convention, it has become customary to say that a result is statistically significant if the associated p-value is less than 0.05. This assures that erroneous rejection of a true null hypothesis will occur only 5% of the time. The merits of this approach are debatable (Salsberg 1985); however, the practice will probably continue. The main point is that statistical significance only indicates that the obtained result was not likely under the null hypothesis. It does not mean that the magnitude of the result has significant implications for agroecology.

Agroecological Significance

Once an effect is found to be statistically significant, its agroecological significance can be evaluated. This is done by focusing on the magnitude of the effect, rather than on its statistical significance. The latter is closely tied to the number of samples taken. Even though an effect is significantly greater than that which would have been due to chance alone, it may still be so small that in a practical sense it is unimportant. Agroecological or biological significance is what the experiment is all about. Statistical significance is simply a protection against inferring an effect that does not really exist. What is agroecological significance, then? How is it measured? This will depend on the system being investigated and will include factors such as strength of effects from other variables, economic considerations, relative benefits of enhanced biological control to yield, environment, health, society, etc.

Conclusion

Common sense and clear, logical thinking are the two most powerful tools in designing and implementing experimental designs. Replication of treatments, use of controls, awareness of confounding factors, use of exploratory data analysis, etc., are actually rigorous, formalized common sense. Statistical techniques often bring a new level of power and discrimination to detecting effective methods for the enhancement of biological control. Computer programs are increasingly becoming available to conduct most types of data analysis on an inexpensive computer. Many good books are available that explain statistics and experimental design at a level that does not require formal training. Biocontrol research will benefit from a systematic adoption of good experimental design and thoughtful analysis.

Acknowledgments

We thank Michael Pitcairn and Andrew Corbett for reviewing the manuscript and providing helpful suggestions.

References

Andow, D. A. & K. Hidaka. 1989. Experimental natural history of sustainable agriculture: syndromes of production. Agric. Ecosyst. Environ. 27:447-462.

Bugg, R. L., F. L. Wäckers, K. E. Brunson, J. D. Dutcher & S. C. Phatak. 1991. Cool-season cover crops relay intercropped with cantaloupe: influence on a general predator, *Geocoris punctipes* (Hemiptera: Lygaeidae). J. Econ. Entomol. 84:408-416.

Carmer, S. G. & W. M. Walker. 1988. Significance from a statistician's viewpoint. J. Prod. Agric. 1:27-33.

Dixon, W. J. 1988. BMDP statistical software manual. University of California Press. Berkeley, CA.

Edens, T. C. & D. L. Haynes. 1982. Closed system agriculture: resource constraints, management options, and design alternatives. Ann. Rev. Phytopathol. 20:363-395.

Eisenhardt, C. 1947. Some assumptions underlying the analysis of variance. Biometrics 1:1-21.

Fienberg, S. E. 1985. The analysis of cross-classified data. (2nd Ed.). MIT Press Cambridge, MA.

Green, R. G. 1979. Sampling design and statistical methods for environmental biologists. Wiley & Sons, New York.

Gurevitch, J. & S. T. Chester, Jr. 1986. Analysis of repeated measures experiments. Ecology 67:251-255.

Heads, P. A. & J. H. Lawton. 1983. Studies on the natural enemies complex of the holly leaf-miner: the effects of scale on the detection of aggregative responses and the implication for biological control. Oikos 40:267-276.

Hedges, L. V. & I. Olkin. 1985. Statistical methods for meta-analysis. Academic Press, New York.

Hosmer, D. W. & S. Lemeshow. 1989. Applied logistic regression. John Wiley & Sons, New York.

Hurlbert, S. H. 1984. Pseudoreplication and the design of ecological field experiments. Ecol. Monogr.. 54:187-211.

Karandinos, M. G. 1976. Optimum sample size and comments on some published formulae. Bull. Entomol. Soc. Amer. 22:417-421.

Kareiva, P. 1990. The spatial dimension in pest-enemy interactions, pp. 213-227. *In* M. Mackauer, L. E. Ehler & J. Roland [eds.], Critical issues in biological control. Intercept Ltd., Andover, England.

Kemp, W. P., T. M. Kalaris & W. F. Quimby. 1989. Rangeland grasshopper (Orthoptera: Acrididae) spatial variability: macroscale population assessment. J. Econ. Entomol. 82:1270-1276.

Kogan, M. & D. C. Herzog [eds.] 1980. Sampling methods in soybean entomology. Springer-Verlag, New York.

Leius, K. 1967. Influence of wild flowers on parasitism of tent caterpillars and codling moth. Can. Entomol. 99:444-446.

Li, C. C. 1975. Path analysis—a primer. Boxwood Press, Pacific Grove, Calif.

Liebhold, A. M., X. Zhang, M. E. Hohn, J. S. Elkington, M. Ticehurst, G. L. Benzon & R. W. Campbell. 1991. Geostatistical analysis of gypsy moth (Lepidoptera:Lymantriidae) egg mass populations. Environ. Entomol. 20:1407-1417.

Mack, T. P., A. C. Appel, C. B. Backman & P. J. Trichilo. 1988. Water relations of several arthropod predators in the peanut agroecosystem. Environ. Entomol. 17:778-781.

MacRae, R. J., S. B. Hill, J. Henning, & G. R. Mehuys. 1989. Agricultural science and sustainable agriculture: a review of the existing scientific barriers to sustainable food production and potential solutions. Biol. Agric. Hortic. 6:173-219.

Mann, C. 1990. Meta-analysis in the breech. Science 249:476-479.

Milliken, G. A. & D. E. Johnson. 1984. The analysis of messy data, Volume I: Designed experiments. Van Norstrand, Reinhold, N.Y.

Mitchell, R. J. 1992. Testing evolutionary and ecological hypotheses using path analysis and structural equation modelling. Functional Ecol. 6:123-129.

Nelson, L. A. 1989. A statistical editor's viewpoint of statistical usage in horticultural science publications. HortScience 24:53-57.

Neter, J., W. Wasserman & M. Kutner. 1990. Applied linear statistical models: regression, analysis of variance, and experimental designs (3rd Ed.). Irwin, Homewood, Ill.

O'Brien, R. G. 1986a. Power analysis for linear models, pp. 915-922. In Proceedings, 11th Annual SAS Users Group International Conference, Atlanta, Ga., February 9-12., SAS Institute, Cary, N.C.

1986b. Using the SAS system to perform power analysis for log linear models, pp. 778-784. In Proceedings, 11th Annual SAS Users Group International Conference, Atlanta, Ga., February 9-12. SAS Institute, Cary, N.C.

Odeh, R. & M. Fox. 1991. Sample size choice. Marcel Dekker Inc. New York.

Robertson, G. P. 1987. Geostatistics in ecology: interpolating with known variance. Ecology 68:744-748.

Rossi, R. E., D. J. Mulla & E. H. Franz. 1992. Geostatistical modeling and interpreting ecological spatial dependence. Ecol. Monogr. 62:277-314.

Rosenthal, R. 1991. Meta-analytic procedures for social research. (Revised edition). Sage publications, Newbury Park, California.

Ruesink, W. G. 1980. Introduction to sampling theory, pp. 61-78. In M. Kogan & D. C. Herzog [eds.], Sampling methods in soybean entomology. Springer-Verlag, New York.

Rzewnicki, P. E., R. Thompson, G. W. Lesoing, R. W. Elmore, C. A. Francis, A. M. Parkhurst & R. S. Moomaw. 1989. On-farm experimental design and implications for locating research sites. Am. J. Alternative Agric. 3:168-173.

Salsburg, D. S. 1985. The religion of statistics as practiced in medical journals. Amer. Stat. 39:220-223.

Schoenig, S. E. & L. T. Wilson. 1992. Patterns of spatial association between spider mites (Acari:Tetranychidae) and their natural enemies on cotton. Environ. Entomol. 21:471-477.

Schotzko, D. J. & L. E. O'Keeffe. 1989. Geostatistical description of the spatial distribution of *Lygus hesperus* (Heteroptera: Miridae) in lentils. J. Econ. Entomol 82:1277-1288.

Sokal, R. R. & F. J. Rohlf. 1981. Biometry. W. H. Freeman & Co., New York.

Southwood, T. R. E. 1978. Ecological methods. Chapman & Hall, London.

Steel, R. D. G. & J. H. Torrie. 1980. Principles and procedures of statistics: a biometrical approach. McGraw-Hill, New York.

Toft, C. A. and P. J. Shea. 1983. Detecting community-wide patterns: estimating power strengthens statistical inference. Am. Nat. 122:618-625.

Tukey, J. 1977. Exploratory data analysis. Addison-Wesley, Reading, Massachusetts.

Wilson, L. T., W. L. Sterling, D. R. Rummel & J. E. DeVay. 1989. Quantitative sampling principles in cotton IPM, pp. 85-105. In E. Frisbee, K. M. El-Zik & L. T. Wilson [eds.], Integrated pest management systems and cotton production. R. Wiley and Sons, New York.

Zar, J. H. 1984. Biostatistical analysis. Prentice-Hall. Englewood Cliffs, New Jersey.

Measuring the Impact of a Natural Enemy Refuge: The Prune Tree / Vineyard Example

BROOK C. MURPHY, JAY A. ROSENHEIM & JEFFREY GRANETT—Department of Entomology, University of California,One Shields Avenue, Davis, CA 95616

CHARLES H. PICKETT—Biological Control Program, California Department of Food & Agriculture, 3288 Meadowview Road, Sacramento, CA 95832

ROBERT V. DOWELL—Pest Detection/Emergency Projects, California Department of Food & Agriculture, 1220 N Street, Sacramento, CA 95814

Many studies have demonstrated that by diversifying monocultures one can potentially stabilize and/or suppress pest populations (Russell 1989; Andow 1991). As a result, many researchers now believe that by identifying essential habitat and selectively integrating these components back into the agroecosystem, some pest species can be regulated solely through the action of their natural enemies (Altieri & Letourneau 1984; Russell 1989; Lewis et al. 1997). A number of studies have investigated the efficacy of habitat management techniques as pest control tactics (Letourneau 1987; Altieri 1983; Russell 1989 & references therein). To date, however, most evaluations have been primarily qualitative in nature or too poorly replicated to represent rigorous tests of the tactic. Furthermore, these studies are rarely conducted under commercial agricultural conditions. However, some of these proposed tactics may be both practical and economical for adoption by growers (van Emden 1990). An emerging challenge for researchers is to develop appropriate experimental designs to evaluate the pest control potential of a habitat management tactic within the economic and management constraints of contemporary agricultural production (Herzog & Funderburk 1985; van Emden 1990). As a case study, we present our work on the ability of prune (*Prunus domestica*) tree refuges to enhance the biological control of the grape (*Vitis vinifera*) leafhopper in grape vineyards.

Prune Refuge / Grape Vineyard System

The grape leafhopper (*Erythroneura elegantula* [Osborn], Homoptera: Cicadellidae) is the most important pest of grapes in many of the grape growing areas of the western United States. If left untreated, leafhopper feeding can result in economic losses due to cosmetic damage to grape berries from leafhopper frass, fruit damage from excess sun exposure, and reduced vine vigor from heavy leaf loss. In addition, high densities of adult leafhoppers can disrupt manual harvest by flying into the eyes, noses and mouths of field laborers.

The most important natural enemy of the grape leafhopper is *Anagrus epos* (Girault) (Hymenoptera: Mymaridae) (Williams 1984), an important egg parasite of many leafhopper species throughout North America (Gordh & Dunbar 1977). Within commercial vineyards, *A. epos* often inflicts major mortality on grape leafhopper populations in the western United States. A key factor associated with the density and effectiveness of *A. epos* populations in grape vineyards is the presence of habitat to support alternate hosts (Doutt & Nakata 1973). *Anagrus epos* overwinters within leafhopper eggs. Because the grape leafhopper overwinters as an adult, a source of alternate leafhopper host eggs is required for *A. epos* to overwinter.

Doutt and Nakata (1965) observed that vineyards located downwind from riparian habitats had higher grape leafhopper egg parasitism and lower grape leafhopper numbers compared to vineyards located distant from riparian habitat. Blackberry brambles (*Rubus* spp.), abundant in riparian and other native habitats, support *Dikrella californica* (Lawson), blackberry leafhopper, a year-round host of *A. epos*. Each spring, prior to the first generation of grape leafhoppers, *A. epos* disperses from blackberry brambles into nearby vineyards. A shorter immigration distance results in earlier establishment of *A. epos* at higher densities, and thereby decreases the likelihood that chemical control of the grape leafhopper will be required (Doutt & Nakata 1973). As a result of these findings, experimental plantings of blackberry were established in central California vineyards distant from riparian habitat to enhance *A. epos* populations. Unfortunately, wild blackberry brambles did not maintain abundant populations of *A. epos* when removed from natural habitats and planted near grape vineyards. Flaherty et al. (1985) attempted to mimic the natural surroundings of blackberries by providing brambles with shade. Although lath-shaded plants produced more leafhoppers than non-shaded plants, the production of parasites was not higher, and further efforts involving blackberries were abandoned.

The greater number of *A. epos* reported downwind of riparian habitat may reflect, in part, the development of parasites on leafhopper eggs associated with plants other than blackberry. *Anagrus epos* attacks several species of leafhoppers, including some associated with rosaceous woody plants found in many natural habitats (Pickett et al. 1990). Kido et al. (1984) found that commercial plantings of prune trees support prune leafhopper, *Edwardsiana prunicola* (Edwards) (Homoptera: Cicadellidae), which can serve as a year-round primary host for *A. epos* because it overwinters in the egg stage. Emergence of overwintering *A. epos* from these trees was found to coincide with the egg-laying period of first generation grape leafhoppers in vineyards, indicating that prunes could potentially serve as useful overwintering habitat for *A. epos* populations. Studies in both commercial and experimental plantings of prune trees demonstrated that populations of *A. epos* can be supported year-round, even when the prunes are located far from riparian habitats (Kido et al. 1984; Pickett et al. 1990). Additional studies have also demonstrated that larger *A. epos* populations and greater grape leafhopper egg parasitism occurred in proximity to prune tree refuges (Pickett et al. 1990).

Over the twenty-five years since the relationship between overwintering refuge and the effectiveness of *A. epos* populations was first hypothesized, the biological rationale behind the use of overwintering refuges to enhance *A. epos* populations has been clearly established. Although the basic framework of the grape leafhopper *A. epos* interaction has been identified, there has not been a rigorous test of the system to determine the level of parasitism that could be produced, whether enhanced parasitism could result in economic suppression of leafhopper populations, or under what geographical or cultural conditions the technique would be effective.

Experimental Design

Our study had three main objectives: to determine (1) if prune trees significantly increase mean parasitism levels of grape leafhopper eggs in adjacent vineyards; (2) if parasitism associated with prune refuges results in economic suppression of grape leafhoppers; and (3) if the impact of refuges varies significantly over a wide geographical area.

Most arthropod biological control systems have three components: a host plant, an herbivore, and one or more natural enemies. Our study had five components: prune trees and grape vines, prune and grape leafhopper populations, and the parasite *A. epos*. The complexity of the system creates potentially novel ecological interactions and more potential sources of variation in pest and natural enemy densities. For example, the number, planting density, age, and vigor of prune trees varies among sites and each of these variables can affect the number of prune leafhoppers on trees. This, in turn, can have a major effect on the number of parasites and their subsequent impact in nearby vineyards. We were also concerned about the impact of area-wide populations of *A. epos*. There are numerous plants outside the prune tree/vineyard system that support leafhoppers parasitized by *A. epos*;

parasites from these external sources may migrate into vineyards and affect the relative contribution of parasites from nearby prune refuges. Lastly, grape cultivar can affect grape leafhopper production, which can affect the abundance of parasites. To have the statistical power to detect treatment effects in the face of these numerous sources of variation, a large number of replicates was needed. Because the impact of the refuge might vary between regions of the state and between years, study sites were selected across northern and central California grape-growing regions, and the study was planned to be conducted over two seasons.

Our experimental design used paired vineyard plots of the same grape cultivar, one with a refuge and the other lacking a prune refuge. Ten pairs of vineyard plots were established in commercial vineyards located in central and northern California. A distance of at least 0.4km between paired vineyard sites was maintained to minimize interaction of hosts or parasites between treatments. Two of the prune refuges used in the study were planted specifically for leafhopper pest control, three were commercial prune orchards adjacent to the vineyard plots, and the remainder were remnants of commercial prune orchards. Given the nature of the prune refuge habitat being tested, we were unable to ensure true randomization of treatments among plots. True randomization would have required us to randomly assign refuges to plots, and would have required a 3- to 5-year delay until the prune trees could become established and mature.

For prune refuge sites, leafhopper egg parasitism and leafhopper nymph densities were monitored by sampling leaves from two transects (parallel to, and downwind from, the refuge), the first at 50 meters and the second at 125 meters downwind from the prune refuge. In the comparison vineyard lacking a prune tree refuge, leaves were sampled randomly across vineyard blocks without using transects. In all vineyards, leaves were visually examined in the field to count leafhopper nymphs and then returned to the laboratory and examined for egg parasitism. Leaves were scored for egg parasitism using two methods. The first method determined egg parasitism for unhatched eggs only. Unhatched eggs were examined under dissecting microscopes for the presence of a developing parasite or developing leafhopper nymph embryo, and the percentage of parasitized eggs calculated. Unhatched eggs provided a point estimate of parasitism coinciding with the sample date. In addition, leaves were examined for the distinctive scars left by leafhopper eggs. The egg scars were examined and scored as one of three types: (1) scars with a round exit hole, indicating parasite emergence; (2) scars with a small tear in the egg and leaf epidermis, indicating a healthy nymph had emerged and (3) scars with no signs of emergence. Egg scars without signs of emergence were dissected for evidence of parasitism and recorded as mortality from parasitism or causes other than parasitism. Egg scars remain recognizable on leaves all season and were therefore used to estimate cumulative parasitism from the beginning of the season until the date samples were taken.

Three yellow sticky traps were placed in each prune refuge to capture emerging parasites. Trap capture data were used to estimate parasite production for each refuge by multiplying the mean number of *A. epos* captured per trap prior to the completion of the first leafhopper generation by the number of trees in the refuge.

Data were statistically analyzed using a split-plot analysis of variance test, where the main factor tested was geographical region and the sub-factor was presence or absence of a prune refuge. Each of the paired vineyard plots represented a block within the analysis. Regional differences were tested by nesting blocks according to the viticultural climatic region of the block's location. Viticultural climatic regions were defined by the cumulative degree days occurring in each region from April through October (Winkler et al. 1974). Comparisons were made between cooler regions (<3,000 degree-days) and warmer regions (>3,000 degree-days). Analyses revealed that percent parasitism was dependent on the density of eggs on leaves. Thus, egg density was used as a covariate within the analysis. Details of the density-dependent relationship will be reported elsewhere.

Results
Cumulative Egg Mortality

Using the number of parasite or nymph emergence holes and the number of dead nymph embryos, we quantified the importance of egg parasitism by *A. epos* relative to all other sources of egg mortality. Parasitism by *A. epos* was the dominant factor affecting leafhopper egg mortality for all vineyards with or without a refuge (Figure 1). Parasitism by *A. epos* was six times larger than other sources of egg mortality during the first generation, and the cumulative mortality due to parasitism increased progressively during the succeeding generations. All other egg mortality combined accounted for less than ten percent of the eggs killed. The cause of death from "other sources" of egg mortality was uncertain, although laboratory and field observation suggested that some mortality might have been due to high temperatures experienced during the early season when vine canopies were small and therefore providing little protection for eggs.

Figure 1. Percent of grape leafhopper egg mortality due to *A. epos* parasitism or other mortality factors; hatched egg data. Estimates (± 1 SE) reflect cumulative mortality across the first leafhopper generation (June sample), across the first and second generations (August sample), and across all three leafhopper generations (October sample).

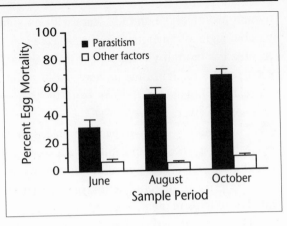

Cumulative Egg Parasitism

Levels of leafhopper egg parasitism were consistently higher in refuge-enhanced vineyards (Figure 2). Egg parasitism measured at the completion of the third leafhopper generation (October) was 13 percent greater at 50 meters downwind of the refuge and 11 percent greater at 125 meters downwind in refuge-enhanced versus the paired non-refuge vineyards. Differences in parasitism for both the 50 and 125 meter transects were significant (F = 4.94; df = 1, 27; P = 0.015). No significant differences were detected among vineyard growing regions (P > 0.05). There was also no significant interaction between refuges and growing regions. Egg parasitism during the second generation was significantly elevated at the 50 meter transect compared to the control (F = 4.91; df = 1,18; P = 0.014), but not at the 125 meter transect (P > 0.05). There were again no significant differences among growing regions. The lack of difference seen for the 125 meter transect may be due to insufficient replication (n = 5), because five of the vineyards were not large enough to include a 125 meter transect. No significant differences in egg parasitism occurred during the first leafhopper generation (P > 0.05), even though the mean parasitism rate in refuge treatments was nearly twofold larger than in control vineyards (see Figure 2).

Point Estimates of Parasitism

Examination of the unhatched eggs provided a measure of egg parasitism during the two-week period prior to the leaf sampling date. Mean differences in egg parasitism between control and refuge-enhanced vineyards were largest during the first leafhopper generation and became smaller with each succeeding generation (Figure 3). Significantly higher parasitism rates

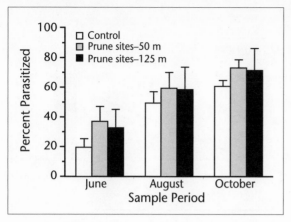

Figure 2. Percent of grape leafhopper eggs parasitized by *A. epos* in control vineyards that lack prune refuges, and the paired prune refuge vineyards at 50 and 125 meters downwind of prune trees; hatched egg data. Estimates (± 1 SE) are cumulative across sample periods.

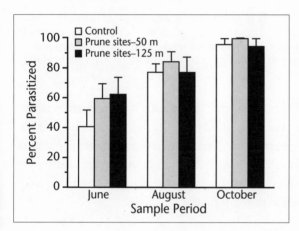

Figure 3. Point estimates of grape leafhopper egg parasitism during the first leafhopper generation (June), the second generation (August), and the third generation (October). Point estimates of egg parasitism use unhatched leafhopper egg data and are not cumulative across sample periods.

during the first generation were detected (F = 4.45; df = 1,18; P = 0.025), with no significant regional effect or interaction between region and prune refuge treatment. Differences in parasitism rates between refuge and non-refuge vineyards and across growing regions were not significant for either the second or third generations (P > 0.05).

Prune Refuge Production

The prune refuges used in the study varied with respect to tree number, tree spacing, grower management practices, and possibly other factors that could affect the density of overwintered parasites from prune trees. Therefore, the relationship between overwintering *A. epos* numbers and the difference in parasitism between each of the paired vineyard plots was analyzed for the

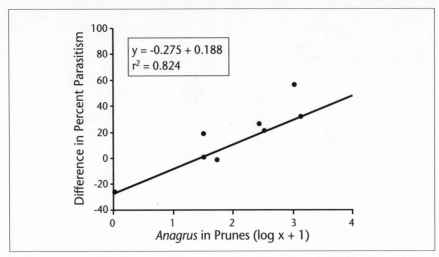

Figure 4. Relationship between early season *A. epos* numbers caught on sticky cards in prune trees (independent variable), and the difference in cumulative egg parasitism mortality through June between the adjacent prune refuge vineyard and the paired control vineyard (dependent variable). Values greater than zero indicate higher parasitism in prune refuge vineyards.

first generation of grape leafhopper eggs (Figure 4). A significant positive relationship was found between the number of emerging parasites and the degree of difference in egg parasitism during the first leafhopper generation ($n = 8$, $P < 0.002$). These results indicated that variation in grape leafhopper egg parasitism between paired vineyards was related to the contribution of overwintered *A. epos* from the prune refuges.

Leafhopper Nymph Densities

The densities of nymphs on grapes in control plots were compared to densities on grape leaves in plots with refuges. Nymph densities were highest during the first generation and became progressively lower during the remainder of the season (Figure 5). Nymph data were analyzed in two ways. The first analysis compared single sample dates during the season and found no significant differences in nymph numbers among plots. The second analysis compared cumulative nymph densities across generations, and also failed to detect significant differences. Leafhopper nymph densities averaged well below the action threshold of 20 nymphs per leaf, where chemical control would have been necessary to reduce vine damage.

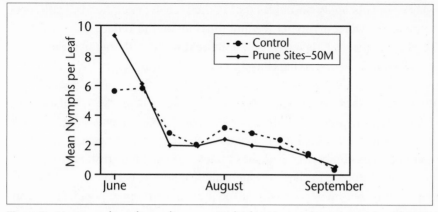

Figure 5. Mean number of nymphs per grape leaf in prune refuge vineyards and in the paired control vineyards.

Discussion

Doutt and Nakata (1973), Flaherty et al. (1985) and Settle and Wilson (1989) reported *A. epos* to be a key natural enemy attacking grape leafhopper eggs in commercial vineyards and also suggested that the parasite could potentially control leafhopper populations without pesticide applications. Their observations were based primarily on examining point estimates of egg parasitism in vineyards located within a limited geographical area. However, single point estimates determine parasitism for only a short time interval and may not estimate the true impact on the egg population (Van Driesche 1983). In our study, egg scars were used to quantify cumulative parasitism across leafhopper generations, which estimated actual parasitism mortality for the leafhopper egg population present over the lifetime of the leaf (Bellows et al. 1992). Our analysis confirmed that *A. epos* was the primary mortality factor for leafhopper eggs across all geographical areas and leafhopper generations examined within the study. Furthermore, egg parasitism increased as the season progressed while all other mortality factors acting on leafhopper eggs remained essentially unchanged. Thus, our results confirm that *A. epos* is a key mortality agent of leafhopper eggs and therefore appears to be a suitable focus for attempts to enhance biological control.

Anagrus epos parasitism furthermore appears to be a significant egg mortality factor whether prune refuges are present or not. In fact, for some vineyards, 100 percent of eggs are attacked by *A. epos* (see Figure 3). However, parasitism often does not reach these high levels until the third leafhopper generation, which is too late to provide economic suppression.

Many growers apply pesticides during the second generation of leafhopper nymphs to reduce nymphal feeding pressure and to reduce the number of third generation adults interfering with field workers during harvest.

A number of researchers have proposed that leafhopper outbreaks could be prevented by enhancing *A. epos* populations using prune refuges planted adjacent to vineyards (Doutt & Nakata 1973; Kido et al. 1984; Wilson et al. 1989). They hypothesized that overwintering parasites emerging from the prune refuges move rapidly into nearby vineyards, ensuring early establishment of large numbers of parasites relative to vineyards distant from overwintering habitats. Earlier parasite establishment was hypothesized to enhance early-season egg parasitism and prevent leafhopper populations from exceeding the economic threshold.

Our first analysis tested the hypothesized association between prune refuges and higher seasonal parasitism rates in adjacent vineyards. We documented significant differences in cumulative egg parasitism between refuge and non-refuge vineyards, confirming the hypothesized association between the presence of the overwintering refuges and enhanced impact of *A. epos* on grape leafhopper egg populations (Doutt & Nakata 1973; Kido et al. 1984; Wilson et al. 1989). Furthermore, we found that the effect is consistent among different growing regions. Differences in parasitism were most pronounced during the first leafhopper generation using either the cumulative or point estimate parasitism data. These results support the hypothesized role of refuges in providing earlier *A. epos* establishment and therefore higher early-season egg mortality (Kido et al. 1984; Wilson et al. 1989).

We also wished to determine if the size of the overwintering *A. epos* populations in prunes influenced parasitism rates in the adjacent vineyards. The majority of the prune refuges used in the study were not planted to manage leafhopper populations. Instead, most of the refuges were remnants of commercial prune orchards. Large differences that existed between prune refuges in the number, age, and management of trees may have affected parasite densities overwintering within trees. Sticky card trap samples showed that densities of parasites varied dramatically among the refuges. Our analysis also revealed a clear positive relationship between parasite production and parasitism rates in nearby vineyards (Figure 4). This relationship suggests that the number of trees planted and their management could potentially be optimized for pest control. Trees with the highest parasite densities were shown to enhance parasitism approximately 40 percent over non-refuge sites.

From a pest management perspective, the value of prune refuges depends on whether nymph densities are maintained below levels that result in economic losses to the grower. Comparison of nymphal densities failed to detect significant differences among the prune refuge and non–prune refuge vineyards. However, none of the plots sustained densities that required pesticide applications (about 20 nymphs per leaf). The low nymph densities may have been the result of unusually cool winter and spring temperatures during 1991. Nevertheless, the trends in leafhopper nymphal densities observed during the first two leafhopper generations do suggest some enhanced suppression in vineyards adjacent to prunes (Figure 5). In the refuge vineyards, nymph densities declined to less than 25% of initial densities (a 75% decline), compared to a 50% decline from initial densities in non-refuge vineyards. Quantifying the role of prune refuges in enhancing suppression of nymph populations is a key objective of ongoing research.

Conclusion

The results of our evaluation of the prune refuge system have thus far been encouraging. We have shown that prune trees can enhance levels of grape leafhopper egg parasitism, and that the method may be effective in many viticultural regions of California. In addition, we have quantified the relationship between the production of *A. epos* in refuges and the enhancement of parasitism in associated vineyards. *Anagrus epos* production is itself a function of the density of parasites per prune tree and the number of trees in the refuge. By managing prune trees to enhance prune leafhopper populations and *A. epos*, and by controlling the number of prunes planted into the refuge, the production of parasites can potentially be adjusted to match pest management needs.

Our results underscore the need to understand the relationship between habitat structure, including the size and spatial arrangement of refuges, and the effectiveness of natural enemies. In many traditional experimental designs for assessing the efficacy of biological control agents, the influence of the experimental treatments on natural enemy abundance has been well defined (Luck et al. 1988). For instance, when evaluating the introduction of classical biological control agents, pest mortality and pest density in plots with the natural enemy present are contrasted with pest mortality and density in plots with the natural enemy absent; the treatment effect is presence versus absence of the natural enemy. Similarly, when evaluating the augmentative releases of commercially available natural enemies, the treatments may include a variety of predetermined release rates. In both cases, tight control over natural enemy densities is maintained within the experimental design. In habitat management studies, however, there are a large number of potential treatment combinations, including the size, spatial arrangement, and cultural management of the habitat, and any particular treatment that can have highly variable influences on natural enemy densities. Control of natural enemy abundance is necessarily indirect, via the influence of providing an alternate habitat. Thus, studies evaluating habitat diversification need to characterize the relationship between the nature of the refuge provided, the degree to which natural enemy densities are altered, and the resultant effect on the level of biological control. This information will be critical in developing and improving habitat management recommendations.

References

Altieri, M. A. 1983. Vegetational designs for insect-habitat management. Environ. Man. 7:3-7.

Altieri, M. A. & D. K. Letourneau. 1984. Vegetation diversity and insect pest outbreaks. CRC Critical Rev. Plt. Sci. 2:131-169.

Andow, D. A. 1991. Vegetational diversity and arthropod population response. Annu. Rev. Entomol. 36:561-586.

Bellows, T. S. Jr. 1992. Life-table construction and analysis in the evaluation of natural enemies. Annu. Rev. Entomol. 37:587-614.

Doutt, R. L. & J. Nakata. 1965. Overwintering refuge of *Anagrus epos* (Hymenoptera: Mymaridae). J. Econ. Entomol. 58:586.

Doutt, R. L. & J. Nakata. 1973. The Rubus leafhopper and its egg parasitoid: an endemic biotic system useful in grape-pest management. Environ. Entomol. 2:381-386.

Flaherty, D. L., L. T. Wilson, V. M. Stern & H. Kido. 1985. Biological control in San Joaquin valley vineyards, pp. 501-520. *In* D. C. Herzog & M. A. Hoy [eds.], Biological control in agricultural IPM systems. Academic Press, Orlando, Florida.

Gordh, G. & D. M. Dunbar. 1977. A new *Anagrus* important in the biological control of *Stephanitis takeyai* and a key to the North American species. Fla. Entomol. 60:85-95.

Herzog, D. C. & J. E. Funderburk. 1985. Plant resistance and cultural practice interactions with biological control, pp. 67-88. *In* D. C. Herzog & M. A. Hoy [eds.], Biological control in agricultural IPM systems. Academic Press, Orlando, Florida.

Kido, H., D. L. Flaherty, D. F. Bosch & K. A. Valero. 1984. French prune trees as overwintering sites for the grape leafhopper egg parasite. Amer. J. Enol. Vitic. 35:156-160.

Letourneau, D. K. 1987. The enemies hypothesis: tritrophic interactions and vegetational diversity in tropical agroecosystems. Ecology 68:1616-1622.

Lewis, W. J., J. C. van Lenteren, S. C. Phatak, and J. H. Tumlinson, III. 1997. A total system approach to sustainable pest management. Proc. Natl. Acad. Sci. 94:12243-12248.

Luck, R. F., B. M. Shepard & P. E. Kenmore. 1988. Experimental methods for evaluating arthropod natural enemies. Annu. Rev. Entomol. 33:367-391.

Pickett, C. H., L. T. Wilson & D. L. Flaherty. 1990. The role of refuges in crop protection, with reference to plantings of French prune trees in a grape agroecosystem, pp. 151-165. In N. J. Bostanian, L. T. Wilson, & T. J. Dennehy [eds.], Monitoring and integrated management of arthropod pests of small fruit crops. Intercept Publishing, Great Britain.

Russell, E. P. 1989. Enemies hypothesis: a review of the effect of vegetational diversity on predatory insects and parasitoids. Environ. Entomol. 18:590-599.

Settle, W. H. & L. T. Wilson. 1989. Invasion by the variegated leafhopper and biotic interactions: parasitism, competition, and apparent competition. Ecology 71:1461-1470.

Van Driesche, R. G. 1983. Meaning of "percent parasitism" in studies of insect parasitoids. Environ. Entomol. 12:1611-1622.

van Emden, H. F. 1990. Plant diversity and natural enemy efficiency in agro-ecosystems, pp. 63-80. In M. Mackauer, L. E. Ehler & J. Roland [eds.], Critical issues in biological control. Intercept Ltd., Andover, England, U.K.

Williams, D. W. 1984. Ecology of the blackberry-leafhopper-parasite system and its relevance to California grape agroecosystems. Hilgardia 52:1-33.

Wilson, L. T., C. H. Pickett & D. L. Flaherty. 1989. French prune trees: refuge for grape leafhopper parasite. Calif. Agric. March-April:7-8.

Winkler, A.J., J. A. Cook, W. M. Kliewer & L. A. Lider. 1974. General viticulture. University of California Press, Berkeley.

Spiders and Vineyard Habitat Relationships in Central California

WILLIAM ROLTSCH—Biological Control Program, California Department of Food & Agriculture, 3288 Meadowview Road, Sacramento, CA 95832

RACHID HANNA—Department of Entomology, University of California, One Shields Avenue, Davis, CA 95616

FRANK ZALOM—Department of Entomology, University of California, One Shields Avenue, Davis, CA 95616

HARRY SHOREY—University of California, Kearney Agricultural Center Parlier, CA 93648

MARK MAYSE—California State University, Department of Plant Science Fresno, CA 93740

Introduction

As noted by Whitcomb (1980), spiders have generally been overlooked in agroecosystems. When spider counts are taken during faunal surveys, species are seldom recognized (i.e., they are simply listed as "spiders"). In attempting to determine why spiders, until recently, have received little attention as biological control agents in agricultural systems, one discovers traits and requisites making spiders one of the most interesting, albeit challenging, subjects for research.

Spiders are easily neglected in agroecosystems. In part, this is due to their species diversity within a system, coupled with highly varied life histories. There are two distinct modes of prey capture: hunting and capture in webs. Many species show a temporal-dependent pattern of activity over a day. A few species, such as two lynx spiders, *Oxyopes salticus* (Hentz) and *Peucetia viridans* (Hentz) (Araneae:Oxyopidae), show little diel (i.e., 24hr) periodicity (LeSar & Unzicker 1978; Nyffeler et al. 1987a). Although web remnants of nocturnal web-building spiders remind us of their presence, the nocturnal hunting species can be cryptic during daytime.

In contrast to their insect predator and parasitoid counterparts, spiders may be considered less capable natural enemies because they lack wings to facilitate colonization of agricultural lands. However, ballooning through the use of silken threads readily allows dispersal of immatures of many spider species and adults of small species (Tolbert 1977; Gertsch 1979).

Most ballooning spiders are 1–2mm long, but larger individuals may also use this mode of dispersal. (Dean & Sterling 1985, 1990; Plagens 1986; Greenstone et al. 1987).

Several other issues are linked to the neglect of spiders as valuable natural enemies. Issues include difficulty in identification, their diminished presence due to extensive use of pesticides in agroecosystems, and the traditional lack of importance attributed to generalist natural enemies in the theory of biological control (Riechert & Lockley 1984).

The goals of this chapter are to provide general information about spiders and their potential importance as natural enemies in agroecosystems. Also, we suggest ways for increasing their abundance in perennial agricultural systems through habitat management, using the grape vineyard agroecosystem as an example. Last, we try to provide a perspective that will generate productive research hypotheses. Although this introduction covers material found in reviews of spider literature by Turnbull (1973), Lucsak (1979), and Riechert and Lockley (1984), we add more recent information. The importance of doing this is underscored by the paucity of literature on such fundamental topics as prey selection and consumption.

General Life History Characteristics of Spiders

There is little information on spider reproductive potential. In noting that some spider species lay as many as 3,000 eggs during their lifetimes while certain very small species lay as few as twelve, Gertsch (1979) generalized that the average level of fecundity per female is approximately 100 eggs. Leigh and Hunter (1969) studied the fecundity of six spider species found in cotton, representing the families Dictynidae, Thomisidae (2 spp.), Lycosidae (2 spp.), and Micryphantidae. Fecundity by member species of this complex, calculated from published mean numbers of egg cases and eggs per case, is estimated to range from 145 to 572 eggs per female. Muniappan and Chada (1970) reported that a crab spider, *Misumenops celer* (Thomisidae) (Hentz), had an average reproductive potential of over 200 offspring. These mean egg counts are well above those estimated by Gertsch, and are more in line with estimates by Turnbull (1973).

Prey consumption by spiders is difficult to quantify, particularly for hunting spiders. Late instars and adults of *M. celer* ate from 5 to14 early instar fall armyworm (*Spodoptera frugiperda* [J. E. Smith], Lepidoptera: Noctuidae) or corn earworm (*Helicoverpa zea* [Boddie], Lepidoptera: Noctuidae) larvae per day under laboratory conditions (Muniappan & Chada 1970). However, similar levels of extensive feeding observed for *Oxyopes salticus* under laboratory conditions were considered unrealistically high (Nyffeler et al. 1987b). Prey consumption studies on hunting spiders common in cotton (*Gossypium* sp.) suggest that consumption ranges from less than one to three prey each day, depending on the relative sizes of the prey and spider predator (McDaniel et al. 1981; Nyffeler et al. 1986; Nyffeler et al. 1987a, b; Dean et al. 1987; Breene et al. 1988).

Quantifying prey consumption by spiders can be complicated by variation in the rate of feeding during particular periods in their life (Miyashita 1968). To help resolve the issue of prey consumption by hunting spiders, consumption rates and dietary effects on development and reproduction should be assessed under various thermal and prey-restriction regimes. This approach should provide a close approximation of the minimum food requirements needed by species to reach maturity and sustain reproduction.

As noted by Turnbull (1973), characterizing prey selection by hunting spiders is particularly difficult. Laboratory studies may suggest unrealistically broad prey ranges; however, several field studies of web-building and hunting spiders have demonstrated broad host ranges (Nyffeler et al. 1987a, b; Nyffeler et al. 1988; Nyffeler et al. 1989; Agnew & Smith 1989). Nonetheless, spiders are far from being indiscriminate predators, and can be selective. For example, several spider species reject larval and adult coccinellids (*Hippodamia parenthesis* [Say], Coleoptera: Coccinellidae), presumably due to low palatability, and adult alfalfa weevils (*Hypera postica* [Gyllenhal], Coleoptera: Curculionidae), apparently due to hardness (Howell & Pienkowski 1971).

Prey range can be determined for many web spiders in the field by repeatedly collecting prey captured in the webs. However, some species remove prey remains from their webs after feeding (Putman 1967). Nyffeler et al. (1989) discussed prey type for several orb-weaver (Araneidae) species, and concluded that selectivity is largely a result of web size, strength, orientation and location in the environment. A recent review characterizes prey range and prey consumption rates for various types of web and hunting spiders (Nyffeler et. al. 1994).

The diel rhythms of spiders can affect research. Feeding studies and sampling surveys show that some species are distinctly diurnal or nocturnal. Kiritani et al. (1972) found that several species in rice (*Oryza sativa*) fed during the day, whereas others fed mainly at night. Several surveys (see below) of populations entailing day and night sampling indicated similar periodicity of spider activity.

Spiders in Agroecosystems

In reviewing the literature on spider species diversity in several row crops in Europe, and alfalfa (*Medicago sativa*) in Europe and North America, Luczak (1979) noted that species from two to five families composed 80% of the spider community within each field. However, most of these studies were limited to sampling the canopy spider fauna, and only one study evaluated the ground fauna using pitfall traps. A review of faunal surveys of spiders in field crops of the United States determined that five of the 48 families in North America contain 61% of the species collected. (Young & Edwards 1990).

Composition of epigeal and foliar spider fauna can be quite different. In soybean (*Glycine max*), Culin and Rust (1980) found 48 species in the ground community and 105 species in the foliage community. Only two species were collected in both. In cotton, Dean et al. (1982) identified 23 species in D-vac samples (Dietrick et al. 1959), 35 in pitfall traps, and 67 spp. by direct observation. Only 10 spp. were collected by all three methods.

Spiders can affect population densities of insect pest and other arthropod species, yet this impact has been assessed to varying degrees in only a few cases (Kiritani et al. 1970; 1972; Mansour et al. 1983; Mansour & Whitcomb 1986; Corrigan & Bennett 1987; Knutson & Gilstrap 1989; Riechert & Bishop 1990; Oraze & Grigarick 1989; Costello & Daane 1995). Spiders may be most appropriately viewed in terms of the impact of species assemblages (Riechert & Bishop 1990). Size and age-class variation and foraging modes are probably important. Although spiders are generally not believed to regulate prey populations, Riechert and Lockley (1984) suggested that as a species assemblage they may prevent outbreaks. This suggests that predation by spiders can provide a form of "environmental resistance" (Chapman 1928; Price 1984), which in combination with other biotic and abiotic factors prevents pest populations from exceeding economic injury levels. This could occur through a sustained reduction of a pest's geometric population growth potential.

Agroecosystem stability may influence species composition. Luczak (1979) stated that few if any populations of spider species are able to become permanent residents in annual agroecosystems, due to highly disruptive farming practices. Species inhabiting annual systems must be highly capable colonizers. Bishop and Riechert (1990) demonstrated that long-distance migrants can largely determine species composition in an annual system.

By contrast, perennial systems subject to minimal disturbance are likely to have more predictable assemblages of species characterized by those that are predominantly resident species (i.e., populations perpetuating on-site from one year to the next). In short, a less disrupted habitat is likely to meet the long-term needs of many potential inhabitants. In comparing the spider fauna in an annual soybean system with that of alfalfa, Culin and Yeargan (1983a, b) demonstrated that spider populations were more similar between years in alfalfa than in soybean.

Increasing plant diversity through cover cropping or weed management can increase the abundance of beneficial arthropods that help control crop pests (Herzog & Funderburk 1986). Cover cropping has been implicated in enhancing the abundance of generalist predators such as spiders, which in turn may lead to a reduction in pest abundance (Letourneau & Altieri 1983; Altieri & Schmidt 1985; Settle et al. 1986).

Effects of habitat manipulations on spider fauna in agricultural settings are little known. Nentwig (1988, 1989) evaluated the role of strip cropping in wheat (*Triticum* spp.) to enhance border sources of spiders. Although species composition differed, spider densities and impact remained unchanged. In contrast, Riechert and Bishop (1990) found that the added habitat provided through a mulch cover in experimental vegetable gardens had increased densities and reduced insect herbivore damage. Although conservation was central to both studies, the former involved creating potential source populations of spiders in an annual crop, whereas the latter dealt with modifying the immediate habitat within which spiders interact on a daily basis. In another study concerning immediate habitat, Ali and Reagan (1986) found that the spider fauna differed little in sugarcane plots containing broadleaf plants and/or grass between rows compared to plots where weedy ground cover was controlled. Sugarcane (*Saccharum officinarum*) was grown as a ratoon crop of three years duration.

Information addressing the impact of pesticides on spiders comes from a few controlled, replicated studies and from repeated monitoring studies lacking true replication. Based on pre- and post-application sampling, applications of malathion in citrus (*Citrus* sp.) had little impact on spider populations dominated by Cheiracanthium *mildei* (L.) (Clubionidae) and *Theridion* spp. (Theridiidae) (Mansour & Whitcomb 1986). By contrast, an application of formothion + carbaryl dramatically decreased spider densities.

Although the impact of several insecticides (endosulfan, azinphos-methyl, methidathion) and an acaricide (cynexatin) did vary, all depressed the spider fauna on apples (*Malus pumila*) (Mansour et al. 1981). In a replicated study in an apple orchard, an organophosphate insecticide (chlorpyrifos) decreased spider densities whereas several synthetic pyrethroid insecticides had an even greater negative impact (Mansour 1987). In the laboratory, the same insecticides were found to be highly toxic to *C. mildei*, one of the predominant species found in several cropping systems in Israel. Several acaricides had no observable effect on mortality, whereas several fungicides and herbicides caused limited mortality. Densities of spiders in cotton and apple appeared to be unaffected by sulfur application (Herne & Putman 1966; Shepard & Sterling 1972), but sulfur caused approximately 40% mortality to *C. mildei* in a laboratory study (Mansour 1987).

Prolonged use of certain insecticides can result in resistant spider populations. Redmond and Brazzel (1968) demonstrated that cotton field populations of striped lynx spider (*Oxyopes salticus*) were more resistant to methyl parathion than populations occurring away from areas of intensive cotton production. Similarly, Mansour (1984) found a malathion-tolerant strain of *C. mildei* from citrus and cotton.

A series of studies pertaining to insecticide-induced secondary pest outbreaks was conducted in the Japanese rice system by Ito et al. (1962), Kiritani et al. (1970, 1972), and Kiritani and Kawahara (1973). Following extensive use of a systemic chlorinated hydrocarbon insecticide (benzene hexachloride, BHC) to control the striped rice borer (*Chilo suppressalis* [Walker], Lepidoptera: Pyralidae), green rice leafhopper (*Nephotettix cincticeps* [Uhler], Homoptera: Cicadellidae), became an important pest. Several investigations, including a population dynamics study, suggested the importance of spiders as generalist natural enemies. Detrimental effects of BHC upon the spider community were observed even when a granular formulation was used. Although the insecticide did not kill green rice leafhoppers, which had fed on treated rice plants, these leafhoppers were lethal to spiders that preyed on them.

In summary, these and other studies indicate that spiders are susceptible to many classes of insecticides, and to other pesticides as well (Laster & Brazzel 1968; Bostanian et al. 1984; Whitford et al. 1987; Hilburn & Jennings 1988). Between-species variation in susceptibility can be expected as a result of fundamental physiological and behavioral differences. However, history of pesticide exposure can also be important, as indicated by within-species variation in susceptibility.

Spider Sampling and Evaluation of Several Techniques in San Joaquin Valley Grape Vineyards

The diversity of spider species encountered in any given system makes sampling a challenge. Associated with the problem of diversity are diel rhythms and other life history and behavior characteristics. Although some diel rhythms of spider activity are known, their impacts on sampling are not well understood.

Howell and Pienkowski (1971) found differences in sweep-net samples taken in alfalfa over a 24-hour period. Jumping (Salticidae) and crab (Thomisidae) spiders were most abundant in samples taken from late morning until early evening. Of two orb weaver species (Araneidae), one was most frequently caught in sweep net samples from mid-morning until early evening, whereas the other was caught from approximately 2100 until 0600hrs. Similar patterns of activity occurred when evaluating the apple limb-beating method of sampling (McCaffrey et al. 1984). Although the results were not definitive, data suggested once again that salticid and thomisid spiders were more common in diurnal samples, but that clubionid hunting spiders were more frequently collected at night. By sampling designated strata within the soybean canopy, LeSar and Unzicker (1978) showed that some spider species have a cyclic recurrence in various parts of the canopy through the course of the day.

As previously noted, in addition to habitat partitioning within the canopy, epigeal and canopy-dwelling spider faunae can be strikingly different (Turnbull 1960; Culin & Rust 1980; Agnew & Smith 1989). This further complicates the development of a sampling protocol.

Although the use of multiple sampling techniques ensures that few species are missed, limitations remain in making quantitative comparisons among species. To compare species abundance, the sampling efficiency of each method must be known for species sampled by a given method. Unfortunately, comparative studies relating sample methods to a so-called absolute sample estimate have rarely been done (McCaffrey et al. 1984, Costello & Daane 1997). In fact, the relative efficiencies of two or more sampling techniques have seldom been studied (Howell & Pienkowski 1971; LeSar & Unzicker 1978). In evaluating the relative abundance of spiders in the alfalfa canopy using a sweep-net and D-vac, Howell and Pienkowski (1971) generally found a high degree of similarity between methods. However, the family Linyphiidae and Lycosidae were relatively uncommon in sweep-net samples, whereas the former was ranked most abundant in D-vac samples.

Vineyard Spider Sampling

We have used primarily three techniques to sample spiders in the canopy of grapevines (i.e., D-vac, shake-cloth and leaf sampling). A D-vac sample was obtained by walking several vine-rows and randomly inserting the vacuum orifice into the canopy for a total of 50 suctions. The total number of samples varied from 1 to 3 per site. For the purpose of comparing methods, D-vac spider counts were converted to numbers per 100 leaves. The number of leaves sampled by each D-vac suction was estimated as a result of evaluating the air velocity under field conditions. Results indicated an effective canopy sampling depth of 10 to 15cm. Within a canopy of average density, it was determined that approximately 17 leaves are sampled in the 40cm diameter ring within a canopy depth of 12cm.

The leaf sample method involved the removal and inspection of two leaves from each of the basal, middle, and terminal regions of two shoots per vine. This leaf selection was repeated on several vines per row. The shake-cloth method utilized a cloth funnel (Figure 1). The unit was held under a vine row, and a trellis wire was given eight vigorous shakes to dislodge spiders. The spiders and debris that fell into the sampler were funneled into a plastic zip-lock bag, placed in a cooler and sorted at a later date.

In 1991 we evaluated the relative efficiency of the three methods detailed above, and a strip sampling approach. This new method, intended to approximate an "absolute" method, involved suspending a PVC frame covered with white denim fabric from the canopy, using stiff wire hooks (60 cm in length) that were attached to each corner of the frame. The cloth-covered frame had the same measurements as the top of the funnel shaped shake-cloth sampler. Once in place, all leaves and new canes above the frame were stripped from the vine and placed in a large plastic bag. The white sheet supported by the frame was used as a catch cloth. It was constantly observed for the presence of falling spiders, which were collected and placed in the bag. Due to the bulky samples and limited cold storage space, we took only ten such samples in each vineyard. This "strip-sample" approach was used because we believed it to be close to an absolute sample, and it could therefore provide a benchmark for other sampling methods. Nonetheless, the strip-sample method was imperfect because some initial disturbance occurred during setup. That is, the foliage was not removed instantaneously and the vine trunk was not sampled.

We sampled vines on 30–31 July 1991 from 0730 to 1500hrs (PDT). All

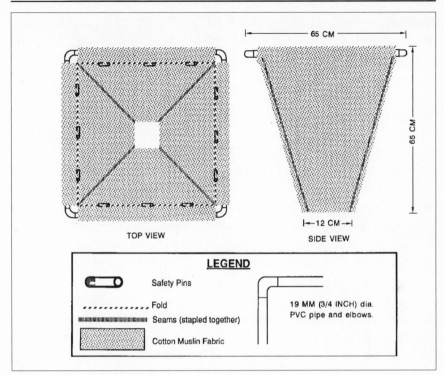

Figure 1. Shake-cloth sampler.

methods were used in each of six vineyards in Fresno County (WF, MAI, MAS, SM2, SM3, RIF) located in San Joaquin Valley, California. Within a 17-row section of each vineyard five rows were flagged, each separated from one another by three unflagged rows. Leaf samples (144 to 180 leaves per vineyard) followed by shake-cloth samples were taken on the inner three of five flagged rows. D-vac samples were taken on the middle row of each unflagged three-row section. Approximately seven of us completed all sampling within a vineyard over a time interval of about 90 minutes.

Sample materials obtained by the D-vac (three samples per vineyard) were placed in half-liter glass jars containing 70% ethanol. The 30 plastic bag samples per vineyard resulting from the shake-cloth method were held in large ice chests until returning to the laboratory, at which time they were stored at 6° C. Within a 12-day period, these samples were poured into a large plastic tray for sorting. During this period, the vast majority of specimens remained alive. Following sorting, all specimens obtained by each method of sampling were placed in vials containing 70% ethanol.

Relative to the strip method, sampling efficiency varied among methods and species (Table 1). *Hololena nedra* (Chamberlin and Ivie) (Agelenidae) was sampled well using the shake-cloth and leaf sampling. *Trachelas pacificus* (Chamberlin and Ivie) (Clubionidae) was effectively sampled only by the shake-cloth method. *Cheiracanthium inclusum* (Hentz) (Clubionidae) was sampled reasonably well by both shake-cloth and leaf sampling. Although the shake-cloth method only removed approximately 50% of this species, the relative sampling efficiency was consistent. A determination of each method's effectiveness in sampling jumping spiders (Salticidae) was inconclusive because they were not collected efficiently in the strip samples and populations were low in all vineyards. However, it appears that all three methods are capable of detecting a consistent percentage of the population. Lynx spiders (Oxyopidae) were not always sampled well by the strip sample method. The shake-cloth and D-vac seemed to be useful sampling devices for this group. The capture of both salticid and oxyopid spiders at higher levels by the shake-cloth and D-vac methods indicates a weakness in capturing these families with the strip sample method. The characteristic jumping behavior of both salticid and oxyopid spiders may have allowed them to escape from the strip sample while the foliage was being removed. Sampling of the small *Theridion* species (Theridiidae) was also inconclusive because of low densities in most vineyards evaluated. However, this group appears to be sampled best by the leaf sample method.

In addition, orb weaver spiders (principally Araneidae) were seldom collected except by the D-vac. Population estimates using the D-vac were 0.35 and 0.51 spiders per 100 leaves in the two vineyards where they were visually apparent. All other methods (including the strip method) found them detectable at only low densities.

In summary, this study indicated that no sample method worked well for all vineyard spider species. Many species of hunting spiders including the funnel-web weaver *H. nedra* were sampled efficiently by the shake-cloth method. *Theridion* cobweb species were most efficiently sampled by leaf sampling. Orb weavers appeared to be best sampled by D-vac sampling. A visual web count could be a useful alternative for sampling orb weavers, although we have not evaluated this method. A similar sampling study was conducted by Costello and Daane (1997). They sampled a single vineyard repeatedly from May to September and included an additional drop-cloth method. In contrast, we included numerous vineyards each sampled once during mid-summer, at a time when each species was well represented by adults. Overall findings of the two studies were similar. While the selection of a sampling technique for *Theridion* spp. was inconclusive in our study, largely due to low population densities, Costello and

Table 1. Percent catch of each of three sampling techniques relative to strip samples.

	Vineyards Sampled						
	WF	MAI	MAS	SM2	SM3	RIF	Mean %
AGELENIDAE *Hololena nedra*							
Shake-cloth	81	112	200[b]	—	100[b]	—	123
D-Vac	10	4	—	—	85[b]	—	33
Leaf Sample	132	205	—	—	—	—	
Strip Sample Mean[a]	3.4	3.7	0.2	0	0.1	0	
CLUBIONIDAE *Trachelas pacificus*							
Shake-cloth	83	71	300	143	—	—	149
D-Vac	—	2	—	122	—	—	
Leaf Sample	—	—	—	—	—	—	
Strip Sample Mean	1.1	1.6	0.6	0.2	0.05	0.05	
Cheiracanthium inclusum							
Shake-cloth	38	50	58	19	57[b]	—	44
D-Vac	—	—	14	—	—	—	
Leaf Sample	110	107	—	275	600[b]	—	273
Strip Sample Mean	1.0	1.3	1.3	0.4	0.1	0	
SALTICIDAE							
Shake-cloth	—	170[b]	200[b]	115[b]	170[b]	—	164[b]
D-Vac	—	85[b]	176[b]	200[b]	1065[b]	—	382[b]
Leaf Sample	—	—	600[b]	300	700[b]	—	533[b]
Strip Sample Mean	0	0.1	0.1	0.2	0.1	0	
THERIDIIDAE *Theridion* spp.							
Shake-cloth	12[b]	8[b]	77[b]	23[b]	34[b]	—	31[b]
D-Vac	9[b]	40[b]	300[b]	75[b]	28[b]	—	90[b]
Leaf Sample	200[b]	—	10,100[b]	3,000	700[b]	—	3,500[b]
Strip Sample Mean	1.1[c]	0.1	0.1	0.2	0.1	0	
OXYOPIDAE *Oxyopes salticus* & *Oxyopes scalaris* Hentz							
Shake-cloth	86	45	—	766	258[b]	—	289
D-Vac	108	—	—	780	307[b]	300[b]	374
Leaf Sample	—	—	—	—	—	—	—
Strip Sample Mean	1.0	0.6	0	0.1	0.2	0.05	

— No spiders collected (i.e., 0 count).
[a] Mean no. spiders per 100 leaves.
[b] Based on very small spider counts.
[c] Most specimens (11 of 15) found in single sample.

Daane (1997) indicated that a shake-cloth (i.e., funnel) technique is effective. However, it should be noted that the small and delicate *Theridion* spp. can be easily destroyed and overlooked among the sample litter during sorting.

Vineyard Spider Populations and the Influence of Habitat Characteristics

Patterns of Seasonal Occurrence

During 1990, seven vineyard sites were surveyed every 2–4 weeks from May to October to determine species abundance and seasonal distribution of vineyard canopy spiders (Njokom 1991). In 1991, all but one of the 1990 sites were surveyed from June to September at monthly intervals. The shake-cloth method was used as the method of sampling in both years. With the exception of one vineyard site (CSC), all were restricted to the use of "soft" pesticides such as soaps, *Bacillus thuringiensis* products, and cryolite. The CSC site received an application of dimethoate during June of 1990. This site was a raisin vineyard and within 300m of a table grape vineyard (CSM). In contrast to 1990, only soft insecticides were used on the former site in 1991.

Spider populations increased rapidly from late spring to early summer (Figure 2). From late June until well into September, population densities remained relatively high (approximately 5–15 spiders per shake-cloth) in all sites except at the CSC site where dimethoate had been used. Total spider densities were compared in each of seven vineyard sites in 1991 versus 1990. The average population density during the mid to late summer period in 1991 was correlated (r = 0.73) with the 1990 spider densities obtained during a similar time period (Figure 3). Several points are illustrated in Figure 3 by the linear relationship of spider densities between years. Spider densities declined from 1990 to 1991 in all but one vineyard. During the period of peak spider abundance among vineyards, densities in 1991 were approximately 30%–50% of those during 1990. The reasons for these differences are unknown, but the colder winter and summer of 1991 may have reduced spider densities.

Only one study site was treated with dimethoate and only in 1990 (Figure 3), which may have suppressed spider populations in this vineyard to very low levels. It is suggested that the change in insecticide usage at this site from 1990 to 1991 reversed the spider density pattern between years relative to that found in other vineyards surveyed, including the adjacent table grape vineyard.

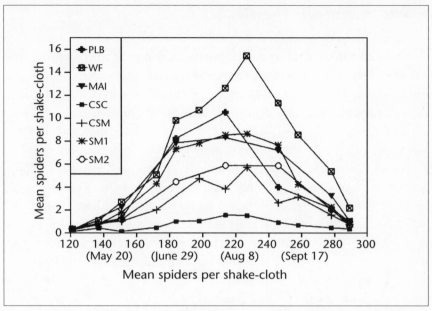

Figure 2. Mean spider counts per shake-cloth in seven vineyards during 1990. The legend reflects codes for the various vineyards.

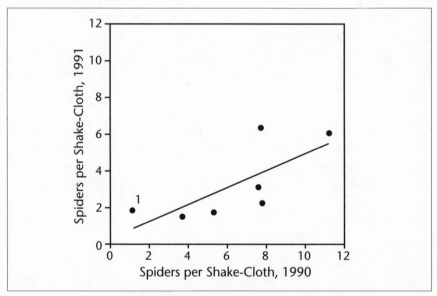

Figure 3. Comparison of relative spider densities in 1991 versus 1990. Values represent mean densities during mid-summer. [1] Treated with dimethoate in June of 1990. Correlation coefficient (r)=0.73.

Species Assemblage in Vineyards

Using the shake-cloth method, we evaluated spider species composition for 11 vineyard sites, including data from the sample method comparison and surveys. Three vineyard sites (SM1, SM2, and SM3) were part of one vineyard, and two sites were in another vineyard system (CSC and CSM). All other sites were widely distributed throughout Fresno County.

During late July 1991, when spider species abundance and diversity were relatively stable, most of the canopy-dwelling spiders represented five families: Agelenidae (1 sp.), Clubionidae (2 spp.), Oxyopidae (2 spp.), Salticidae (4 or more spp.), and Theridiidae (2 or more *Theridion* spp.) (Figures 4 and 5), (Figure 6). Based on the sampling efficiency study results, the proportion of *C. inclusum* in these vineyards is underestimated by approximately 50%. Because of inefficiencies in sampling *Theridion* spp. and orb weavers with the shake-cloth method, their relative abundance was probably several times greater than illustrated in Figure 5. Our results are similar to those of Costello and Daane (1995).

Among the vineyards sampled, SM1-3 vineyards sites were located adjacent to riparian habitat on the east side of the San Joaquin Valley. SM1 and SM3 were farther from the river, and more elevated, than SM2. SM1 differed from the other two table grape sites in that the variety was 'Thompson Seedless' (gable trellis), while the other two sites contained 'Ruby Seedless' (two line trellis). In all three sites, overall spider abundance was relatively low among the vineyards surveyed (Figure 4). Despite the proximity of the three sites (<800 m apart), there were differences in the spider fauna. Most notably, *H. nedra* was common in SM1 and undetectable at SM2. Oxyopid densities were low in SM1 and relatively high in SM2 and SM3. *T. pacificus* densities were similar among sites.

In contrast to the SM1-3 sites, the PLB vineyard was located on the west side of the San Joaquin Valley where vineyards are uncommon. Therefore it was relatively isolated from other vineyards. Its spider fauna featured almost exclusively and at high density the hunting spider, *T. pacificus*.

Canopy and Covercrop Spider Activity

A study was conducted in 1990 to survey the arthropod fauna in vineyard canopies and associated ground cover composed of naturally occurring resident plant species. The D-vac was used to sample the canopy as previously outlined. The ground cover was sampled by positioning the vacuum orifice straight down upon the vegetation.

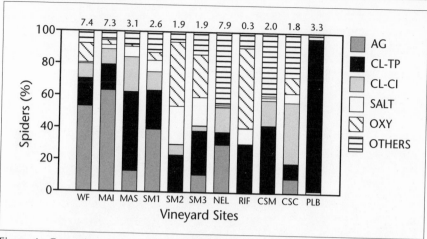

Figure 4. Grapevine canopy spider composition in Fresno County vineyards. Results are based on shake-cloth sampling during mid-summer of 1991. Values above upper X-axis represent mean number of total spiders per shake-cloth (N=30) for each vineyard at the time of sampling. AG: Agelenidae, *Hololena nedra*; CL-TP: Clubionidae, *Trachelas pacificus*; CL-CI: Clubionidae, *Cheiracanthium inclusum*; SALT: Salticidae; OXY: Oxyopidae, *Oxyopes salticus* & *O. scalaris*; OTHERS: other taxa (see Figure 5).

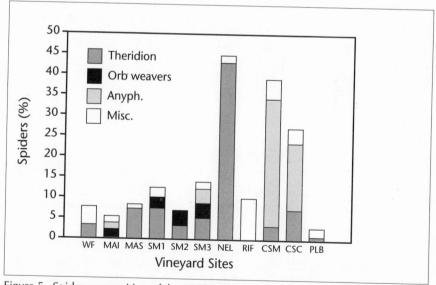

Figure 5. Spider composition of the "OTHERS" group presented in Figure 4. THERIDION: Theridiidae, *Theridion* spp.; Orb weavers: primarily Araneidae; Anyph: Anyphaenidae, *Aysha incursa*; Misc: miscellaneous species of low occurrence.

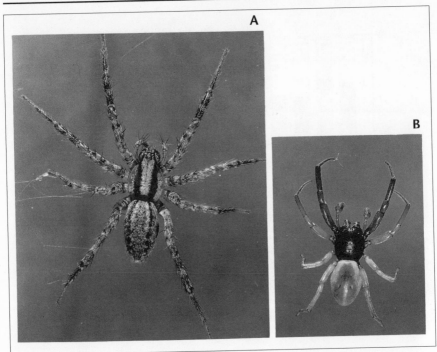

Figure 6. A) Funnel-web spider (technically a hunting spider), *Hololena nedra* (Chamberlin & Ivie) (Agelenidae). Female specimen, body length is 11 mm. Body coloration includes a tan background, reddish-brown markings throughout, and two longitudinal gray stripes on abdomen. Flat web with funnel retreat is located on outer canopy. B) Sac spider (hunting spider), *Trachelas pacificus* (Chamberlin & Ivie) (Clubionidae). Male specimen, body length is 7mm. Body coloration includes a dark reddish-brown carapace (consisting of head and thorax) and forelegs, and cream to light brown abdomen and hind legs.

Agelenids, clubionids, salticids, oxyopids and araneids that are common in the canopy also occur in ground cover habitats (Figure 7A–D and G). In contrast, linyphiids, which are noted for having extremely large populations in various habitats, have low populations in the canopy relative to what is found in the ground cover (Figure 7E). Lycosids were largely restricted to the ground cover (Figure 7F).

Although these data suggest a greater abundance in the ground cover for all species discussed, it is important to address several potentially important biases. The D-vac sampler apparently catches a disproportionate number of small spiders. Furthermore, it is unlikely that the D-vac has equal effectiveness in sampling in ground cover versus a canopy, which is devoid of a background surface. Because linyphiid spiders are very small even as adults, sample counts in the canopy and ground cover may closely approximate the

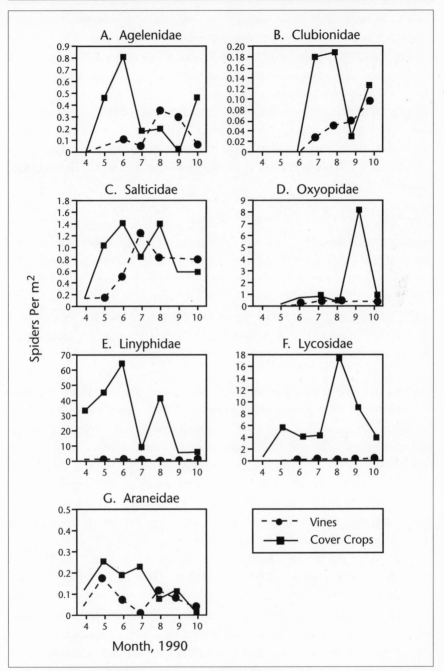

Figure 7. Relative comparison of spider population seasonal trends in vineyard canopies and associated cover crops in San Joaquin Valley, California 1990. D-vac sampler used in both canopy and cover crop.

relative abundance of this group in these two habitats. However, the relative efficiency of D-vac in the two zones may have been appreciably different for the remaining spider groups, which have a far greater body size as late instars and adults. As a result, sample counts may have overestimated densities in the ground vegetation relative to densities in the vine canopy. Nevertheless, these data suggest that population dynamics of spider species commonly found in the canopy may be directly affected by ground cover habitat. Because only young spiders were found in the canopy and ground cover samples, it is unknown if there is an age class–specific interaction between vegetational strata.

Influence of Vineyard Habitat on Spiders

A study was initiated in 1990 to evaluate the impact of a polyculture cover crop on the vineyard arthropod community. The insect pest of greatest concern was the variegated leafhopper (*Erythroneura variabilis* Beamer, Homoptera: Cicadellidae), a key pest in San Joaquin Valley vineyards. The study site was a 35-year-old 'Thompson Seedless' vineyard located near Madera, California. This vineyard had not been sprayed for leafhoppers during the previous two seasons. The cover-crop was composed of a 40:40:20 (by weight of seeds) mixture of common vetch (*Vicia sativa*), purple vetch (*Vicia benghalensis*) and oat (*Avena sativa*). The cover crop was planted in alternating row-middles on 15 November 1989. During the following spring, all but four cover-cropped row-middles were disced and incorporated into the soil in the first week of March 1990. The remaining four cover-cropped row-middles were approximately 300m in length and separated by 15 rows of vines. They were used as experimental plots for determining the effect of cover crops on pest and beneficial arthropods. The cover crop was mowed shortly before bud break to delay maturity. These methods were followed during the 1990–91 season, as well.

Arthropod populations were sampled in the cover crop from 19 April until when it was mature on 21 June, and in the vine canopy from 4 April to 12 September 1990. Sampling was conducted in plots located toward the center of each cover crop middle. Sweep-net sampling was used to evaluate arthropods in the cover. In the vine canopy next to the cover crop, and in rows of vines midway between those adjacent to cover-cropped middles (7–8 rows away), arthropods were sampled by direct observation of individual leaves and with sticky traps.

Spiders were the predominant generalist natural enemy component. The agelenid (*H. nedra*) and theridiid (*Theridion*) spiders were more abundant

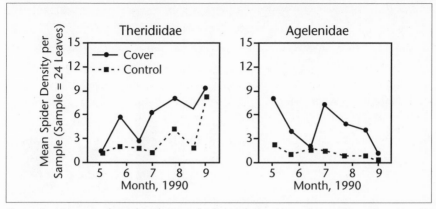

Figure 8. Mean number of spiders per leaf sample in vineyard canopies adjacent (COVER) and distant (CONTROL) to cover crop, 1990.

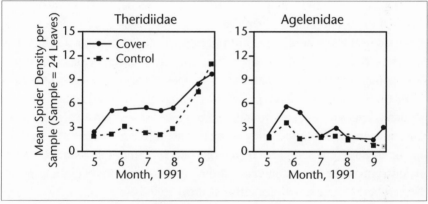

Figure 9. Mean number of spiders per leaf sample in vineyard canopies adjacent (COVER) and distant (CONTROL) to cover crop, 1991.

during 1990 in the vine canopy next to the cover crop than in those vine-row canopies that were distant from the cover crop (Figures 8 and 9). Similar relationships were found for the *Theridion* spp. during 1991.

Densities of variegated leafhopper showed a corresponding but inverse relationship (Figures 10 and 11). In 1990, the second and third generation leafhopper populations were larger in locations distant from the cover crop, whereas in 1991 a similar relationship occurred only in the second generation population. The 1991 season was particularly cool, and could have contributed to lower abundance in the third generation. In this study, agelenid densities fluctuated more than was indicated by vineyard survey data (summarized earlier), which illustrated a steady increase of total spider

Figure 10. Mean number of variegated leafhopper nymphs per leaf in
canopies adjacent (COVER) and
distant (CONTROL) to cover crop
for each of three generations, 1990.
Samples were taken during two
consecutive weeks during peak
abundance for each generation.
Standard error represented by vertical lines

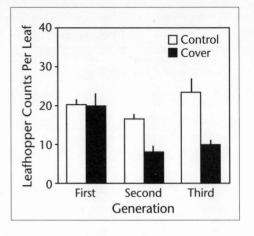

densities through spring and early summer. The late June decline of
agelenids in the vine canopy next to cover crops suggests an interaction.
From mid to late June, cover crop senescence was accompanied by an
increase in the agelenid population in the cover crop prior to discing. This
suggests that agelenids were dispersing from vines to senescent cover crops.
The population was not identified to species because it was composed of
immatures at this time. The species was probably *H. nedra* because no other
agelenid species has been found in any of these vineyards to date. Adult
agelenids in San Joaquin Valley vineyards are rarely found in early summer,
being absent from samples taken from late June through mid September.
Adults appear be most abundant in the vineyard canopy beginning in late
September, as was shown for citrus (Carrol 1980).

Additional Evidence for a Connection
Between Spiders and Leafhopper Abundance

As previously noted, the 1990 survey included one vineyard site that received an early summer application of dimethoate insecticide. Although
leafhopper densities were similar during late spring in the dimethoate treated
(CSC) and adjacent (CSM) sites, late summer densities were dramatically
different (Figure 12). In fact, leafhopper abundance in all sites except for
the CSC vineyard remained relatively low during late summer. Conversely,
spider abundance was lowest in the CSC vineyard throughout most of the
season (Figure 2).

Overall, these survey findings in combination with results from the
cover crop study suggest a direct relationship among spider abundance,
variegated leafhopper abundance, and cover-cropped vineyards. While the

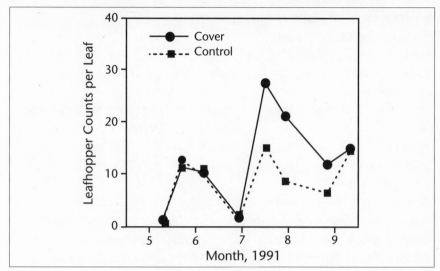

Figure 11. Mean number of variegated leafhopper nymphs per leaf in canopies adjacent (COVER) and distant (CONTROL) to cover crop throughout the season, 1991.

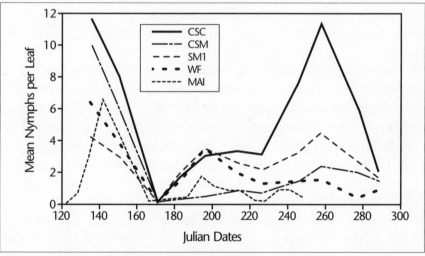

Figure 12. Mean number of variegated leafhopper nymphs per leaf in vineyards surveyed in 1990. CSC vineyard was treated with dimethoate in June of 1990.

potential importance of other biological control agents on leafhopper abundance cannot be dismissed, spiders were by far the most apparent and consistent predators found. Although the combined findings in the cover crop study and survey are not a substitute for detailed life-table analysis, they strongly suggest the important ecological roles played by spiders in vineyard agroecosystems.

A Perspective on Habitat Management in
Woody Perennial Systems

When understory vegetation is present, woody perennial systems such as grape vineyards can provide a highly complex habitat containing multiple strata. The cover crop stratum in vineyards contains spider species found in the vine canopy. Therefore, a continuum of species activity may occur between the vines and other vegetation in vineyards. This suggests that the spider fauna in the vineyard canopy stratum has population-sustaining linkages with other strata.

Cover crop study results suggested that the agelenids moved to the cover crop during June. Additional observations of agelenids in another vineyard strongly suggest the positive potential importance of interactions between spiders and strata. In that vineyard, the area immediately below the vine row (i.e., berm) contained an undisturbed grass cover that grows in the spring and remains dormant throughout the remainder of the year. At various times during the year, many agelenids have webs within the grassy berm. Webs are also found on gnarled vine trunks and constructed between trunks and vine stakes. The importance of these microhabitats in the context of spider survivorship and pest management is unknown.

Another field observation underscores the importance of timely habitat availability, and suggests that understory habitat elimination is not necessarily destructive to spider populations when properly timed. During 1991, the diverse vegetative understory present in the WF vineyard was clean cultivated in late May. Nevertheless, the spider density was among the greatest for vineyards surveyed during 1990 as well as 1991 (Figures 2 and 4). Furthermore, species composition and densities were comparable to those found in the MAI vineyard, which had undisturbed berms containing resident grasses that became dormant during early summer. The key to preserving the spider community may have been the result of having a well developed grapevine canopy present when the cover crop was eliminated by plowing. Therefore, it is hypothesized that a sink (i.e., the canopy) was available for spiders that would have been otherwise lost had plowing occurred earlier in the spring. This also suggests that water-consuming cover crops need not be present at all times to be of value in pest management.

Last, we have observed that corrugated cardboard bands placed around the trunks of grapevines and walnut trees during winter often contain numerous spiders. This reflects the importance of suitable overwintering habitat, as shown for spider populations in corn (*Zea mayze*) residue in a cultivated annual cropping system (Plagens & Whitcomb 1986). Overwintering habitat may be particularly scarce when vines and trees are young and have smooth bark.

Woody perennial crops afford the opportunity to meet the needs of the spider community in space and time. This could include timely manipulation of the understory vegetation on berms and between rows. Such an approach would support spider populations during the growing season and provide overwintering habitat to ensure carryover to the subsequent year.

Summary and Future Research Needs

The spider assemblages in San Joaquin Valley vineyard canopies are dominated by species from seven families. Two clubionid species (*T. pacificus* and *C. inclusum*) were common in the majority of vineyards using "soft" pesticides. Agelenid and oxyopid spiders are very common in approximately 40% of such vineyards. Because oxyopid populations consisted primarily of 1–3mm juveniles through the entire season, the family appears to represent a highly ephemeral component. Its potential value as a natural enemy in vineyards seems questionable despite reputed importance in many other agroecosystems (Young & Lockley 1985). By contrast, other commonly found species in vineyard canopies demonstrate the presence of a progressively maturing population through the course of the summer and fall. The salticids and orb weavers were numerous in a relatively small portion of vineyards throughout the region. The regional occurrence of orb weavers and species of *Theridion* in vineyard canopies has not been well established, but they can be very abundant. Most sampling methods do a poor job of sampling araneids while *Theridion* spp. are perhaps best sampled by direct observation of grape leaves.

Future studies should concern the effects of pesticides, cover cropping, and overwintering habitat on seasonal and long-term population dynamics of the principal spider species in vineyards. Little is known about their biology and ecology; for instance, immigration and emigration, prey selection, overwintering, and the influence of temperature.

Many questions need to be answered. What species are good colonizers, and which are long-term residents requiring maintenance throughout the year? What role do cover crops play in maintaining the predominant canopy spiders? What is the collective impact of spiders on the vineyard insect pest community? To what extent are other natural enemies affected by spiders and vice versa? Does feeding during the cold winter period occur and is it of significance to pest populations (Aitchison 1984)? Answers to these and other ecological questions will provide information necessary to determine the potential value of spiders as biological control agents and are required for developing effective habitat management recommendations.

Acknowledgments

We thank Marjorie Moody for her invaluable help in identifying numerous specimens for our reference collection. The diligent assistance of Curtis Sisk and Mbah Njokom was also greatly appreciated. We also thank Jack Kelly Clark for photographs, and personnel at CSU–Fresno California Agricultural Technology Institute for the shake-cloth sampler illustration.

References

Agnew, C. W. & J. W. Smith, Jr. 1989. Ecology of spiders (Araneae) in a peanut agroecosystem. Environ. Entomol. 18:30-42.

Aitchison, C. W. 1984. Low temperature feeding by winter-active spiders. J. Arachnol. 12:297-305.

Ali, A. D. & T. E. Reagan. 1986. Influence of selected weed control practices on araneid faunal composition and abundance in sugarcane. Environ. Entomol. 15:527-531.

Altieri, M. A. & L. L. Schmidt. 1985. Cover crop manipulation in northern California orchards and vineyards: effects on arthropod communities. Biol. Agric. Hortic. 3:1-24.

Bishop, L. & S. E. Riechert. 1990. Spider colonization of agroecosystems: mode and source. Environ. Entomol. 19:1738-1745.

Bostanian, N. J., C. D. Dondale, M. R. Binns & D. Pitre. 1984. Effects of pesticide use on spiders (Araneae) in Quebec apple orchards. Can. Entomol. 116:663-675.

Breene, R. G., W. L. Sterling & D. A. Dean. 1988. Spider and ant predators of the cotton fleahopper on wooly croton. Southw. Entomol. 13:177-183.

Carroll, D. P. 1980. Biological notes on the spiders of some citrus groves in central and southern California. Entomol. News 91:147-154.

Chapman, R. N. 1928. The quantitative analysis of environmental factors. Ecology 9:111-122.

Corrigan, J.E. & R.G. Bennett. 1987. Predation by *Cheiracanthium mildei* (Araneae, Clubionidae) on larval Phyllonorycter blancardella (Lepidoptera, Gracillariidae) in a greenhouse. J. Arachnol. 15:132-134.

Costello, M.J. & K.M. Daane. 1995. Spider (Araneae) species composition and seasonal abundance in San Joaquin Valley grape vineyards. Environ. Entomol. 24:823-831.

_____ 1997. Comparison of sampling methods used to estimate spider (Araneae) species abundance and composition in grape vineyards. Environ. Entomol. 26:142-149.

Culin, J. D. & R. W. Rust. 1980. Comparison of the ground surface and foliage dwelling spider communities in a soybean habitat. Environ. Entomol. 9:577-582.

Culin, J. D. & K. V. Yeargan. 1983a. Comparative study of spider communities in alfalfa and soybean ecosystems: foliage-dwelling spiders. Ann. Entomol. Soc. Am. 76:825-831.

Culin, J. D. & K. V. Yeargan. 1983b. Comparative study of spider communities in alfalfa and soybean ecosystems: ground-surface spiders. Ann. Entomol. Soc. Am. 76:832-838.

Dean, D. A., W. L. Sterling & N. V. Horner. 1982. Spiders in eastern Texas cotton fields. J. Arachnol. 10:251-260.

Dean, D. A. & W. L. Sterling. 1985. Size and phenology of ballooning spiders at two locations in eastern Texas. J. Arachnol. 13:111-120.

Dean, D. A. & W. L. Sterling. 1990. Seasonal patterns of spiders captured in suction traps in eastern Texas. Southw. Entomol. 15:399-412.

Dean, D. A., W. L. Sterling, M. Nyffeler & R. G. Breene. 1987. Foraging by selected spider predators on the cotton fleahopper and other prey. Southw. Entomol. 12:263-270.

Dietrick, E. J., E. I. Schlinger & R. van den Bosch. 1959. A new method for sampling arthropods using a suction collecting machine and modified Berlese funnel separator. J. Econ. Entomol. 52:1085-1091.

Gertsch, W. J. 1979. American spiders. Van Nostrand Reinhold Co., New York, N.Y.

Greenstone, M. H., C. E. Morgan & A. Hultsch. 1987. Ballooning spiders in Missouri, USA, and New South Wales, Australia: family and mass distributions. J. Arachnol. 15:163-170.

Herne, D. H. C. & W. L. Putman. 1966. Toxicity of some pesticides to predacious arthropods in Ontario peach orchards. Can. Entomol. 98:936-942.

Herzog, D. C. & J. E. Funderbrk. 1986. Ecological bases for habitat management and pest cultural control, pp. 217-250. *In* M. Kogan [ed.], Ecological theory and integrated pest management practice. Wiley & Sons. New York.

Hilburn, D. J. & D. T. Jennings. 1988. Terricolous spiders (Araneae) of insecticide-treated spruce-fir forests in west-central Maine. Great Lakes Entomol. 21:105-114.

Howell, J. O. & R. L. Pienkowski. 1971. Spider populations in alfalfa, with notes on spider prey and effect of harvest. J. Econ. Entomol. 64:163-168.

Ito, Y., K. Miyashita & K. Sekiguchi. 1962. Studies on the predators of the rice crop insect pests, using the insecticidal check method. Japanese J. Ecol. 12:1-11.

Kiritani, K., N. Hokyo, T. Sasaba & F. Nakasuji. 1970. Studies on population dynamics of the green rice leafhopper, *Nephotettix cincticeps* Uhler: regulatory mechanism of the population density. Res. Popul. Ecol. 12:137-153.

Kiritani, K., S. Kawahara. 1973. Food-chain toxicity of granular formulations of insecticides to a predator, *Lycosa pseudoannulata,* of *Nephotettix cincticeps.* Botyu-Kagaku 38:69-75.

Kiritani, K., S. Kawahara, T. Sasaba & F. Nakasuji. 1972. Quantitative evaluation of predation by spiders on the green rice leafhopper, *Nephotettix cincticeps* Uhler, by a sight-count method. Res. Popul. Ecol. 13:187-200.

Knutson, A.E. & F.E. Gilstrap. 1989. Direct evaluation of natural enemies of the southwestern corn borer (Lepidoptera: Pyralidae) in Texas corn. Environ. Entomol. 18:732-739.

Laster, M. L. & J. R. Brazzel. 1968. A comparison of predator populations in cotton under different control programs in Mississippi. J. Econ. Entomol. 61:714-719.

Leigh, T. F. & R. E. Hunter. 1969. Predacious spiders in California cotton. Calif. Agric. 23:4-5.

LeSar, C. D. & J. D. Unzicker. 1978. Soybean spiders: species composition population densities, and vertical distribution. Ill. Nat. Hist. Surv. Biol. Notes No. 107.

Letourneau, D. K. & M. A. Altieri. 1983. Abundance patterns of a predator, *Orius tristicolor* (Hemiptera: Anthocoridae), and its prey, *Frankliniella occidentalis* (Thysanoptera: Thripidae): habitat attraction in polycultures versus monocultures. Environ. Entomol. 12:1464-1469.

Luczak, J. 1979. Spiders in agrocoenoses. Pol. Ecol. Stud. 5:151-200.

Mansour, F. 1984. A malathion-tolerant strain of the spider *Cheiracanthium mildei* and its response to chlorpyrifos. Phytoparasitica 12:163-166.

Mansour, F., D. Rosen, H. N. Plaut & A. Shulov. 1981. Effect of commonly used pesticides on *Cheiracanthium mildei* and other spiders occurring on apple. Phytoparasitica 9:139-144.

Mansour, F., D. B. Richman & W. H. Whitcomb. 1983. Spider management in agroecosystems: habitat manipulation. Environ. Mgmt. 7:43-49.

Mansour, F. & W. H. Whitcomb. 1986. The spiders of a citrus grove in Israel and their role as biocontrol agents of *Ceroplastes floridensis* (Homoptera: Coccidae). Entomophaga 31:269-276.

Mansour, F. 1987. Effect of pesticides on spiders occurring on apple and citrus in Israel. Phytoparasitica 15:43-50.

McCaffrey, J. P., M. P. Parrella & R. L. Horsburgh. 1984. Evaluation of the limb-beating sampling method for estimating spider (Araneae) populations on apple trees. J. Arachnol. 11:363-368.

McDaniel, S. G., W. L. Sterling & D. A. Dean. 1981. Predators of tobacco budworm larvae in Texas cotton. Southw. Entomol. 6:102-108.

Miyashita, K. 1968. Quantitative feeding biology of *Lycosa T-insignita* Boes. et Str. (Araneae: Lycosidae). Bull. Nat. Instit. Agric. Sci. 22:329-344.

Muniappan, R. & H. L. Chada. 1970. Biology of the crab spider, *Misumenops celer.* Ann. Entomol. Soc. Am. 63:1718-1722.

Nentwig, W. 1988. Augmentation of beneficial arthropods by strip-management. I. Succession of predacious arthropods and long-term change in the ratio of phytophagous and predacious arthropods in a meadow. Oecologia 76:597-606.

Nentwig, W. 1989. Augmentation of beneficial arthropods by strip-management II. Successional strips in a winter wheat field. J. Plant Dis. Protection. 96:89-99.

Njokom, M. S. 1991. Abundance, distribution and ecological role of spiders as natural enemies in grape agroecosystems. M.S. Thesis, Calif. State Univ., Fresno, CA.

Nyffeler, M., D. A. Dean & W. L. Sterling. 1986. Feeding habits of the spiders *Cyclosa turbinata* (Walckenaer) and *Lycosa rabida* Walckenauer. Southw. Entomol. 11:195-201.

1987a. Feeding ecology of the orb-weaving spider *Argiope aurantia* [Araneae: Araneidae] in a cotton agroecosystem. Entomophaga 32:367-375.

1987b. Evaluation of the importance of the striped lynx spider, *Oxyopes salticus* (Araneae: Oxyopidae), as a predator in Texas cotton. Environ. Entomol. 16:1114-1123.

1988. Prey records of the web-building spiders *Dictyna segregata* (Dictynidae), *Theridion australe* (Theridiidae), *Tidarren haemorrhoidale* (Theridiidae), and *Frontinella pyramitela* (Linyphiidae) in a cotton agroecosystem. Southw. Entomol. 33:215-218.

1989. Prey selection and predatory importance of orb-weaving spiders (Araneae: Araneidae, Uloboridae) in Texas cotton. Environ. Entomol. 18:373-380.

Nyffeler, M., W. L. Sterling, & D. A. Dean. 1994. How spiders make a living. Environ. Entomol. 23:1357-1367.

Oraze, M. J. & A. A. Grigarick. 1989. Biological control of aster leafhopper (Homoptera: Cicadellidae) and midges (Diptera: Chironomidae) by *Pardosa ramulosa* (Araneae: Lycosidae) in California rice fields. J. Econ. Entomol. 82:745-749.

Plagens, M. J. 1986. Aerial dispersal of spiders (Araneae) in a Florida cornfield ecosystem. Environ. Entomol. 15:1225-1233.

Plagens, M. J. & W. H. Whitcomb. 1986. Corn residue as an overwintering site for spiders and predaceous insects in Florida. Florida Entomol. 69:665-671.

Price, P. W. 1984. Insect ecology (2nd ed.). Wiley & Sons. New York, N.Y.

Putman, W. L. 1967. Prevalence of spiders and their importance as predators in Ontario peach orchards. Can. Entomol. 99:160-170.

Redmond, K. R. & J. R. Brazzel. 1968. Response of the striped lynx spider, *Oxyopes salticus,* to two commonly used pesticides. J. Econ. Entomol. 61:327-328.

Riechert, S. E. & T. Lockley. 1984. Spiders as biological control agents. Annu. Rev. Entomol. 29:299-320.

Riechert, S. E. & L. Bishop. 1990. Prey control by an assemblage of generalist predators: spiders in garden test systems. Ecology 71:1441-1450.

Settle, W. H. & L. T. Wilson, D. L. Flaherty & G. M. English-Loeb. 1986. The variegated leafhopper, an increasing pest of grapes. Calif. Agric. 40:30-32.

Shepard, M. & W. Sterling. 1972. Effects of early season applications of insecticides on beneficial insects and spiders in cotton. Tex. Agric. Exp. Stn. Misc. Publ. 1045.

Tolbert, W. W. 1977. Aerial dispersal behavior of two orb weaving spiders. Psyche 84:13-27.

Turnbull, A. L. 1960. The spider population of a stand of oak (*Quercus robus* L.) in Wytham Wood, Berks., England. Can. Entomol. 92:110-124.

Turnbull, A. L. 1973. Ecology of the true spiders (Araneomorphae). Annu. Rev. Entomol. 305-348.

Whitcomb W. H. 1980. Sampling spiders in soybean fields, pp. 544-558. *In* M. Kogan & D.C. Herzog [eds.], Sampling methods in soybean entomology. Springer-Verlag, New York, N.Y.

Whitford, F., W. B. Showers & G. B. Edwards. 1987. Insecticide tolerance of ground- and foliage-dwelling spiders (Araneae) in European corn borer (Lepidoptera: Pyralidae) action sites. Environ. Entomol. 16:779-785.

Young, O. P. & T. C. Lockley. 1985. The striped lynx spider, *Oxyopes salticus* (Araneae: Oxyopidae), in agroecosystems. Entomophaga 30:329-346.

Young, O. P. & G. B. Edwards. 1990. Spiders in United States field crops and their potential effect on crop pests. J. Arachnol. 18:1-27.

Farmscaping in California: Managing Hedgerows, Roadside and Wetland Plantings, and Wild Plants for Biointensive Pest Management

ROBERT L. BUGG—University of California Sustainable Agriculture Research & Education Program, University of California, One Shields Avenue Davis, CA 95616-8716

JOHN H. ANDERSON—Hedgerow Farms, 21740 County Road 88 Winters, CA 95694

CRAIG D. THOMSEN— Department of Agronomy and Range Science, University of California, One Shields Avenue, Davis, CA 95616

JEFF CHANDLER—Cornflower Farms, P.O. Box 896, Elk Grove, CA 95759

Introduction

Biointensive integrated pest management relies mainly on biological and cultural controls (Frisbie & Smith 1989) to attain what might be termed "agroecosystem resistance" to pests. Permanent plantings along the borders of fields, orchards, and vineyards can play a major role in this, because they can enhance both biological and cultural pest control. Techniques for developing such plantings may be viewed in the context of farmscaping. Farmscaping is the modification of agricultural settings, including management of cover crops, field margins, hedgerows, windbreaks, and specific vegetation growing along roadsides, catchments, watercourses, and adjoining wildlands. The goal is usually more sustainable, ecologically based agricultural systems. This may be achieved by improving esthetics, controlling erosion, reducing wind, providing weed control through competition, and by affording habitat for wildlife (Mineau & McLaughlin 1996), including non-pest butterflies (Holl 1995; Loertscher et al. 1995), solitary bees (Kevan et al. 1990; Buchmann & Nabhan 1996), and arthropods that attack agricultural pests.

Various aspects of farmscaping have long been in use. However, in most cases, plantings have not been designed with arthropods in mind. Thus, any contribution to biologically intensive integrated pest management has been left to chance. Based on many years of practical and research experience, we have developed designs using native and introduced trees, shrubs, perennial grasses, and herbaceous perennial and annual broadleaf species to attract and sustain beneficial arthropods and other desirable wildlife.

For discussion of entomological issues surrounding cover crops, we refer readers to reviews by Bugg (1991, 1992) and Bugg and Waddington (1994). We herein present a philosophical and historical overview of various farmscaping features, including hedgerows, roadside plantings, plantings for catchments and watercourses, and wild plants. We then present practical management issues in farmscaping. Finally, we discuss specific insect associations with non-crop vegetation that have implications for California farms.

The Elements of Farmscaping:
Historical and Philosophical Overview

Hedgerows

A hedgerow is usually thought of as a row of shrubs or trees. Since ancient times, hedgerows have played key roles in agriculture. In ancient Europe and the British Isles, hedgerows marked property lines, sheltered farmlands and dwellings from wind, and provided people with food (game animals, hazelnut, walnut, quince) and fodder (acorns). Coppicing tree species were repeatedly harvested for both structural and fuel wood (Pott 1989). Windbreaks are plantings or structures designed to reduce undesirable effects of wind. Tree windbreaks are hedgerows of special types. During the 1930s and 1940s, the United States Department of Agriculture Soil Conservation Service (S.C.S.—now termed the Natural Resource Conservation Service [N.R.C.S.]) supervised the planting of thousands of acres of windbreaks throughout the midwestern states, to reduce wind erosion (Van Eimern 1964). This program emphasized shelterbelts, which are broad, multi-rowed, multi-tiered windbreaks.

California agriculture has long used both exotic and native trees for windbreaks in areas where wind is a persistent problem. Coastal and desert areas have especially benefited from wind shelter provided by Arizona cypress (*Cupressus glabra*), athel (*Tamarix aphylla*), blue gum (*Eucalyptus globulus*), Lombardy poplar (*Populus nigra* var. *italica*), Monterey cypress (*Cupressus macrocarpa*), and Siberian elm (*Ulmus pumila*). However, these species were used in single-species windbreaks, without considering value to beneficial arthropods or other desirable wildlife.

Relatively few of the dust bowl–era shelterbelts and the prehistoric hedgerows remain. Their loss was due in part to the push in the U.S. and in Europe during the 1960s and 1970s to plant arable crops from fencerow to fencerow in order to "feed the world." The trend towards "clean farming" entails indiscriminate herbiciding or disking of non-cropped ground in order to avoid weed problems, and use of so-called selective herbicides once crop seedlings have emerged. Aerial application of broad-spectrum herbicides, an imprecise technique that often damages adjoining crops and other non-target vegetation, continues unabated in most of the United States. This practice has been banned in some countries. In California, paradoxically, continual disturbance of agricultural field margins by disking, blading, and herbicides promotes colonization by invasive introduced weeds. The weeds that are favored by these conditions are typically worse pests of agriculture than the native plants displaced. This puts the farmer on an "herbicide treadmill."

Together, these agricultural techniques and trends have led to a great loss of non-invasive wild, and often native, vegetation that once persisted in roadsides and along sloughs. As a result, biodiversity has been greatly reduced on farmlands. However, recent changes in the marketplace and reduced federal crop subsidies have lessened the collective urge to farm and otherwise disturb marginal lands. Furthermore, there is an increasing awareness of the need to preserve and restore native vegetation and wildlife habitat.

To partially offset the loss of biological diversity, we suggest border plantings. We believe that hedgerows and windbreaks can be diversified to enhance beneficial arthropod activity. This could help counteract the habitat fragmentation commonly encountered in agricultural landscapes, which otherwise may interfere with biological control (Kruess & Tscharntke 1994). These plantings provide habitat for other desirable wildlife, including amphibians, reptiles, birds, and mammals. There is a need for research to document the wide range of possible outcomes, which may be positive, negative, or neutral with respect to production agriculture (see reviews by Van Eimern 1964; Van Emden 1965; Altieri and Letourneau 1982). We believe that through proper planning and management, the rural landscape will be beautified by adding native trees, shrubs, and herbs, and that restoration ecology and production agriculture can dovetail through ecologically based farming.

Our beliefs are based both on the scientific literature and our extensive field background, including nearly 20 years of experience at Hedgerow Farms (owned by J.H. Anderson). This site near Winters, Yolo County,

Table 1. Birds observed at Hedgerow Farms (Yolo County Roads 27 and 88, between Winters and Esparto, Yolo County, California) (R. Jones and J.H. Anderson, personal communications).

Order	Common Names
Ciconiiformes	American Bittern, Great Egret, Snowy Egret, Green Heron, Great Blue Heron
Anseriformes	Mallard, Cinnamon Teal
Falconiformes	Turkey Vulture, White-tailed Kite, Northern Harrier, Golden Eagle, Red-tailed Hawk, Swainson's Hawk, Rough-legged Hawk, American Kestrel, Merlin, Prairie Falcon
Galliformes	Ring-necked Pheasant, California Quail
Charadriformes	Killdeer, Longbilled Curlew
Columbiformes	Mourning Dove, Rock Dove
Strigiformes	Great Horned Owl, Common Barn Owl, Burrowing Owl
Apodiformes	Anna's Hummingbird, Black-chinned Hummingbird
Picciformes	Northern Flicker, Redbreasted Sapsucker, Nuttal's Woodpecker, Downy Woodpecker, Acorn Woodpecker
Passeriformes	Western Kingbird, Ash-throated Flycatcher, Black Phoebe, Say's Phoebe, Horned Lark, Barn Swallow, Cliff Swallow, Northern Rough-winged Swallow, Scrub Jay, Yellow-billed Magpie, Common Raven, American Crow, Plain Titmouse, Bush-tit, Redbreasted Nuthatch, Bewick's Wren, House Wren, Ruby Crowned Kinglet, Hermit Thrush, Swainson's Thrush, American Robin, Western Bluebird, Mountain Bluebird, Northern Mockingbird, American Pipit, Cedar Waxwing, Loggerhead Shrike, European Starling, Warbling Vireo, Yellow-rumped Warbler, Yellow Warbler, Wilson's Warbler, Orange-crowned Warbler, Black-throated Grey Warbler, Townsend's Warbler, Hermit Warbler, Western Tanager, Black-headed Grosbeak, California Brown Towhee, Rufous-sided Towhee, Savannah Sparrow, Song Sparrow, Dark-eyed Junco, White-crowned Sparrow, Golden-crowned Sparrow, Fox Sparrow, Western Meadowlark, Red-winged Blackbird, Brewer's Blackbird, Brown-headed Cowbird, Northern Oriole, American Goldfinch, Lesser Goldfinch, House Finch, Purple Finch, House Sparrow

California, features roadside and hedgerow plantings that by a conservative estimate comprise 14 native and 9 introduced species of perennial grasses, 9 native and 6 introduced genera of shrubs, and 6 native and 5 introduced tree species. Many of these plants have strong associations with beneficial insects and other desirable wildlife. At least eighty-eight species of birds have been observed at the site (Table 1), reflecting the richness of the habitats provided. The knowledge developed at Hedgerow Farms is now being applied and adapted in many parts of California. Farmers throughout California are using multispecies hedgerows and other plantings to attract beneficial insects and other desirable wildlife, in attempts to promote biologically intensive integrated pest management.

Roadside Plantings

In much of California, conventional management of agricultural roadsides leads to domination by weeds. This occurs because weeds colonize gaps created by repeated blading and herbicide applications. Roadside plantings of native perennial grasses and associated species have been used in Iowa and other midwestern states to suppress weeds, reduce the need for herbiciding and other maintenance tools, and restore wildlife. Similar practices are now being explored in California (Stromberg & Kephart 1996). Established stands of California native perennial grasses reduce growth by some weeds (Bugg et al. 1997). Availability of seed for various native grasses is increasing, as is public interest in their restoration, providing opportunities for large-scale plantings. Complexes of native grasses and forbs can be established using either herbicide-intensive or non-herbicidal techniques, which are summarized in Table 2. Once these herbaceous plants are established in vigorous mixed stands, herbicides typically are used only for spot application. Rural roadsides typically contain several topographic zones or niches with varying environmental conditions and management requirements. Growth form, height, and environmental optima and tolerances of the various perennial grass species dictate the niches in which they are suitable (Figure 1, adapted from Bugg et al. 1997).

Plantings for Catchments and Watercourses

California agriculture owes its productivity in part to the ability of farmers to apply water where and when it is needed. Because most of the state's intensive production occurs in arid to semi-arid conditions, this ability is contingent on a vast and intricate network of reservoirs, canals, small catchments (structures for collecting or draining water), and ditches, as well as sophisticated irrigation systems. This infrastructure presents an array of moisture gradients and zones that give rise to a wide range of farmscaping opportunities. Moist zones vary in their seasonal water supply, their tendency to retain water once it is supplied, their associated soil types, and their management needs, but nearly all afford niches for wetland plantings, including moist prairies, marshes, riparian corridors, and seasonal or intermittent ponds. To some extent, these artificial wetlands mimic natural

Table 2. Weed-control procedures for enhancing establishment of California native perennial grasses in rights-of-way.

Objective	Year	Months	Herbicide Procedure[1]	Alternative Procedure
Reduce weed densities and seed reservoir prior to seedbed preparation	1–2 years before seeding	Feb.–March April–May	Apply an appropriate broad-spectrum contact herbicide.	Establish summer fallow by disking weeds before seed has matured. This reduces the weed seed bank; the practice can be used for one or more years prior to seeding a site with native perennial grasses. In the final year, it can simultaneously prepare a seed bed for fall seeding of native perennial grasses. Disking perpetuates the disturbed conditions under which weeds prosper. On some sites, scraping is better, because it will remove much of the weed seed reservoir. Mowing before seed has matured can also be effective against some species of weeds, but others, such as annual ryegrass and introduced annual wild barleys can produce seed heads low to the ground following mowing.
Prepare seed bed	1	Sept.–Jan.	…	Harrowing with disk to break up clods and/or other tillage implements to create seedbed of good tilth. This process also kills weeds.
Reduce weed population after seedbed preparation	1	Sept.–Jan.	After winter-annual weeds have germinated and sprouted in response to rain or preirrigation, apply glyphosate. This is after seed-bed preparation but prior to seeding native grasses.	Flame-kill emerged weed seedlings after rains or preirrigation, but prior to seeding native perennial grasses. Very shallow tillage (e.g., spike-tooth harrowing) can also be effective. Shallowness is essential to avoid bringing more weed seed to the surface.

Seeding	1	...	...	Drill or broadcast seed; if broadcast, incorporate to about 0.5 inch depth with spike-tooth harrow or other suitable implement. Relatively high seeding rates may lead to more rapid attainment of vegetative cover, and better weed suppression.
Mulching	1	Sept.–Jan.	...	Mulching with barley or oat straw reduces soil crusting and may improve establishment of native grasses.
Irrigation or rain	1	Sept.–Jan.	...	Prompt irrigation or rain after seeding promotes rapid development and establishment of grass seedlings.
Reduce weed population after seedbed preparation but prior to seeding native grasses. Winterannual weeds have germinated and sprouted in response to rain or pre-irrigation.	1	Sept.–Nov.	Apply an appropriate broad-spectrum contact herbicide.	Use propane flamer to kill emerged weed seedlings. Very shallow tillage (e.g., disking or spike-tooth harrowing) can also be effective. Shallowness is essential to avoid bringing more weed seed to the surface.
Reduce weed densities after seeding but before emergence of native grass seedlings	1	Sept.–Nov.	Apply an appropriate broad-spectrum contact herbicide.	Use propane flamer to kill cool-season annuals. Timing is critical, because any emerged native grasses will be killed, as well.
Reduce weed competition and densities during establishment of native grasses	1	Feb.–April	Apply an appropriate contact herbicide that is selective for broadleaf weeds.	Mowing, haying, or grazing by livestock can remove the taller-growing and (often) more-palatable winter annuals during late winter and early spring. This improves the competitive posture of the native perennial grasses.

Continued on following page.

Table 2. continued

Objective	Year	Months	Herbicide Procedure[1]	Alternative Procedure
Reduce weed competition, densities, and seed production during first season of native grass growth		March–mid-April	Wick application of an appropriate broad-spectrum contact herbicide can be made to selectively kill annual ryegrass, ripgut brome, wild oats, and other weedy annual grasses that are over-topping lower-growing native grasses. After April, natives are usually tall enough to be susceptible. Also, by this time, many weeds will have already gone to seed, so the value of the technique would be reduced.	High mowing (13 cm minimum) can be used to remove seed heads of winter-annual grasses, and thus reduce seed bank of these weeds in subsequent years. Most native grasses produce seed later, and can recover from high mowing at this time.
Reduce weed densities and seed production during first season of native grass growth		April–June	Apply an appropriate contact herbicide that is selective for broadleaf weeds.	If stands of natives are vigorous and producing seed, mowing or grazing of native perennial grasses should be avoided during this period, to allow seed maturation. By contrast, if introduced grasses are dominant and have set seed, hay can be made by mowing, raking, and baling. The hay can be fed to livestock. Thus, the weed seed and excessive residue will be removed, and native stands can be augmented by further seedings during the subsequent autumn.

Reduce warm-season grasses.	May–Sept.		Spot-application of an appropriate broad-spectrum contact herbicide can be used against warm-season annual and perennial grasses (e.g., bermuda grass, johnson grass, barnyard grasses). Dust on the plants reduces efficacy of herbicide: under dusty conditions, increase use of adjuvants.	Mechanical control of weeds through hoeing, weed whipping, mowing, or digging can be used to remove, or prevent reseeding by patches of warm-season annual and perennial weeds.
Inhibit germination of winter-annual weed seeds	Sept.–Nov.	2	Apply an appropriate contact herbicide that is selective for broadleaf weeds.	Seed native Californian annual and perennial forbs, particularly in areas where native perennial grass growth is sparse, or where control burns have removed leaf litter. Supplemental seeding with native perennial grass seed is also an option.
Selective control of remaining weed infestations	Feb.–Sept.	2	Spot-application of appropriate broad-spectrum or selective contact herbicides can be used against yellow starthistle (*Centaurea solstitialis*), prickly lettuce (*Lactuca serriola*), cheeseweeds (*Malva* spp.) and other weeds.	Mow as mentioned for the first year of establishment. Spot use of hoe or weed whip can remove patches of annual and perennial weeds.
Maintenance of mature stands of native grasses	Feb.–Sept.	3	Spot-application of appropriate broad-spectrum or selective contact herbicides can be made as needed.	Mow as mentioned for the first year of establishment. Spot use of hoe or weed whip can remove patches of annual and perennial weeds. Control burns may be used about once in 4 years to remove old residues and permit increased reproduction by annual forbs.

[1]Standard safety precautions should be observed in all use of herbicides. Restricted herbicides must be used in accordance with county, state, and federal regulations. For information regarding permits and usage, contact county agricultural commissioners' offices.

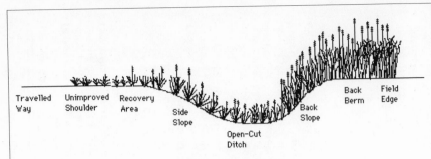

Native Grass Complex #1

Unimproved shoulder: Pine bluegrass (*Poa secunda* ssp. *secunda*)
Recovery area: Red fescue (*Festuca rubra*) and California brome
(*Bromus carinatus*).
Side slope: Meadow barley (*Hordeum brachyantherum* ssp.
brachyantherum).
Open-cut ditch: Meadow barley.
Back slope: Blue wildrye (*Elymus glaucus*).
Back berm and field edge: Creeping wildrye (*Leymus triticoides*).

Native Grass Complex #2

Unimproved shoulder: Pine bluegrass, California barley (*Hordeum
brachyantherum* spp. *californicum*), and Idaho fescue (*Festuca
idahoensis*).
Recovery area: Pine bluegrass, Idaho fescue, purple needlegrass
(*Nassella pulchra*), and squirreltail (*Elymus multisetus*).
Side slope: Meadow barley, Idaho fescue, purple needlegrass,
nodding needlegrass (*Nassella cernua*), California oniongrass (*Melica
californica*).
Open-cut ditch: Meadow barley, purple needlegrass, saltgrass (*Distichlis
spicata* var. *nana*) and spikerush (*Eleocharis* spp.).
Back slope: Blue wildrye, purple needlegrass, slender wheatgrass (*Elymus
trachycaulus* ssp. *majus*).
Back berm: Blue wildrye, California brome, creeping wildrye, and slender
wheatgrass
Field edge: Creeping wildrye.

Figure 1. Schematic diagram of roadside topographic zones and appropriate native perennial grasses (From Bugg et al. 1997).

features of the Sacramento Valley, yet there are important differences. For example, the seasonal abundance of water may differ profoundly: many artificial wetlands contain water intermittently and during seasons when water may not be available in nearby natural systems. Mixtures of wetland and upland plant species can be planted along standpipes, valves, reservoirs, irrigation catchments, and tailwater ponds. Tailwater ponds are located in the low ends of agricultural fields and receive drainage water resulting from rains or irrigation. Recirculating tailwater ponds are those that include pumps to return accumulated water to the tops of fields, potentially conserving water and reducing erosion and downstream sedimentation. Wetland plantings in the sites just mentioned receive supplemental water as a normal consequence of crop irrigation, and may support a diverse array of insects. Some small moist sites may accommodate only a few plants; larger, more diverse sites can contain many species, planted in multiple tiers and appropriately established along moisture gradients and given proper topographic relief and exposure to light. Table 3 presents a list of plant species that we have used in these situations.

Wild Plants

Wild plants play a paradoxical role in farming systems. Some species, which may be termed weeds, compete excessively for water and nutrients, harbor other pests or pathogens, or sustain ice-nucleating bacteria, thereby increasing frost damage to tree fruit. However, weeds have also been termed "the guardians of the soil" (Cocannauer 1950), which is hardly an exaggeration in light of their tendency to quickly colonize and cover disturbed areas, thereby reducing soil erosion. Weed status of a plant is contextual and dependent on the management system involved. Therefore, we believe that the objective of managing weeds should be to limit their deleterious effects, rather than to kill them. The field known as "weed science" is still dominated by an agrichemical agenda, and is directed mainly toward the use of herbicides to destroy weeds. Jordan (1996, 1997) has called for the development of a weed science that is oriented more toward alternative tactics and collaboration with new partners, including environmental advocacy groups.

Many wild plants that occur in roadsides, waste areas, or wildlands have very limited impacts on field or row crops in adjoining farmlands. In the understories of orchards and vineyards, wild plants can be managed as cover crops, and thereby become assets. When so managed, the term "resident vegetation" is usually applied, rather than "weeds." Other things being equal, such resident species should be tolerated. The timing of mowing or tillage can influence the prevalent weed complex (Altieri & Whitcomb 1978/1979).

Practical Implementation

We distinguish here among management of hedgerows, roadsides, wetlands, and resident vegetation, yet these categories actually grade into one another, and include plant materials and management options in common. For example, plantings around a tailwater pond—typically located in the

Table 3. Native California plants currently used in hedgerows and tailwater pond plantings in the Sacramento Valley of California.

Tier	Plants
Understory Plants	Aster (*Aster chilensis*)
	Blue-eyed grass (*Sisyrinchium bellum*)
	California goldenrod (*Solidago californica*)
	Creeping wildrye (*Leymus triticoides*)
	Deergrass (*Muhlenbergia rigens*)
	Dutchman's pipe (*Aristolochia californica*)
	Gumweed (*Grindelia camporum*)
	Hedge-nettle (*Stachys ajugoides*)
	Heliotrope (*Heliotropium curassivicum* var. *oculatum*)
	Meadow barley (*Hordeum brachyantherum* ssp. *brachyantherum*)
	Narrow-leaved milkweed (*Asclepias fascicularis*)
	Purple needlegrass (*Nassella* [*Stipa*] *pulchra*)
	Slender wheatgrass (*Elymus trachycaulus*)
	Spike-rushes (*Heleocharis* spp.)
	Western goldenrod (*Euthemia occidentalis*)
	White-root sedge (*Carex barbarae*)
	Wild licorice (*Glycyrrhiza lepidota*, Fabaceae)
	Yarrow (*Achillea millefolium* [=*Achillea borealis*])
Shrubs	Brewer saltbush (*Atriplex lentiformis*)
	California bladderpod (*Isomeris arborea*)
	California buckwheat (*Eriogonum fasciculatum*)
	California coffeeberry (*Rhamnus californica*)
	California fuchsia (*Zauschneria californica*)
	California rose (*Rosa californica*)
	Cleveland sage (*Salvia clevelandii*)
	Coyote brush (*Baccharis pilularis*)
	Holly-leaf cherry (*Prunus ilicifolia*)
	Mexican elderberry (*Sambucus mexicana*)
	Mule fat (*Baccharis viminea*)
	Saint Catherine's lace (*Eriogonum giganteum*)
	Toyon or Christmasberry (*Heteromeles arbutifolia*)
Trees	Fremont cottonwood (*Populus fremontii*)
	Gray pine (*Pinus sabiniana*)
	Valley oak (*Quercus lobata*)
	Willows (*Salix* spp.)

low corner of a laser-leveled field—may be used to "anchor" a more extensive hedgerow, providing water and harboring an enriched flora, and thereby enrich the fauna of the area.

Propagation techniques for California native plants are given by Lenz (1977), Schmidt (1980), Emery (1988), and Young & Young (1990). Farmscaping projects may employ from one to more than 50 plant species, and from relatively few to thousands of individual plants. Propagation methods vary according to several factors, but generally involve division, which includes both roots and shoots; cuttings, which include only shoot material that produces roots after being inserted into the soil; or seed. Division is commonly used for grasses (Poaceae), rushes (Juncaceae), and sedges (Cyperaceae); cuttings are used for many woody perennial plants (e.g., trees and shrubs); and seeds can be used for almost all species, although with varying degrees of difficulty.

Most propagation by division is quite easy, and can be performed in the field. Poplars (*Populus* spp.) and willows (*Salix* spp.) can be propagated quite easily by direct sticking of whips (cuttings) during late fall or winter. For many oaks (*Quercus* spp.), acorns can be direct-seeded in the field with some protection from herbivores. Those species that are easy to propagate also can be container grown and are available from many native plant nurseries. However, most shrubs and trees and many wetland herbaceous species require precise protocols and specialized facilities. Propagation of these plants is easier left to the nurseries.

For large projects, the ultimate source of the plant materials may be a concern as regards establishment and long-term viability. Moreover, on farms with adjoining wildlands, it is desirable to preserve the genetic integrity of wild plant populations. However, site-specific propagation of plant materials necessitates collection of propagules one year prior to planting. If no wild populations are located nearby, using regionally obtained materials would be the next-best option.

For large projects, we typically assign one or more experienced practitioners the role of determining where various plant species belong. This person traverses the area and inserts color-coded survey flags at appropriate sites. In order to install the proper plants, the laborers who follow have only to match color-coded flats, potted plants, or seed mixes to the corresponding flags.

Hedgerows

Various hedgerow designs have been used: (1) bands of perennial grasses with other herbaceous and occasional woody plants interspersed; (2) single-row windbreaks with or without understory plants; and (3) multirow, multitier shelterbelts. Wind is not a major problem for agricultural production in Yolo County. Therefore, to minimize the shading of adjacent farmlands, we have avoided the use of continuous lines of tall trees and focused instead on clusters of shrubs and occasional trees set amid stands of perennial grasses and perennial and annual forbs. A list of plant species used in various projects is presented in Table 3.

For large-scale plantings, we use standard farming implements and practices to propagate and establish shrubs and trees amid stands of native perennial grasses, with phased establishment periods. In one planting at Hedgerow Farms (1992), the hedgerow was 4.58 m (180") wide, and comprised 3, 1.52-m (60") beds, separated by irrigation furrows. All beds were seeded to native grasses; the central bed contained shrubs as well. Seed of the perennial grasses was drilled to a depth of 1.27 cm (0.5") during October. The outer edges of each of the outer beds were seeded to creeping wildrye (*Leymus triticoides*), because our observations showed that it can recover rapidly from inadvertent disking or herbicide application. The remainder of each of the two outer beds and the entire central bed were seeded with mixes of other grass species. Sections were assigned to mixtures of blue wildrye (*Elymus glaucus*), California brome (*Bromus carinatus*), meadow barley (*Hordeum brachyantherum* ssp. *brachyantherum*), and slender wheatgrass (*Elymus trachycaulus* ssp. *majus*) (by-weight 1:1:1:1 blend of seeds) or of meadow barley and purple needlegrass (*Nassella pulchra*) (1:1). Grasses were seeded using a Truax® native grass seed drill. Shrubs and herbaceous perennial plants (12–18-month-old seedlings) were planted from November 1992 through March 1993 in the central bed.

In general, the use of fertilizers to accelerate plant growth, compost to augment soil fertility and improve structure, irrigation, mechanical or chemical weed control, and protection from vertebrate and grasshopper damage may be necessary during the first year. During that time, shrubs and trees usually require the local removal of potentially competing herbaceous plants. Weed control may involve combinations of cultivation; herbicides; straw, wood, or plastic mulches; and selective mowing or removal of weeds. By the beginning of the second growing season, woody plants can usually tolerate an understory of herbaceous plants. However, if burning is used as a management tool, low-growing plants may carry a ground fire to sensitive woody species. Where fire-sensitive perennial forbs, shrubs, and trees are used, fire must be avoided or managed with great care. Moreover, fire may destroy habitat for beneficial insects, e.g., bunchgrass tussocks that harbor masses of diapausing lady beetles (e.g., *Hippodamia convergens* Guerin-Meneville, Coleoptera: Coccinellidae). Therefore, to avoid wholesale destruction, we recommend small- rather than large-scale burns. Fire should be used in roadside plantings only with caution and appropriate control and re-routing of traffic.

Many native broadleaf herbaceous perennial plants have low seedling vigor, and are more successfully established as plugged transplants during spring. Some of the perennial forbs we have used include yarrow (*Achillea millefolium*), narrow-leaf milkweed (*Asclepias fascicularis*), aster (*Aster chilensis*), grass-leaved goldenrod (*Euthamia occidentalis*), wild licorice (*Glycyrrhiza lepidota*), gumweed (*Grindelia camporum*), heliotrope (*Heliotropium curassivicum* var. *oculatum*), California goldenrod (*Solidago californica*), hedge-nettle (*Stachys ajugoides* var. *ajugoides*), and blue-eyed grass (*Sisyrinchium bellum*). Transplanting may be postponed until year 2, especially by practitioners who use blanket application of broadleaf herbicides during perennial grass establishment. Removal of other, competing vegetation is important to ensure establishment of broadleaf herbaceous perennial plants.

A schedule of operations and approximate costs for hedgerow establishment are presented in Table 4. The estimated cost of establishment per 0.8-km (1/2-mile) strip is US$1,378.00.

Other protocols for establishing hedgerows are certainly possible, including the initial establishment of trees and shrubs in bare ground, followed by seedbed preparation and seeding of grasses and forbs. As suggested above, the simultaneous establishment of grasses and forbs limits the options for herbicide use. However, diverse stands of herbaceous plants that include both rhizomatous and bunching species and both grasses and forbs probably lead to greater long-term suppression of herbaceous weeds.

Plantings for Catchments and Watercourses

Plantings for catchments and watercourses present challenges in terms of unreliable seasonal access, steep slopes, similarly steep moisture gradients (temporally as well as spatially), variety of exposures, and competitive resident vegetation. Watercourse plantings may be at risk through erosion or sedimentation, and these same factors can lead to highly variable soils that affect plant establishment and long-term growth. Invasion of canal banks and tailwater ponds by some species of agricultural weeds is a liability, because weed propagules may be dispersed by water to previously uninfested irrigated fields. This points up the importance of maintaining high proportions of canopy cover by non-invasive species in order to reduce establishment of weedy plants. It also underscores the value of thorough and ongoing monitoring and spot treatment of weed problems.

Wetlands may accumulate pesticide residues from farmlands, presenting another type of problem. These residues may poison both flora and fauna in farm ponds and ditches. For example, J.H. Anderson (personal observation) has observed death of bullfrog (*Rana catesbeiana*) larvae where insecticide from a treated alfalfa field apparently flowed with irrigation runoff into a tailwater pond. This observation is consistent with toxicological

Table 4. Establishment schedule for hedgerows.

Operation	Time Period	Cost/Acre (½-mile long strip)
1. Till soil and form 60" beds	September–October	$ 90.00
2. Preirrigate or wait for early rains to germinate weeds	September 20– October 20	0.00–50.00
3. Eliminate early-germinated weeds with herbicides, flamer, or shallow tillage. (Steps 2 and 3 may not be necessary if weed seed bank is depleted, or weed complex is not overly competitive.)	October	40.00
4. Using the Truax® native grass seed drill, seed the native grass mixture into the 3 beds	October–November	300.00
5. Plant woody vegetation at approximately 10' centers in center bed. Costs include include those for the plants, fertilizer tabs, protective devices (milk carton, stakes, weed nets). By this planting arrangement, a 1/2 mile-long hedgerow would comprise about 264 plants at $1.00 per. If plants are arranged in islands or groups, reduce plant number to 132.	November–March	528.00
6. Broadleaf weed control: herbicides; shields protect woody vegetation.	January–March	40.00
7. Vegetation control around individual woody plants.	February–May	40.00
8. Irrigation every 2–4 weeks for the first year.	June–January	150.00
9. Summer spot weed control, using herbicide or hoe.	June–September	40.00
10. Drill or broadcast mixes of forb seeds into outer beds: toothpick ammi, fennel yarrow, blazing star, bee phacelia, etc.	September-October	100.00
Total		$1,378.00

data collected from other tailwater ponds in the area (Kathleen Groody personal communication 1997). R.L. Bugg (personal observation) has noted reduced establishment and growth by meadow barley (*Hordeum brachyantherum* ssp. *brachyantherum*) in ditch beds where the pre-emergence herbicide oryzalin had been sprayed. Nearby untreated ditches did not show this problem. Oryzalin and some other pre-emergence herbicides are often transported with sediment and thus concentrate in ditch beds (Clyde Elmore personal communication 1993). Oryzalin reduces growth of meadow barley (Cynthia S. Brown & Robert L. Bugg unpublished data; but see data from Lanini et al. [1996]).

Wetland plantings may comprise a wide range of species, including all the trees, shrubs, and herbs mentioned for hedgerows, as well as aquatic and emergent forms. Given the steep moisture gradients and differing optima and tolerances of various plant species, a great diversity can be concentrated in a small area. For example, two small tailwater pond sites at Dixon Ridge Farms (organic walnut orchards, in Solano County near the town of Winters) currently include plantings of 26 of the species listed in Table 3. Many native Californian plants undergo summer dormancy, and are susceptible to soil-borne fungal pathogens if the root masses remain waterlogged during warm weather. Proper elevation and drainage are essential in maintaining such senstive plants, which include toyon (*Heteromeles arbutifolia*), wild buckwheats (*Eriogonum* spp.), and wild lilacs (*Ceanothus* spp.).

When establishing wetland stands of grasses and forbs by direct seeding, seed-bed preparation and control of early weeds are again essential. To avoid contamination of water, herbicides should be used with caution and in accord with label recommendations. With mechanized operations, lightweight, stable vehicles with low centers of gravity (e.g., some 4-wheeled all terrain vehicles) are used where there is sloping terrain and a likelihood of compacting moist soils.

Many wetland species establish well by vegetative propagation: cuttings of willows (*Salix* spp.) and poplars (*Populus* spp.) are often used, as are divisions of rhizomatous plants, such as creeping wildrye (*Leymus triticoides*), hedge nettle (*Stachys* spp.), and spikerushes (*Eleocharis* spp.). In many cases, propagules of these species may be simply inserted into the mud, amid standing water. To promote genetic diversity, we suggest using multiple locally obtained clones of the same species for source material. Maintenance of wetland plantings requires personnel trained in field identification of desirable and undesirable plant species. Otherwise, weed contol measures quickly become non-selective.

Roadsides

Roadside plantings may be designed and managed like hedgerows, with the stipulations that design and management should not interfere with motorists and cyclists. The remainder of this section is based on an account by Bugg et al. (1997).

Rural roadsides typically comprise several topographic zones (Fig. 1): (1) unimproved shoulder; (2) recovery area; (3) side slope; (4) open-cut ditch (drainage); (5) back slope; (6) back berm; and (7) field edge. These zones present a range of environmental conditions and management options and requirements, and may require varied plant materials. Different native perennial grasses have diverse environmental optima and tolerances and differing growth habits, and thus may lend themselves to different topographic zones. In general, it is a reasonable goal to use plant species that lead to reduced total above-ground biomass yet retain high proportions of perennial vegetative groundcover. Other things being equal, this should reduce the threat of flooding, wildfire, and soil erosion. Our experience with large-scale roadside plantings has provided information on the suitability and bases for assigning various perennial grasses to the various topographic zones.

Low-statured bunchgrasses, e.g., Idaho fescue (*Festuca idahoensis*), pine bluegrass (*Poa secunda* ssp. *secunda*) and low-growing forms of California barley (*Hordeum brachyantherum* ssp. *californicum*) are perhaps most suitable for the unimproved shoulder, because they do not obstruct motorists' vision, are unlikely to break up pavement, and tolerate but do not require close mowing. Creeping red fescue (*Festuca rubra*), California brome, and lower-growing forms of blue wildrye are intermediate in height and may be used in the recovery area, through which motor vehicles occasionally travel. Perennial grasses like meadow barley, California barley, California brome, California oniongrass (*Melica californica*), purple needlegrass, and nodding needlegrass (*Nassella cernua*) are candidates for both the recovery area and the side slope. The moderate biomass production of these grasses makes them unlikely to interfere with motor vehicles that occasionally use the area. Also, these grasses are drought tolerant and thus adapted to these two zones that may receive less water than do either the unimproved shoulder or the open-cut ditch. Meadow barley is tolerant of intermittent flooding and has moderate stature. Thus, it would tolerate conditions encountered in most open-cut ditches and be unlikely to block the flow of water. If the open-cut ditch contains water for extended periods, spike-rushes (*Eleocharis* spp.) would be better adapted. For the back slope and back berm, tall-statured grasses, such as California brome, blue wildrye, slender wheatgrass, and purple needlegrass are suitable. If mowing is frequent, these species could also be used on the back slope and in intermittently-flooded ditch beds. The field edge is subject to inadvertent damage by herbicides and agricultural implements. Therefore, the rhizomatous creeping wildrye (*Leymus triticoides*) is probably appropriate. This species is tall, recovers rapidly from mechanical damage, and is believed by some workers to show resistence to glyphosate, a commonly-used broad-spectrum contact herbicide (J. H. Anderson personal observation).

Following preparation of a seedbed and control of early-emerging weed seedlings by contact herbicides, flaming, or shallow cultivation, cool-season perennial grass seed may be broadcast and incorporated or drilled to a depth of 1.27 cm. Seeding dates range from mid-October to as late as mid-January. Protocols are presented in Table 2.

As mentioned earlier for hedgerow management, fire should be used in roadside plantings with caution and appropriate control and re-routing of traffic.

Wild Plants

Resident wild plants are the most important factor limiting the restoration of perennial native vegetation, yet they also include native species, and collectively provide important resources to beneficial arthropods, as noted earlier. Since their status as weeds is contextual, wild plants that require control in one setting may be tolerable, or even useful, in another. Therefore, control programs should be thoughtfully crafted and implemented with restraint.

As mentioned, herbaceous annual or perennial vegetation should be controlled during the early establishment phases of perennial woody plants or native perennial grasses. Many wild plants may be tolerated in non-cropped areas thereafter, if they have minimal impacts on adjoining farming systems. Thus, aquatic or emergent plants like swamp smartweed (*Polygonum coccineum*) and spikerushes (*Eleocharis* spp.), both natives, are weeds in rice, but can be tolerated or even encouraged in recirculating tailwater ponds that serve other crops. Likewise, the introduced apiaceous weed toothpick ammi (*Ammi visnaga*) germinates only in cold weather, is rarely a problem in field crops (occurring only occasionally in overwintered sugar beet fields), and could be encouraged along field margins. Fennel (*Foeniculum vulgare*), an apiaceous herbaceous perennial plant, is an important wildlands weed of the Californian North Coast, but is almost never a problem in fields or orchards, and can be tolerated in borders.

Dense, established stands of perennial grasses reduce recruitment by many weedy species (Fenner 1978). Prickly lettuce (*Lactuca scariola*) is one of the few winter-germinating weeds that is able to establish amid dense stands of California native perennial grasses. However, even small gaps in such stands may be colonized aggressively by wild oat (*Avena fatua*) and yellow starthistle (*Centaurea solstitialis*). None of these three annual weeds is of great value to parasitic or predatory insects, but yellow starthistle is heavily used as a nectar source by honeybee (*Apis mellifera*) and by various native butterflies and solitary bees.

In-field application of weed-control measures should eliminate the need for wholesale treatment of field borders. Selective control of undesirable wild plants in farmscaped areas can be achieved by hand removal, cultivating, mowing, or spot treatment with herbicides by hand-held sprayer or wick.

Arthropods Associated with Farmscaping Features

Perennial non-crop vegetation, including that contained in hedgerows, harbors some arthropods that do not venture much into surrounding arable crops ("hard-edge species" in the usage of Duelli et al. 1990). For example, Bishop and Riechert (1990) showed that there were few species in common between the spider fauna of vegetable gardens and those of a nearby wooded area. Similar results were obtained by Maelfait and De Keer (1990) when they compared the spider faunae of intensively-grazed pastureland and its border areas. On the other hand, some "soft-edge species" may show increased—or decreased—abundance in the arable lands adjoining perennial vegetation. There can be negative or positive consequences for such patterns.

On the negative side, hedgerow vegetation might include plants that serve as reservoirs for insect-borne plant-pathogenic viruses, bacteria, and mycoplasmas, or as overwintering sites for boll weevil (*Anthonomus grandis grandis* Boheman, Coleoptera: Curculionidae) (Slosser et al. 1984) and other pests. Poplars and cottonwoods (*Populus* spp.) are required alternate hosts for various root aphids (*Pemphigus* spp., Homoptera: Eriosomatidae) that attack beet (*Beta vulgaris*) and lettuce (*Lactuca* spp.) (Davidson & Lyon 1986). In South Africa, windbreaks of silky oak (*Grevillea robusta*) may predispose for problems with citrus thrips (*Scirtothrips aurantii* [Moulton], Thysanoptera: Thripidae) (Grout & Richards 1990). Windbreaks may also cause local infestations of aphids (Homoptera: Aphididae) and thrips (Thysanoptera: Thripidae) because the winged colonizing forms settle in sheltered areas (Lewis 1965a, 1965b). Weeds and wild plants within or outside orchards or vineyards can be important reservoirs for pest or beneficial arthropods (see reviews by Altieri & Letourneau 1982 and Andow 1988). *Lygus* spp. (Hemiptera: Miridae) are well known to emanate from weed hosts (Fye 1980), and under some conditions may damage orchard crops. Certain other Miridae cause similar problems.

Forest, chaparral (scrub), and other plant complexes may harbor pest arthropods. For example, the causal organism of Pierce's disease of grapes, the bacterium *Xylella fastidiosa* and its vector the bluegreen sharpshooter (*Graphocephala atropunctata* [Signoret], Homoptera: Cicadellidae), are hosted by certain wild native or introduced perennial plants. Likewise, eleven-spotted cucumber beetle (*Diabrotica undecimpunctata* ssp. *undecimpunctata* Mannerheim) emanates from wild plants that may adjoin farmlands.

On the positive side, hedgerows and other non-crop vegetation may harbor important beneficial arthropods. Much of the relevant research has been done in England and Germany (see Nentwig; Wratten et al.; Helenis this volume). Ground beetles (Coleoptera: Carabidae) may occupy hedgerows during winter, then disperse to adjoining field crops with the advent of spring, as was suggested for *Platynus dorsalis* Pontoppidan (Coleoptera: Carabidae) in an unreplicated study of an intensively run farm near Kiel, Germany (Knauer & Stachow 1987). In England, willows (*Salix* spp.) and an alder (*Alnus glutinosa*) can harbor predatory true bugs (Hemiptera: Anthocoridae, Miridae) in the spring. These predators later disperse to orchards and attack codling moth (*Cydia pomonella* [L.], Lepidoptera: Tortricidae) or pear psylla (*Psylla pyricola* Foerster, Homoptera: Psyllidae) (Solomon 1981; Gange & Llewellyn 1989). Entomophagous arthropods may also benefit from weeds and other wild plants that provide shelter, pollen, alternate hosts or prey, honeydew, or nectar.

Shelter

Wind shelter (Lewis 1965a; Pollard 1971) and flowers (Bowden & Dean 1977) provided by hedgerows determine the distribution of hoverflies (Diptera: Syrphidae) that attack aphid pests of crops (Beane & Bugg this volume). In California, perennial bunchgrasses provide sites for aestival-hibernal diapause for convergent lady beetle (*Hippodamia convergens* Guerin-Ménévile, Coleoptera: Coccinellidae) (Hagen 1962, 1974).

Pollen As Food

In South Africa, facultatively pollinivorous predatory mites (*Euseius addoensis addoensis* [van der Werwe & Ryke] Acari: Phytoseiidae) can be particularly abundant in citrus (*Citrus sinensis*) sheltered by windbreaks of pollen-producing Monterey pine (*Pinus radiata*) (Grout & Richards 1990). Another predatory mite *Euseius tularensis* (Congdon) (Acari: Phytoseiidae), which attacks various pests of citrus in California, also feeds on pollens of various plants (Kennett et al. 1979). The mite uses pollen of numerous species of grasses, including native perennial bunchgrasses, domestic cereal grains used as cover crops, and several annual grasses that commonly occur as weeds (Ouyang et al. 1992). None of the grass pollens thus far tested are sufficient diets to sustain reproduction at peak levels for more than one generation (Ouyang et al. 1992). The thick exine coats of grass pollens may render them less suitable as foods than are other pollens (e.g., apple pollen) (Ouyang et al. 1992). Table 5 summarizes flowering periods for several cool-season annual weedy grasses, domestic cereal grains, and native perennial grasses, based on data presented by Munz (1973).

Alternate Hosts And Prey

Anagrus epos Girault (Hymenoptera: Mymaridae) is an important egg parasite of grape leafhopper (*Erythroneura elegantula* Osborn, Homoptera: Cicadellidae) and variegated leafhopper (*Erythroneura variabilis* Beamer, Homoptera: Cicadellidae). The parasite overwinters on eggs of various leafhoppers. These hosts include *Dikrella californica* (Lawson) (Homoptera: Cicadellidae) on wild blackberrry (*Rubus* spp.) and wild grape (*Vitis californica*) (Doutt & Nakata 1965, 1973), and prune leafhopper (*Edwardsiana prunicola* [Edwards], Homoptera: Cicadellidae) on French prune (*Prunus domestica*) (Kido et al. 1984; Pickett et al. 1990). French prune trees are now being planted alongside California vineyards in some winegrape districts, to enhance biological control (Murphy et al. this volume). The ichneumonid wasp *Exochus nigripallipus subobscurus* Townes (Hymenoptera: Ichneumonidae) is a larval-pupal parasite of orange tortrix (*Agrotaenia citrana* [Fernald], Lepidoptera: Tortricidae), itself a lepidopterous pest of grape (*Vitis vinifera*) in the Central Coast region of California. Orange tortrix and *Aristoteliae argentifera* Busck (Lepidoptera: Gelechiidae) may infest coyote brush (*Baccharis pilularis*). Kido et al. (1981) suggested that high levels of parasitism of orange tortrix may occur in vineyards with abundant coyote brush nearby. Orange tortrix may infest various vineyard weeds, including mallow (*Malva parviflora*), curly dock

Table 5. Flowering (pollen-shedding) periods for grasses in California. For many grass species, early flowering may occur if moisture is available during the previous summer. Flowering may also be prolonged on moist sites. Flowering of annuals typically ends with the exhaustion of soil moisture. The flowering periods given were obtained from Munz (1973).

Species	Flowering Period
Annual Ryegrass (*Lolium multiflorum*)	June–August
Barley (*Hordeum vulgare*)	April–July
Blue Wildrye (*Elymus glaucus*)	June–August
California Brome (*Bromus carinatus*)	April–August
Cereal Rye (*Secale cereale*)	May–August
Cultivated Oat (*Avena sativa*)	April–June
Foxtail Fescue (*Vulpia megalura*)	April–June
Meadow Barley (*Hordeum brachyantherum* ssp. *brachyantherum*)	May–August
Rattail Fescue (*Vulpia myuros*)	March–May
Ripgut Brome (*Bromus rigidus*)	April–June
Slender Wild Oat (*Avena barbata*)	March–June
Wild Barley (*Hordeum leporinum*)	April–June
Wild Oat (*Avena fatua*)	April–June

(*Rumex crispus*), filaree (*Erodium* spp.), lupin (*Lupinus* sp.), mustard (*Brassica* sp.), pigweed (*Amaranthus* sp.), and California poppy (*Eschscholtsia californica*) (Kido et al. 1981). It is not clear what role these weeds or various cover crops could play in sustaining parasites.

Several annual weeds sustain aphids or other Homoptera that in turn sustain beneficial arthropods (Table 6). Annual sowthistle (*Sonchus oleraceus*) sustains several species of aphids; some of these are suitable prey for lady beetles, whereas others are toxic.

Nectar

Nectar-bearing plants, including those contained in hedgerows, have long been known to extend the longevity of parasitic and predatory insects, in some cases leading to improved biological control of pests (Van Emden 1962). In California, coyote brush (*Baccharis pilularis*) is heavily visited by various nectarivorous predatory and parasitic insects (Steffan 1997). Various native Californian shrubs attract large numbers of beneficial insects: blue elderberry (*Sambucus caerulea*), coyote brush (Steffan 1997), California coffeeberry (*Rhamnus californica*), California lilacs (*Ceanothus* spp.) (Bugg & Heidler 1981), California buckwheat (*Eriogonum fasciculatum*) (Swisher 1979), holly-leaved cherry (*Prunus ilicifolia*), toyon (*Heteromeles*

Table 6. California weeds that harbor alternate hosts or prey of beneficial insects.

Weed	Phytophagous Insects	Seasonal Dynamics of Phytophagous Insects and Notes on Associated Beneficial Arthropods	Source of Information
Annual Sowthistle (*Sonchus oleraceus*)	*Hyperomyzus* (*Nasonovia*) *lactucae* (L.) (formerly *Amphorophora sonchi*); Potato aphid (*Macrosiphum euphorbiae*); Other aphid species	*Hyperomyzus* (*Nasonovia*) *lactucae* (powdery green with inflated cornicles) and *M. euphorbiae* occur from April through July, and are prey to convergent lady beetle (*Hippodamia convergens* Guerin-Ménéville). The two generalist aphidiid wasps *Aphidoletes aphidimyza* Rondani and *Aphidoletes meridionalis* Felt are found in this aphid community. A red aphid species occurring on annual sowthistle is toxic to lady beetles.	K.S. Hagen and H. Lange, personal communications
Burclover (*Medicago polymorpha*)	Pea aphid (*Acyrthosiphon pisum*)	*Acyrthosiphon pisum* can be abundant from March through early May, and sustains reproduction by various lady beetles and syrphid flies. The plant also sustains *Lygus hesperus*, a pest of some orchard crops.	R.L. Bugg, personal observation
Common Knotweed (*Polygonum aviculare*)	*Aphis avicularis*; *Aphalara curta* (Psyllidae)	*Aphis avicularis* occurs from August through November and sustains reproduction by the ladybeetles *Scymnus* sp., *Hippodamia convergens*, and *Coccinella novemnotata*; and the syrphid *Paragus tibialis*. The host-specific psyllid *A. curta* may also be an important prey item to generalist predators.	Bugg et al., 1987; R.L. Bugg, personal observation

Host Plant	Aphid Species	Notes	Source
Pineapple Weed (*Matricaria matricarioides*)	*Brachycaudus helichrysi*; Bean aphid (*Aphis fabae*); Green Peach Aphid (*Myzus persicae*)	*Brachycaudus helichrysi* is the predominant aphid. It sustains reproduction by convergent lady beetle (*Hippodamia convergens*) and other Coccinellidae	K.S. Hagen, personal communication; H. Lange, personal communication
Mayweed (*Anthemis cotula*)	*Brachycaudus helichrysi*; Bean aphid ; Green peach aphid (*Myzus persicae*)	*Brachycaudus helichrysi* is the predominant aphid. It sustains reproduction by convergent lady beetle (*Hippodamia convergens*) and other Coccinellidae. *Lygus* sp. can be abundant on mayweed.	K.S. Hagen, personal communication; H. Lange, personal communication; R.L. Bugg, personal observation
Wild Barley (*Hordeum leporinum*)	Bird cherry—Oat aphid (*Rhopalosiphum padi* L.); English Grain Aphid (*Macrosiphum avenae*); Other aphid species	*Rhopalosiphum padi* predominates. Aphid populations build in February. With flowering, aphids can become extremely abundant in the panicles. Wet weather or heat can devastate populations. Mild weather can permit aphids to survive into May, when wild barley senesces. Various lady beetles and syrphid flies reproduce on these aphids, as does the generalist aphidiid wasp *Diaeretiella rapae*.	M. Van Horn, personal communication
Wild Oat (*Avena fatua*)	Bird cherry—Oat aphid (*Rhopalosiphum padi*); English Grain Aphid (*Macrosiphum avenae*); Other aphid species	*Rhopalosiphum padi* predominates. Aphid populations build in February. Wet weather or heat can devastate populations. Mild weather can permit aphids to survive into May, when wild oat senesces. Various lady beetles and syrphid flies reproduce on these aphids, as does the generalist aphidiid wasp *Diaeretiella rapae*.	M. Van Horn, personal communication

arbutifolia), and various native willows (*Salix* spp.). Exotic plant species may be frequented by beneficial insects: such exotic insectary plants include soapbark tree (*Quillaja saponaria*) (Bugg 1987), and sweet fennel (*Foeniculum vulgare* var. *dulce*) (Maingay et al. 1991; Patt et al. 1997). In Massachusetts, sweet fennel attracted at least 48 species of parasitic ichneumonid wasps (Hymenoptera: Ichneumonidae) (Maingay et al. 1991) (Table 7). These included parasites that attack larvae of codling moth and grape berry moth. Also observed were 4 species of Sphecidae (solitary wasps) and 4 of Vespidae (social wasps) (Table 8). Related predatory wasps are commonly seen in Californian vineyards, but their agroecological roles have not been documented. Patt et al. (1997) found that sweet fennel is a suitable nectar source for two species of Eulophidae (Hymenoptera), *Edovum puttleri* Grissell (an egg parasite of Colorado potato beetle, *Leptinotarsa decemlineata* Say [Coleoptera: Chrysomelidae]) and *Pediobius foveolatus* Crawford (a larval parasite of Mexican bean beetle [*Epilachna varivestis* Mulsant, Coleoptera: Coccinellidae]). The former two pests are not established in California..

In northern California, two weed species have been suggested as major nectar sources for beneficial entomophagous insects: common knotweed (*Polygonum arenastrum*) (Bugg et al. 1987) (Table 9) and toothpick ammi (*Ammi visnaga*) (Bugg & Wilson 1989) (Table 10). Both weeds commonly occur along rural roadsides, and common knotweed is often abundant within vineyards and orchards as well. Wild carrot (*Daucus carota*) is common in the northern coastal counties of California, and widely recognized as a potent nectar source used by beneficial parasitic and predatory insects (Bohart & Nye 1960; Hirose 1966; Judd 1970).

Poison hemlock (*Conium maculatum*) is common throughout the central and northern coastal areas of California, and its flowers are attractive to many beneficial insects (Bugg, pers. obs.). Common chickweed flowers are also visited by various nectarivorous parasites and predators (Batra 1979; Foster & Ruesink 1984, 1986).

All these weeds commonly occur on vineyard and orchard margins in California, and common knotweed and common chickweed are often found within vineyards and orchards, as well.

Other weedy forbs that we have noted as important nectar sources for predatory and parasitic insects are the following natives: common sunflower (*Helianthus annuus*), swamp smartweed (*Polygonum coccineum*), and turkey mullein (*Eremocarpus strigerus*); and these introduced species: bishop's weed (*Ammi majus*), prostrate spurge (*Euphorbia maculata*),

Table 7. Ichneumonidae (Hymenoptera) collected at the flowers of sweet fennel, *Foeniculum vulgare* Miller var. *dulce* Battandier & Trabut (Apiaceae), Falmouth, Massachusetts, 1985 (from Maingay et al., 1991).

Subfamily	Genus or Species and Authority
Pimplinae	*Zaglyptus varipes incompletus*
	Tromatobia ovivora
	T. rufopectus
	Coccygomimus equalis
	Coccygomimus pedalis
	Coccygomimus disparis
	Coccygomimus annulipes
	Iseropus coelebs
	Scambus (Ateleophadnus) pterophori
	Scambus sp.
	Itoplectis conquisitor
	Theronia atalantae fulvescens
Tryphoninae	*Phytodietus vulgaris*
Cryptinae	*Stibeutes* sp.
	Mastrus sp.
	Atractodes sp.
	Polytribax contiguus
	Polytribax pallescens
	Cubocephalus annulatus
	Acroricnus stylator aequalis
	Aritranis byrsina
	Cryptus moschator iroquois
Ichneumoninae	*Phaeogenes* sp. prob. new
	Colpognathus helvus
	Stenobarichneumon duplicans
	Cratichneumon vescus
	Pterocormus laetus
	Pterocormus creperus
	Pterocormus annulatorius
	Ctenichneumon minor
	Ichneumon sp. poss. new
	Vulgichneumon brevicinctor
	Setanta compta
	Ectopimorpha wilsoni
	Eutanyacra vilissima
Banchinae	*Lissonota* sp.
	Glypta spp.
Campopleginae	*Cymodusa distincta*
	Campoletis sp.
	Campoplex sp.
	Diadegma sp.
	Hyposoter annulipes
	Casinaria eupitheciae
Cremastinae	*Temelucha curtipetiolata*
	Nothocremastus brunneipennis
Anomaloninae	*Barylypa prolata*
	Barylypa fulvescens
	Agrypon prismaticum
	Therion longipes

Table 8. Predatory Hymenoptera observed feeding at the flowers of sweet fennel, *Foeniculum vulgare* Miller var. *dulce* Battandier & Trabut (Apiaceae), Falmouth, Massachusetts, 1985 (from Maingay *et al.*, 1991)

Family	Genus or Species and Authority
Sphecidae	*Sceliphron caementarium*
	Ectemnius lapidarius
	Ectemnius continuus
	Ectemnius stirpicola
Eumenidae	*Monobia quadridens*
	Ancistrocerus campestris
	A. adiabatus
Vespidae	*Polistes fuscatus pallipes*
	Vespula maculata
	Vespula arenaria
	Vespula vidua
Pompilidae	2 unidentified spp.

shepherd's purse (*Capsella bursa-pastoris*), spiny clotbur (*Xanthium spinosum*), white sweetclover (*Melilotus album*), and yellow sweetclover (*Melilotus officinalis*).

Flowering periods are summarized for various wild trees, shrubs, and forbs in Figures 2 and 3.

Conclusion

Wild plants and perennial plantings for hedgerows, roadsides, and wetlands have associated complexes of beneficial and pest animals. Trees and shrubs have long been used in windbreak and farmstead plantings in California, yet only recently has attractancy to beneficial arthropods been used as a selection criterion. Farmers in many parts California are exploring the joint use of these types of plantings in order to provide key resources for predators and parasites. Critical experiments remain to be done on farmscaping to enhance biological control of agricultural pests and provide additional benefits, while avoiding other problems, such as the exacerbation of certain other pests.

Local native plant complexes that have been reduced through agricultural development should be considered for farmscaping projects. Native low-growing herbaceous plants are not as noticeable or visually appealing as trees and shrubs, and in many regions of the world, such inconspicuous plants have been among the first plants to be destroyed in agricultural development. They have also been the last to be restored to hedgerows and other plantings in northern California. Nevertheless, this category of plants in-

Table 9. Insects observed as flower visitors at common knotweed (Polygonum aviculare L., Polygonaceae) in northern California, from 1980–1984.

Order	Family or Subfamily	Genus, species, and authority
Hemiptera	Lygaeidae	Geocoris atricolor
		G. pallens
		G. punctipes
	Miridae	Lygus sp.
	Nabidae	Nabis sp.
	Anthocoridae	Orius tristicolor
Coleoptera	Carabidae	Undetermined spp.
	Melyridae	Collops vittatus
	Coccinellidae	Hippodamia convergens
	Anthicidae	Anthicus spp.
Diptera	Syrphidae	Allograpta sp.
		Eumerus sp.
		Paragus tibialis
		Sphaerophoria sp
		Syritta pipiens
	Muscidae	Musca domestica
	Calliphoridae	Undetermined spp.
	Tachinidae	Archytas californiae
		Undetermined spp.
Hymenoptera	Braconidae	Chelonus sp.
		Undetermined microgastrine spp.
	Ichneumonidae	prob. Compsocryptus sp.
		prob. Pristomerus sp.
	Chalcidoidea	Undetermined spp.
	Dryinidae	Undetermined sp.
	Formicidae	Conomyrma bicolor
		Conomyrma insana
		Iridomyrmex humilis
		Tetramorium caespitum
	Vespidae	Polistes apachus
		Polistes fuscatus
	Eumenidae	Euodynerus sp.
	Pompilidae	Undetermined spp.
	Sphecidae	Ammophila sp.
		Tachytes sp. prob. distinctus
		Undetermined spp.
	Halictidae	Undetermined spp.
	Megachilidae	Undetermined sp.
	Apidae	Apis mellifera

cludes many important nectar sources, butterfly plants, and overwintering sites for predators. Moreover, several vigorous species, including both grasses and broadleaf plants, may prove valuable when used in concert to suppress agricultural weeds.

The same plants we recommend for farmscaping may be used similarly around community gardens, schoolyards, and private residences. The myriad ecological associations may redound not only to improved pest control, but also to enriched educational opportunities and quality of life.

Table 10. Insects feeding at flowers of toothpick ammi, *Ammi visnaga* (L.) Lamarck (Apiaceae), in Yolo County, California, from 1979-1984 (Bugg and Wilson, 1989).

Order	Family	Subordinate taxon and Authority
Hemiptera	Lygaeidae	*Geocoris pallens*
		Geocoris punctipes
	Miridae	*Lygus* sp.
	Nabidae	*Nabis* sp.
	Anthocoridae	*Orius tristicolor*
	Reduviidae	*Zelus renardii*
		Sinea sp.
Coleoptera	Melyridae	*Collops vittatus*
	Coccinellidae	*Coccinella novemnotata*
		Hippodamia convergens
		Guerin-Meneville
Neuroptera	Chrysopidae	*Chrysoperla carnea*
Diptera	Syrphidae	*Allograpta obliqua*
		Eristalis tenax
		Eristalis sp.
		Eumerus sp.
		Eupeodes volucris
		Helophilus sp.
		Prob. *Melanostoma mellinum*
		Toxomerus marginata
		Metasyrphus sp.
		Paragus tibialis
		Platycheirus sp.
		Scaeva pyrastri
		Sphaerophoria sp.
		Syritta pipiens
	Tachinidae	*Archytas apicifer*
		Prob. *Cylindromya* sp.
		Prob. *Gymnosoma fuligonsum*
		Peleteria texensis
		Lepesia archippivora
		Leucostoma aterrimum
		Prob. *Trichopoda plumipes*
	Sarcophagidae	*Senotainia* sp.
		Wohlfahrtia vigil
Hymenoptera	Braconidae	*Microgaster* sp.
		Chelonus insularis
		Chelonus (*Microchelonus*) sp.
	Ichneumonidae	*Biolysia* sp. prob *tristis*
		Charitopes albilabris
		Compsocryptus calypterus
		Cryptus rufovinctus
		Ethylurgus nigriventris
		Prob. *Hyposoter* sp.
		Patrocloides montanus
		Pristomerus spinator
		Pterocormuscupitus
		Pterocormus sp.
		Undet. spp.
	Chalcidoidea	Undet. spp.
	Chalcididae	*Spilochalcis delumbis*
	Dryinidae	Undet. spp.

Table 10. *continued*

Order	Family	Subordinate taxon and Authority
	Formicidae	*Conomyrma bicolor*
		Conomyrma insana
		Formica sp.
		Iridomyrmex humilis
		Solenopsis sp.
	Vespidae	*Polistes apachus*
		Polistes fuscatus
	Eumenidae	Undet. spp.
	Pompilidae	Undet. spp.
	Sphecidae	*Ammophila* sp.
		Undet. Cercerini
		Chlorion sp.
		Philanthus sp.
		Prionyx sp.
		Tachytes sp. prob. *distinctus*
		Undet. spp.

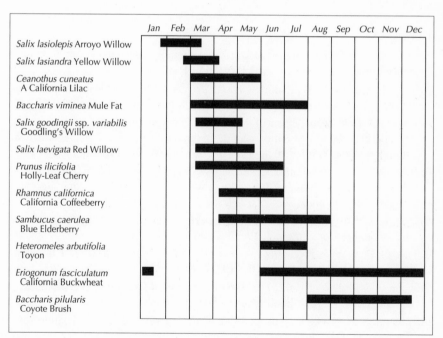

Figure 2. Flowering periods of native insectary trees and shrubs.

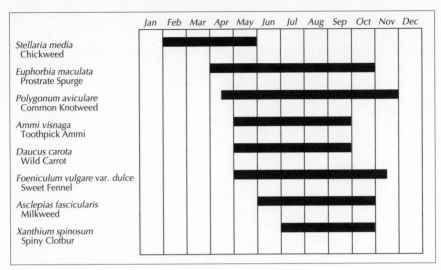

Figure 3. Flowering periods of wild insectary forbs.

References

Altieri, M. A. & D. K. Letourneau. 1982. Vegetation management and biological control in agroecosystems. Crop Protection 1:405-430.

Altieri, M. A. & W. H. Whitcomb. 1978/1979. Manipulation of insect populations through seasonal disturbance of weed communities. Protection Ecol. 1:185-202.

Andow, D. A. 1988. Management of weeds for insect manipulation in agroecosystems, pp. 265-301. *In* M. A. Altieri, & M. Liebman [eds.], Weed management in agroecosystems: ecological approaches. CRC Press, Boca Raton, Florida.

Batra, S. W. T. 1979. Insects associated with weeds in the northeastern United States. III. Chickweed, *Stellaria media*, and stitchwort, *S. graminea* (Caryophyllaceae). J. New York Entomol. Soc. 87:223-235.

Bishop, L. & S. E. Riechert. 1990. Spider colonization of agroecosystems: source and mode. Environ. Entomol. 19:1738-1745.

Bohart, G. E. & W. P. Nye. 1960. Insect pollinators of carrots in Utah. Utah State University Agric. Exp. Stn. Bull. 419.

Bowden, J. & G. J. W. Dean. 1971. The distribution of flying insects in and near a tall hedgerow. J. Appl. Ecol. 14:343-354.

Buchmann, S. L. & G. H. Nabhan. 1996. The forgotten pollinators. Island Press, Washington, D.C.

Bugg, R. L. 1987. Observations on insects associated with a nectar-bearing Chilean tree, *Quillaja saponaria*. Pan-Pacific Entomol. 63:60-64.

Bugg, R. L., 1991. Cover crops and control of arthropod pests of agriculture, pp. 157-163. *In* W.L. Hargrove [ed.], Cover crops for clean water. Proceedings, International Conference, West Tennessee Experiment Station, Jackson, Tennessee, April 9-11, 1991. Soil and Water Conservation Society. Ankeny, Iowa.

Bugg, R. L. 1992. Using cover crops to manage arthropods on truck farms. HortScience 27:741-745.

Bugg, R. L., C. S. Brown & J. H. Anderson. 1997. Restoring native perennial grasses to rural roadsides in the Sacramento Valley of California: establishment and evaluation. Restoration Ecology 5:214-228.

Bugg, R. L., L. E. Ehler & L. T. Wilson. 1987. Effect of common knotweed (*Polygonum aviculare*) on abundance and efficiency of insect predators of crop pests. Hilgardia 55(7):1-53.

Bugg, R. L. & N. F. Heidler 1981. Pest management with California native landscape plants. University of California Appropriate Technology Program, Research Leaflet Series #8-78-28.

Bugg, R. L. & C. Waddington. 1994. Managing cover crops to manage arthropod pests of orchards. Agric. Ecosystems Environ. 50:11-28.

Bugg, R. L. & L. T. Wilson. 1989. *Ammi visnaga* (L.) Lamarck (Apiaceae): associated beneficial insects and implications for biological control, with emphasis on the bell-pepper agroecosystem. Biol. Agric. Hortic. 6:241-268.

Cocannauer, J. 1950. Weeds: guardians of the soil. Devin-Adair, New York.

Davidson, R. H. & W. F. Lyon. 1987. Insect pests of farm, garden, and orchard. John Wiley and Sons, New York, N.Y.

Doutt, R. L. & J. Nakata. 1965. Overwintering refuge of *Anagrus epos* (Hymenoptera: Mymaridae). J. Econ. Entomol. 58:586.

Doutt, R. L. & J. Nakata. 1973. The rubus leafhopper and its egg parasitoid: an endemic biotic system useful in grape pest management. Environ. Entomol. 2:381-386.

Duelli, P., M. Studer, I. Marchand & S. Jakob. 1990. Population movements of arthropods between natural and cultivated areas. Biol. Conservation 54:193-207.

Emery, D. E. 1988. Seed propagation of native California plants. Santa Barbara Botanic Garden, Santa Barbara, California.

Fenner, M. 1978. A comparison of the abilities of colonizers and close-turf species to establish from seed in artificial swards. J. Ecol. 66:953-963.

Foster, M. A. & W. G. Ruesink. 1984. Influence of flowering weeds associated with reduced tillage in corn on a black cutworm (Lepioptera: Noctuidae) parasitoid, *Meteorus rubens* (Nees von Esenbeck). Environ. Entomol. 13:664-668.

Foster, M. A. & W. G. Ruesink. 1986. Impact of common chickweed, *Stellaria media*, upon parasitism of *Agrotis ipsilon* (Lepidoptera: Noctuidae) by *Meteorus rubens* (Nees von Esenbeck). J. Kansas Entomol. Soc. 59:343-349.

Frisbie, R. E. & J. W. Smith, Jr. 1989. Biologically intensive integrated pest management: the future, pp. 151-164. In J. J. Menn & A. L. Steinhauer [eds.], Progress and perspectives for the 21st century. Centennial National Symp. Entomol. Soc. Am., Lanham, Maryland.

Fye, K. E. 1980. Weed sources of Lygus bugs in the Yakima Valley and Columbia Basin in Washington. J. Econ. Entomol. 73:469-473.

Gange, A. C. & M. Llewelyn. 1989. Factors affecting colonisation by the black-kneed capsid *Blepharidopterus angulatus* (Hemiptera: Miridae) from alder windbreaks. Ann. Appl. Biol. 114:221-230.

Grout, T. G. & G. I. Richards. 1990. The influence of windbreak species on citrus thrips (Thysanoptera: Thripidae) populations and their damage to South African citrus orchards. J. Entomol. Soc. South Africa 53:151-157.

Hagen, K. S. 1962. Biology and ecology of predaceous Coccinellidae. Annu. Rev. Entomol. 7:289-326.

Hagen, K. S. 1974. The significance of predaceous Coccinellidae in biological and integrated control of insects. Entomophaga 7:25-44.

Hirose, Y. 1966. Parasitic Hymenoptera visiting the flowers of carrot planted in the truck crop field. Science Bull. Faculty Agric., Kyushu University, Series 2. 22:217-223.

Holl, K. D. 1995. Nectar resources and their influence on butterfly communities on reclaimed coal surface mines. Restoration Ecol. 3:76-85.

Jordan, N. 1996. Weed prevention: priority research for alternative weed management. J. Prod. Agric. 9:485-490.

Jordan, N. 1997. My view—a new strategy for progress in weed science. Weed Sci. 45:191.

Judd, W. W. 1970. Insects associated with flowering wild carrot, *Daucus carota* L., in southern Ontario. Proc. Entomol. Soc. Ontario 100:176-181.

Kennett, C. E., D. L. Flaherty & R. W. Hoffman. 1979. Effect of windborne pollens on the population dynamics of *Amblyseius hibisci* (Acarina: Phytoseiidae). Entomophaga 24:83-98.

Kevan, P. G., E. A. Clark & V. G. Thomas. 1990. Insect pollinators and sustainable agriculture. Am. J. Alternative Agric. 5:13-22.

Kido, H., D. L. Flaherty, C. E. Kennett, N. F. McCalley & D. F. Bosch. 1981. Seeking the reasons for differences in orange tortrix infestations. California Agric. 35: 27-28.

Kido, H., D. L. Flaherty, D.F. Bosch & K.A. Valero. 1984. French prune trees as overwintering sites for the grape leafhopper egg parasite. Am. J. Enol. Vit. 35:156-160.

Knauer, N. & U. Stachow. 1987. Activitäten von Laufkäfern (Carabidae Col.) in einem intensiv wirtschaftenden Ackerbaubetrieb—ein Beitrag zur Agrarökosystemanalyse. J. Agron. Crop Sci. 159:131-145.

Kruess, A. & T. Tscharntke. 1994. Habitat fragmentation, species loss, and biological control. Science 264:1581-1584.

Lanini, W. T., R. F. Long & J. Anderson. 1996. Preemergence herbicides have little effect on vigor of perennial grasses. California Agric. 50(5):38-41.

Lenz, L. W. 1977. Native plants for California gardens. Rancho Santa Ana Botanic Garden. Claremont, California.

Lewis, T. 1965a. The effects of an artificial windbreak on the aerial distribution of flying insects. Ann. Appl. Biol. 55:503-512.

Lewis, T. 1965b. The effect of an artificial windbreak on the distribution of aphids in a lettuce crop. Ann. Appl. Biol. 55:513-518.

Loertscher, M., A. Erhardt & J. Zettel. 1995. Microdistribution of butterflies in a mosaic-like habitat—the role of nectar sources. Ecography 18:15-26.

Maelfait, J.-P. & R. De Keer. 1990. The border zone of an intensively grazed pasture as a corridor for spiders. Araneae. Biol. Conservation 54:223-238.

Maingay, H., R. L. Bugg, R.W. Carlson & N.A. Davidson. 1991. Predatory and parasitic wasps (Hymenoptera) feeding at flowers of sweet fennel (*Foeniculum vulgare* Miller var. *dulce* Battandier & Trabut, Apiaceae) and spearmint (*Mentha spicata* L., Lamiaceae) in Massachusetts. Biol. Agric. Hortic. 7:363-383.

Mineau, P. & A. McLaughlin. 1996. Conservation of biodiversity within Canadian agricultural landscapes: integrating habitat for wildlife. J. Agric. Environ. Ethics 9:93-113.

Munz, P. A. (in collaboration with D.D. Keck). 1973. A California flora and supplement. University of California Press, Berkeley.

Ouyang, Y., E. E. Grafton Cardwell & R. L. Bugg. 1992. Effects of various pollens on development, survivorship and reproduction of *Euseius tularensis* (Acari: Phytoseiidae). Environ. Entomol. 21:1371-1376.

Patt, J.M., G. C. Hamilton & J. H. Lashomb. 1997. Foraging success of parasitoid wasps on flowers: interplay of insect morphology, floral architecture and searching behavior. Entomologia Experimentalis et Applicata 83:21-30.

Pickett, C. H., L. T. Wilson & D. L. Flaherty. 1990. The role of refuges in crop protection, with reference to plantings of French prune trees in a grape agroecosystem, pp. 151-165. *In* N. J. Bostonian, L. T. Wilson & T. J. Dennehy [eds.], Monitoring and integrated management of arthropod pests of small fruit crops. Intercept Ltd., Andover, Hampshire, England.

Pollard, E. 1971. Hedges. VI. Habitat diversity and crop pests: a study of *Brevicoryne brassicae* and its syrphid predators. J. Appl. Ecol. 8:751-780.

Pott, R. 1989. Historische und aktuelle Formen der Bewirtshaftung von Hecken in Nordwestdeutschland. Forstwissenshaftliches Centralblatt 108:111-112.

Schmidt, M. G. 1980. Growing California native plants. California Natural History Guides; 45. University of California Press. Berkeley, California.

Slosser, J. E., R. J. Fewin, J. R. Price, L. J. Meinke & J. R. Bryson. 1984. Potential of shelterbelt management for boll weevil (Coleoptera: Curculionidae) control in the Texas rolling plains. J. Econ. Entomol. 77:377-385.

Solomon, M. G. 1981. Windbreaks as a source of orchard pests and predators, pp. 273-283. *In* J. M. Thresh [ed.], Pests, pathogens and vegetation. Pitman, London.

Steffan, S. A. 1997. Flower-visitors of *Baccharis pilularis* De Candolle subsp. *consanguinea* (De Candolle) C. B. Wolf (Asteraceae) in Berkeley, California. Pan-Pacific Entomol. 73:52-54.

Stromberg, M. R. & P. Kephart. 1996. Restoring native grasses in California old fields. Restoration Manage. Notes 14:102-111

Swisher, R. G. 1979. A survey of the insect fauna on *Eriogonum fasciculatum* in the San Gabriel Mountains, Southern California. M.S. thesis, Dept. of Biology, Calif. State Univ., Los Angeles.

Van Eimern, J. (Chairman). 1964. Windbreaks and shelterbelts. World Meteorol. Assoc. Tech. Note No. 59.

van Emden, H. F. 1962. Observations of the effect of flowers on the activity of parasitic Hymenoptera. Entomologist's Mon. Mag. 98:265-270.

van Emden, H. F. 1965. The role of uncultivated land in the biology of crop pests and beneficial insects. Sci. Hortic. 17:121-136.

Young, J. A. & C. G. Young. 1990. Collecting, processing and germinating seeds of wildland plants. Timber Press, Portland, Oregon.

Within-field and Border Refugia for the Enhancement of Natural Enemies

STEVE D. WRATTEN—Lincoln University, Department of Entomology & Animal Ecology, P.O. Box 84, Canterbury, New Zealand.

HELMUT F. VAN EMDEN—School of Plant Sciences, University of Reading, Whiteknights, Reading, RG6 2AS, U.K.

MATTHEW B. THOMAS—Center for Population Ecology, Imperial College at Silwood Park, University of London, Ascot, Berkshire, SL5 7PY, U.K.

Introduction

The idea that diversification of a crop or of its margins can benefit the natural enemies of pests is intuitively logical and seems to involve common-sense ecological principles. It has become part of gardening folklore that mixing vegetables and flowers or crops of different types, as well as the provision of certain "companion plants" has benefits; it has indeed been practiced by gardeners and their professional advisers for many decades. However the mechanisms behind these purported interactions are usually barely understood. At the farm scale, some of the mechanisms have begun to be studied in some detail and are now being exploited in arable crops around the world. Sometimes, however, the potential role of unmanipulated wild areas adjacent to crops is overlooked.

The field margin definitions in Figure 1 will help in the discussions that follow. There are at least five potential mechanisms involved in the interaction between pest dynamics and within-field or border refugia. These are:

1. The provision of overwintering or aestivation sites.
2. The enhancement of the quantities of pollen and/or nectar available to predators and parasitoids.
3. The provision of alternative prey for predators or alternatively the hosts of parasitoids.
4. The provision of plant food for carnivores that are occasionally phytophagous.
5. A more subtle interaction involving tri-trophic level interactions between the crop and/or the non-crop, the pest and/or the non-pest and the predators and parasitoids.

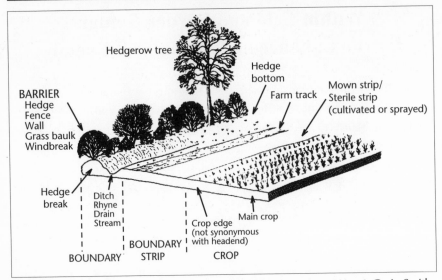

Figure 1. The main components of an arable field margin. (From Way & Greig-Smith, 1987.)

Mechanisms 1 and 2 will be covered in some detail here (see also Gurr et al.in press), while 3 and 4 are relatively minor; 5 was covered recently by van Emden and Wratten (1991).

All these processes have potential negative as well as positive effects on population dynamics of pests, which is why it is important to understand the mechanisms involved, rather than to rush headlong into the intuitive creation of such refugia. Sometimes the improvement or creation of refugia benefits beneficial arthropods other than those with potential in biological control. Conservation of endangered or aesthetically-prized Lepidoptera will be covered later. Nearly all the protocols and ideas given in the following sections share a common theme. They attempt to increase the diversity of the cultivated landscape, thereby returning it to the more complex and structured physical and biological environment assumed present before modern agriculture. It does not follow, however, that making an agricultural landscape more diverse will necessarily lead to greater predator-prey stability, as described by May (1973). Indeed, there is little experimental evidence from cropping systems that crop monocultures cause unstable pest populations (Way 1977). As Perrin (1977) pointed out, in many cases effective pest management can be designed around very simple forms of diversity unique to man-manipulated systems.

Agriculture in the "West," especially in the European Economic Community, is at a crossroads; during the writing of this chapter changes to the Common Agricultural Policy have been accepted, whereby arable farmers will be required to remove 15% of their cropped area because of overproduction of grain within the Community. Virtually the only option available for dealing with this "surplus" land is "set-aside," in which land is left to develop "naturally" but with only cutting, not cropping, as an option. Some ecologists believe that "extensification" of the land (i.e., lower inputs throughout the cropped area, rather than an intensive 85% and an uncared-for 15%) would do more for wildlife and biocontrol (Potts 1991; Sotherton 1991). However, despite an agricultural template set by European politicians rather than by ecologists, useful options still remain for encouraging biocontrol, and wildlife generally, via within-field and border refugia. This chapter covers these options. However, given the "meta-population" (*sensu* Levins 1970) characteristics of much of the fauna of cultivated land, and the likely movement of many groups from refugia to the intensively-managed crop, refugia next to very intensive cultivation may be thought by some to be the agricultural equivalent of establishing a kindergarten next to a busy freeway.

Field Boundaries

The commonest type of field boundary in Britain and northwest continental Europe is the hedge. This may be only a grassy bank but typically is "...a row of closely planted bushes forming a boundary to land" (Pollard et al. 1974). The first reference found by these authors to the existence of hedges in the U.K. was that of one built by Ida of Northumbria, in northeast England, around his new settlement; this was recorded in the Anglo-Saxon Chronicle to have existed in A.D. 547. Hedges a century later protected plants from cattle and are mentioned in the Laws of Ine, King of Wessex. These Crops Laws formed a basis for later ones attributed to King Alfred (Pollard et al. 1974).

The greatest development of hedges in the U.K. began with the Acts of Parliament that enclosed individual parishes in the seventeenth century, and the disappearance of the open or common field system continued via the enclosure, for villagers, of areas designated for meadowland, common grazing, rotational crops, etc. The last major Enclosure Act was in 1903 in Yorkshire, U.K., although most were passed between 1760 and 1820, in the reign of George III (Pollard et al. 1974).

Developments in agriculture in the eighteenth and nineteenth centuries, such as the seed drill and improvements in plough design, led to larger fields and loss of hedges. Wool production declined and arable farming dominated, and hedge losses are presumed to have continued. However, the period from 1950 onwards in the U.K. led to an accelerating rate of hedgerow loss. This continued at least until the 1980s, despite new plantings (Anonymous 1986; Greaves & Marshall 1987). As hedges have an important role in biocontrol (see below) this loss of hedge habitat is important.

Overwintering Sites for Predators

Field boundaries have long been considered important overwintering sites for beetles and other invertebrates. Pollard (1971) demonstrated that a hawthorn (*Crataegus*) hedgerow provided a variety of different habitats. The effect of removing the bottom flora of a hedgerow (by applying a herbicide) was to decrease the abundance of several species of overwintering carabid beetles (Coleoptera: Carabidae) including *Agonum dorsale* Pont., *Harpalus rufipes* (Degeer), *Loricera pilicornis* (F.), subsequently shown to be potential aphid predators (Sunderland & Vickerman 1980). Similarly, Gorny (1970) and Bonkowska (1970) both identified shelterbelts (mature boundary habitats dominated by tree and shrub species and 14-17 m high with an average breadth of 36m in the example of Bonkowska [1970]) as important permanent features of the agricultural environment. These provide alternative habitats for several species of predatory Carabidae.

Luff (1966) showed that the tussock-forming grasses cock's-foot *Dactylis glomerata* and tufted hair grass *Deschampsia caespitosa* harboured large communities of beetles both in summer and winter. It was demonstrated that a temperature of at least −17°C outside the tussock would be needed before the lethal temperature (−8.5°C) of any of the overwintering species was reached inside the tussock. Many species of beetle (as well as other insects) probably have evolved the habit of overwintering in grass tussocks in order to obtain shelter from cold conditions. Desender and D'Hulster (1982) and D'Hulster and Desender (1983), studying the hibernation of staphylinid beetles in field-edge sites, also concluded that dense vegetation, a deep aerated sod and a well developed litter layer provided a buffering of temperature fluctuations and made such a biotope a suitable overwintering site. Furthermore, in a boundary study on the overwintering of the carabid beetles *A. dorsale*, *Bembidion lampros* (Herbst) and *Demetrias atricapillus* (L.), Desender (1982) established positive correlations between the densities of the overwintering predators and the biomass of living and dead grass, and the mean depth of the compact sod layer.

Sotherton (1984, 1985) and Wallin (1985, 1986) showed that many polyphagous predators overwinter almost exclusively in field boundaries. Boundary quality differed markedly between sites, with habitats such as raised banks with rough grass cover (which occur at the bases of hedges and fence lines in many parts of Europe; Greaves & Marshall 1987) being the preferred habitat types (Sotherton 1985).

Coombes and Sotherton (1986) studied the phenology of crop invasion by the boundary overwintering predators in the spring. They revealed that although some species of Staphylinidae and Linyphiidae (Araneae) could invade the crop using aerial dispersal mechanisms, certain species of Carabidae entered the crop by walking only, and therefore significant numbers of individuals of the carabid species studied were not found at the field centres until June (figure 2).

In modern arable systems, the accommodation of increasingly large machinery has in western Europe led to the removal of hedges to produce larger fields (Davies & Dunford 1962; Edwards 1970). This process accelerated rapidly in the late 1940s as a result of government policy and the introduction of grants for hedgerow removal. Over the last 50 years ca. 40% of the hedges that were recorded in lowland Britain in 1940 have disappeared (Anonymous 1991). Despite new plantings, it is estimated that there is still an annual net loss of hedgerows in England and Wales (Anonymous 1986; Greaves & Marshall 1987). Accompanying this, the incidence of spraying herbicides to control weeds in hedgerows has increased (Boatman 1989), reducing the quality of boundaries as sites for overwintering predators. This environmental degradation has two major consequences for pest biocontrol:

1. With increasingly large fields, spring colonisation of field centres by non-flying predators such as many Carabidae could be impaired (Wratten et al. 1984; Coombes & Sotherton 1986; Wratten 1988), leading to reduced predation rates in the field centres during the normal time of colonization (Chambers et al. 1982). Related to this, field-scale applications of broad-spectrum insecticides have been shown to result in aphid resurgence due to reduced predator pressure (a product of limited predator re-colonisation) at the field centre (Duffield & Aebischer 1993).

2. With the small boundary: field area ratio of large fields and reduced availability of non-crop habitats, the overall densities of polyphagous predators in arable ecosystems may become reduced, and their potential to influence pest numbers limited accordingly.

Figure 2. Density of the carabid beetle *Demetrias atricapillus* at different distances from a field boundary in which it overwinters (from Coombes & Sotherton 1986).

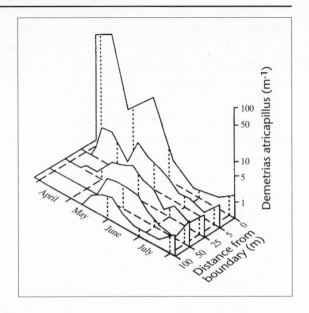

The above outline shows that some of the botanical and environmental factors that could limit the effectiveness of native natural enemies via their overwintering requirements have been identified (also see Bugg and Pickett, this volume). With this knowledge it has become possible to augment their densities within an integrated control programme and thus raise the natural enemy: pest ratio (van Emden 1988; van Emden and Wratten 1991; Gurr et al. in press). This has been carried out via the manipulation of overwintering biology in the U.K. by Thomas et al. (1991) and in Sweden by Chiverton (1989). In the U.K. study, facsimiles of the raised earth bank on which many hedges are planted were created across arable fields in southern England. These banks were 0.4m high, 1.5m wide and 300–400m long and crossed the centres of the fields. The banks did not extend completely to the existing field margins; areas of cultivated crop (20m wide in small fields and about 50m wide in large ones) were left at each end to allow movement of farm machinery from one field side to the other, without damaging the bank.

Following an application of a broad-spectrum herbicide to remove broadleaved weeds that colonized the banks following their establishment, sections of each new bank were hand-sown at commercial sowing rates with various grass species in a linearly randomized block design, with six blocks per bank. Each block contained one replicate of each of eight treatments, each replicate being 6m long. In four of the treatments, grasses were sown in single-species stands. The grasses and seed rates were cock's foot (*Dactylis glomerata*) 3 g m^{-2}, perennial rye-grass (*Lolium perenne*) 3 g m^{-2}, creeping bent (*Agrostis stolonifera*) 8 g m^{-2}, and Yorkshire fog (*Holcus lanatus*) 4 g m^{-2}, and were selected for their qualities of fast growth and potentially good winter cover, requiring little maintenance; they included both matt-forming (*L. perenne* and *A. stolonifera*) and tussock-forming (*D. glomerata* and *H. lanatus*) species. These species are not invasive, aggressive weeds of the crop so that the earth banks could not be considered as foci of pernicious weeds. As well as the single-species treatments, fifth and sixth treatments containing mixtures of three (*A. stolonifera* excluded) and the four species, respectively (to study the effects of seed "cocktails") were sown. The seventh treatment consisted of bare ground controls (maintained by hand-weeding or the use of a herbicide).

Predator densities in the first winter (assessed by surface searching in quadrats) differed significantly among treatments, with *D. glomerata* plots reaching levels of ca. 150 predators/m^2. These were mainly Carabidae, Staphylinidae and Araneae.

In subsequent winters destructive sampling became necessary as surface-searching was inefficient due to the density of the vegetation. Densities reached nearly 1500 predators/m^2, with *D. glomerata* and *H. lanatus* being the most favoured grass species (Table 1).

Hourly temperature recordings were taken automatically from the ridges during winter. Probes were positioned at the base of each grass type to monitor any differences that might exist between the grass species. Bare earth soil-surface and air temperatures (0.3m above ground) were also recorded. Daily means and transformed daily variances were compared.

There were no significant differences between the mean temperatures between micro-habitats on the bank. The transformed variances for the temperatures at each site, however, showed that the within-soil and within-grass refuges provided habitats with the least variable temperature. *D. glomerata* and *H. lanatus* provided the temperature environments that were significantly less variable than those in the other treatments.

The high predator densities achieved in two years were impressive, but for the predators to realise their potential in biocontrol, emigration from the ridges in spring must mimic that from "natural" boundaries. Vacuum-net sampling in transects out from the ridge into the adjacent crop showed that, in spring, there was indeed a wave of emigration, best shown by the carabid *Demetrias atricapillus* (L.), which disperses mainly by walking (Figure 3). These and other data led to a protocol for farmers in which ridges were recommended at ca. 200m intervals across fields.

Table 1. Mean densities (m^{-2}) of groups of polyphagous predators on a within-field refuge (a grass-sown earth ridge). Treatments within a column with the same letter do not differ significantly at $P = 0.05$ (from Thomas et al. 1991).

	Number of Predators					
Treatment	Carabidae Demetrias atricapillus	Total	Staphylinidae Tachyporus hypnorum	Total	Total Araneae	Total Predators
Agrostis stolonifera	8.3 (c)	157.5 (b)	129.2 (a)	160.3 (ab)	170.0 (bc)	487.8 (b)
Dactylis glomerata	922.93 (a)	1112.5 (a)	120.8 (a)	152.5 (b)	222.5 (ab)	1487.5 (a)
Holcus lanatus	662.5 (a)	765.0 (a)	227.1 (a)	272.4 (a)	360.3 (a)	1397.7 (a)
Lolium perenne	64.5 (b)	107.4 (b)	26.3 (b)	50.6 (b)	117.7 (c)	275.7 (b)
Field	0.0	10.0	2.5	3.1	10.0	25.6

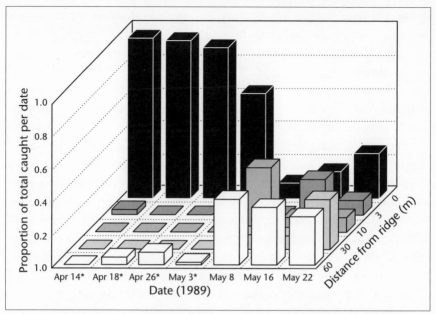

Figure 3. Emigration pattern of the carabid *Demetrias atricapillus* from a mid-field refuge. *Denotes significant between-distance differences at the 5% level. Proportions were used to correct for any between-date variation in sampling efficiency due to abiotic factors (from Thomas, et al. 1991).

The Economics of Ridge Production

The costs of establishing these within-field ridges were outlined in Wratten (1988). Updated to 1998 UK/US prices, the combination of labour costs for bank establishment (1–2 person days) with yield loss due to the land taken out of production (ca. £30/$50 assuming average yield = 6t/ha at £110/ $180/t), together with the cost of grass seed (ridge sown with only *D. glomerata* = £5/$8), would amount to ca. £85/$140 in the first year for a 20 ha winter wheat field. Subsequent costs would comprise gross yield loss at only £30/$50 per year. However, an aphid population kept below a spray threshold by enhanced natural enemy populations could save £300/$495 per annum in labour and pesticide costs for a 20 ha field; alternatively, prevention of an aphid-induced yield loss of 5% could save £660/$1090 for a field of the same size.

Once such ridges (often called beetle banks) are established, the competitive nature of the grass species chosen should exclude most noxious weeds; such species as cleavers (*Galium aparine*) and black grass (*Alopecurus myosuroides*) were occasional on the ridges and occurred only in areas of poor grass establishment. Such sites, however, were a consequence of hand sowing small areas. In a similar study in Sweden it was possible to establish equivalent new habitats by machine drilling, thus improving cover by sown grass (Chiverton 1989). Another potential problem is that autumn or spring applications of grass herbicide to adjacent crops could damage the grass cover; a bare rotovated or residual herbicide-treated strip between ridge and crop would minimise this, although obviously this would increase the area of land taken out of production. Analysis of agrochemical drift, using methods similar to those of Cilgi and Jepson (1995) would quantify this risk. In the absence of such a strip, no such damage to the grasses on the ridges in the present study occurred during three commercial cropping seasons. Furthermore with respect to herbicide drift/management, *Festuca* spp. are not affected by many graminicides (Marshall & Nowakowski 1991) and therefore may be especially appropriate. The role of such grasses as overwintering refuge sites has been quantified in the U.K. by Harwood, working at Southampton University in collaboration with Willmot Conservation. The potential role of *Festuca* spp. warrants further investigation (Harwood et al. 1992).

Beyond the potential for enhancing biocontrol in arable land and thereby reducing insecticide use, within-field ridges may also reduce soil erosion. Wind erosion is not likely to cause great problems in the U.K., but Evans and Cook (1987) concluded that water erosion occurred widely throughout England and was increasing. Evans (1985) estimated that some 40% of arable land in England and Wales was at risk. Soil erosion on slopes, however, is undoubtedly reduced by hedgerows along contours, the hedge underbanks (equivalent to the within-field ridges) being of particular importance in reducing surface run-off (Forman & Baudry 1984). Similarly, from a study in California, Bugg and Anderson (this volume) suggested restoration of native perennial bunch grasses, along roadside verges, could help to control soil erosion. Moreover, the distance between hedgerows is an important factor in inhibiting soil erosion, and Pihan (1976) estimated 40% more erosion with a doubling in distance between hedgerows. Similar patterns of mineral nutrient run-off may be expected (Forman & Baudry 1984). However, from a management perspective, soil erosion is not the only factor governing field size, nor is the number of ridges per field. Several other biological and agronomic factors need to be considered when designing a field system.

Biological Considerations Relating to Arable Land Management

The experimental system described in the previous section is available, with simplification, as a management strategy, but some questions remain unanswered. For example, it is not known from that study whether the introduction of within-field ridges actually contributed to a reduction in aphid numbers. As there can be large variations in aphid numbers between fields (in some cases between adjacent fields on the same farm) (Wratten et al. 1991), no attempt was made to compare aphid populations between control fields and the three fields containing ridges. Dennis and Wratten (1991) showed using small inclusion barriers that enhanced predator densities resulted in reduced aphid numbers. Therefore, enhanced predator densities resulting from the introduction of a ridge could be expected to provide an increase in predator pressure and a subsequent reduction in aphid numbers.

It is not known at this time whether the ridges have an optimum "lifespan" and whether a point may be reached where further successional changes might prove less favourable for agricultural purposes (as suggested in a study on habitat creation by Nentwig (1988 and this volume). Six years' monitoring to date has shown no overall decline in predator populations, however (A. Macleod, unpublished). Similarly, it is unclear whether the creation of predator populations at the field centres is simply a consequence of redistributing existing populations within the field or whether populations are enhanced. That is, is overwintering mortality a "key factor" (Varley & Gradwell 1960) in predator populations and, therefore, does the creation of ridges reduce this mortality? Also, although one ridge may influence predator dispersal patterns in the spring, optimum distance between ridges is still uncertain. Refugia may influence predator ecology only at the field scale, or at a larger, landscape or metapopulation scale as well (Corbett this volume; Opdam 1990).

The spider complexes on the ridges shifted from pioneer r-selected species towards more specialist K-selected species. Similarly, Brown and Southwood (1987) showed that "young fields" of 0–2 years were exploited by opportunistic plants and insects (r-strategists). An "old field" of 6–8 years contained more K strategists and a diverse complex of predators.

Forman and Baudry (1984) suggested that hedgerow species diversity initially increases as birds and wind bring in new species. These early colonists will be largely field and forest-edge species, but as trees develop and the shrub layer becomes more dense, forest interior species will invade. Near established woods, this process probably takes place more rapidly, as appears to be the case for wrens (Williamson 1969), butterflies (Pollard *et al.* 1974), snails (Cameron *et al.* 1980), and shrubs (Helliwell 1975). Interaction among forest, hedgerow, and field habitats has been recorded by Wallin (1985, 1986) for several species of carabid. Forman and Baudry (1984) concluded that hedgerows function as corridors for movement by many plants and animals. However, recent work at the Norwegian Institute of Nature Conservation (NINA) at As (Fry & Main 1992) suggest that field-border habitats that may represent refugia for some invertebrates may be impediments to movement for butterflies. Fry and Robson (1994) at NINA have shown that some Lepidoptera may be "trapped" ecologically by continuous field boundaries (Figures 4, 5 & 6). In this case, intermittent hedges may permit more inter-field movement by butterflies than do more "dense" hedges (Figures 5 & 6). Similarly such hedges may harbour high predator densities (e.g. Sotherton 1984, 1985), but they can restrict inter-field movement by insects such as Carabidae (Figure 7). Their role in delaying recolonisation by predator populations following application of insecticides (Wratten et al. 1988; Sotherton et al. 1987) is under-studied. Many predators are slow dispersers and are late to penetrate to the centres of large sprayed areas, allowing pests to resurge (Duffield & Aebischer 1993; see Bugg and Pickett, this volume). Some boundary types could possibly contribute further to the slow recovery of some predatory groups (see Figure 7), or delay their emigration from fields when populations of their prey have been reduced by insecticides.

The concept of connectivity in the arable landscape involves interpatch distance, density of "stepping stones" and corridors, permeability of the landscape matrix for dispersers, i.e., the landscape characteristics governing the dispersal flow (Opdam 1990). Field size, shape, and position relative to other habitats, are important factors in determining the speed and direction of succession on the ridges. A ridge placed in the centre of a very large field for example (i.e., a large distance from source populations), might undergo succession more slowly than a ridge placed within a smaller field with greater connectivity. Differences in connectivity could be expressed in two ways. First, adoption of the new habitats as overwintering sites by predatory arthropods would be slower in the large field than in the small one because the predators themselves might be slow to disperse and because suitable prey items also could be slow to colonise. Second, if the ridges do indeed reach an optimum successional stage (as discussed above), then the ridge in the small field will have a shorter life than that in the large.

Small fields take proportionally longer to cultivate than large fields,

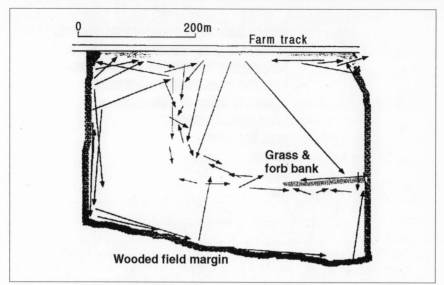

Figure 4. Butterflies are closely associated with edge habitats on arable farms. The field boundaries shown here are a farm track with a roadside verge, a grassy bank, and a conifer forest. Each arrow depicts a single butterfly tracking over 15 minutes. The start, finish and direction of movement are shown by the arrows. (From Fry & Robson 1994.)

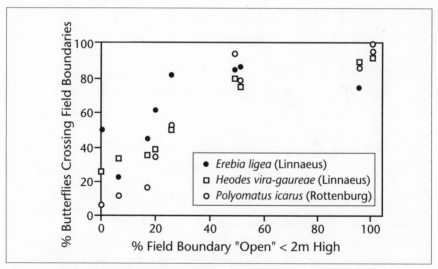

Figure 5. Butterfly (Lepidoptera) movement across landscapes for *Erebia ligea* (Linnaeus) (Lepidoptera: Satyridae), *Heodes vira-gaureae* (Linnaeus) (Lepidoptera: Lycaenidae), and *Polyommatus icarus* (Rottenburg) (Lepidoptera: Lycaenidae). There are clear similarities between these species, which appear to cross boundaries of herbs and shrubs freely but whose movements are restricted by taller tree vegetation. *Erebia ligea* is typical of open forest habitats and crosses wooded boundaries at twice the frequency of the other species. (From Fry & Robson, 1994)

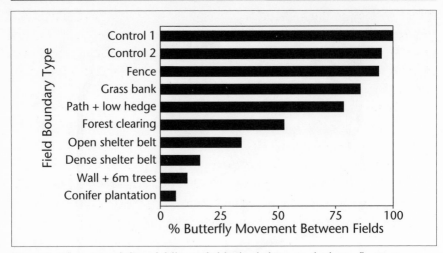

Figure 6. The permeability of different field edge habitats to the butterfly *Heodes viragaureae*. Results are for 100 encounters with each boundary type. Control 1 = random lines across meadows. Control 2 = edges between two crops with no boundary zone. (From Fry & Robson 1994).

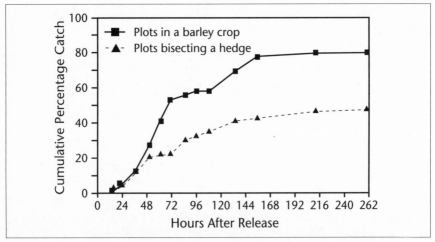

Figure 7. Cumulative percentage recapture of marked carabid beetles (*Pterostichus madidus*) in gutter traps. Traps were at one end of barrier plots and beetles were released at the opposite end. (From Mauremootoo et al. 1995).

because of the proportionally greater time spent turning with smaller farm machinery. Four hectares is the minimum economically viable field size (Le Clezio 1976). Beyond 20 ha, the reasons for larger fields are tenuous (Anonymous 1991). Therefore, within-field ridges could be used to reduce the size of fields to 20 ha or less. However, Forman and Baudry (1984) suggest that mesh size, the area of landscape elements enclosed by the boundary lines, is important in relation to grain size (the area required for a species to feed, reproduce etc.). Thus in Brittany, France, the predatory ground beetle *Poecilus cupreus* L. disappeared where the average mesh size of fields was greater than 4 ha (Deveaux 1976). Of course, it is important that predatory species not only persist within the arable landscape, but penetrate to the field centres in substantial numbers.

Thomas et al. (1991) and Coombes and Sotherton (1986) indicated that the Carabidae were the predators most influenced by field size during the spring emigration period. Staphylinidae and Linyphiidae by contrast can disperse aerially. Reduced field size would allow carabids to colonise field centres during the aphid build-up phase, when their potential for biocontrol is highest (Edwards et al. 1979; Chambers et al. 1982) and prior to the time when application of aphicides might be considered necessary. The current U.K. M.A.F.F. threshold for aphicide application against the grain aphid *Sitobion avenae* F., (Homoptera: Aphididae) is 66% stem infestation at the flowering stage of wheat. Flowering occurs between mid- and end-June in southern U.K. Coombes and Sotherton (1986) showed that of the carabid species studied, all were evenly dispersed by late May to early June. If both field edge and ridge populations contribute to field colonisation, then a distance of 200m between ridge and boundary would result in even field cover at the appropriate time. Thus, a square 20ha field (c 450m x 450m) would require one centrally positioned ridge to achieve uniform predator cover early in the season, the ridge design, with no connection between the ridge ends and the existing margins allowing the field still to be worked as a 20ha unit.

Natural Enemy Population Dynamics

An indication of whether predator populations were enhanced or simply redistributed via ridge creation can be provided by comparing overwintering data from several field boundary habitats with data collected from the boundaries surrounding experimental field systems with ridges. An estimate of the existing population range of a predator group can be calculated (using 95% C.L.) for one field system (ridge + boundary populations). Similarly, an estimate of the population range for a field lacking a ridge can be calculated using data from other typical arable boundaries. If the two predicted population ranges for a particular group do not overlap, then the populations can be considered different, a higher mean value in the existing population indicating enhancement. When two fields were compared in this way for a range of predator groups, although the means of the populations exceed predictions for most groups, no enhancement was indicated (Thomas 1991). However, even if the ridge predator populations are a product of redistribution alone, the enhanced field penetration in spring is still beneficial from a pest management perspective.

Overwintering habitat can also benefit other natural enemies. Parasitoid aphidiid wasps, (Hymenoptera) can reduce grain aphid densities (Vorley & Wratten 1985). Aphidiidae emanating from grasslands colonise late-sown winter wheat and spring wheat fields in the U.K. in the spring; early-sown grain crops may harbour actively-foraging parasitoids throughout mild winters (Wratten & Powell 1991). Fougeroux et al. (1988) found an inverse relationship between early parasitism rates in cereal fields and subsequent peak aphid densities. Thus, early-season colonisation by aphidiids is important, although the spatial dynamics of parasitoids on farmland are poorly understood. Vorley and Wratten (1987) demonstrated that one early-sown winter wheat field produced enough female *Aphidius* parasitoids to colonise approximately twenty-five nearby late-sown winter wheat fields by late May. Simulations suggest that parasitoids can keep grain aphid populations at one-seventh of the level at which they would have been in the absence of parasitoids (Vorley & Wratten 1985).

Parasitoid colonisation could be facilitated by perennial grasses in field margin strips. These should harbour aphids during winter and represent a source of emigrating parasitoids in the spring. There is increasing interest in margin strips of grass in the U.K., as biological barriers to weed egress from the field boundaries, to provide access for people, animals and farm machinery and to "set aside" cultivated land. With regard to parasitoids, research on strips should concern: (1) aphid performance on various grass species/cultivars; (2) parasitoid foraging in grasses of differing structure and growth forms; (3) mowing and cutting regimes, etc; (4) spatial dynamics of adult parasitoids; (5) parasitism rates in the adjacent crop as a function of distance. Parasitoids could also be enhanced by satisfying their spring and summer requirements, as outlined in the next section.

Enhancement of Pollen and Nectar Resources in Refugia with Reference to Syrphidae (Diptera)

Impact of Field Margins on Syrphid Populations

Female Syrphidae (Diptera) feed on pollen to sustain egg production. This essential food may affect syrphid oviposition and consequently their control of aphids. Modern agronomic practices have removed the flowering weeds used by adult syrphids (Boatman 1989).

Von Klinger (1987) found that margin strips of *Sinapsis arvensis* and *Phacelia tanacetifolia* led to higher densities of polyphagous predators in the strips and in adjacent fields than in wheat plots without strips. Syrphid adults attained higher densities in the strips than in the field, presumably because the flies foraged on the *P. tancetifolia* and *S. arvensis*. Impact of different predatory groups on aphid populations was not quantified; however, there was a trend towards lower aphid densities in the field with adjoining strips. Sengonça and Frings (1988) found syrphid adults more abundant in sugar beet (*Beta vulgaris*) plots with *P. tanacetifolia* margin strips than in sugar beet monocultures. The density of syrphid eggs and larvae was highest in control plots where aphid density was highest.

Impact of Flower Diversity on Syrphid Diversity

Along field boundaries, Syrphidae were less diverse and numerous in narrow margins of grass than in wider (1.5–2m) margins of dicotyledonous plants (Molthan & Ruppert 1988). The greatest diversity and density occurred on an 8 m wide margin with an adjacent embankment and a high diversity of flowering plants. Kühner (1988) found that syrphids were more abundant where crop edges comprised a range of flowering plants than where the crop edge was characterised by different species of grasses and few flowering dicotyledons.

Kühner (1988) also found that cereal aphid density was lower in herbicide-free plots than in herbicide-treated plots where naturally occurring boundaries consisted of a range of flowering weeds. However, in crop edges dominated by grasses there was no significant difference in aphid density between two treatments. Felki (1988) found that syrphid larvae were more numerous in border strips treated with herbicide and having higher aphid infestations than in herbicide-free edges. However, carabids, staphylinids, and spiders were more abundant in the herbicide-free edges.

Until recently, only van Emden (1965) and Pollard (1971) had studied the impact of field margin composition on the diversity and density of syrphids in the U.K. However, recent work on southern English field boundaries has quantified hoverfly "preference" for native wild flowers in field margins (Cowgill et al. 1993a). The study ranked native wild flowers in relation to their use by the predatory syrphid, *Episyrphus balteatus* (De Geer). Some plant species were strongly preferred, giving a guide to possible floral manipulation of boundaries via herbicide/cutting regimes (see Harwood et al. 1992).

In practice, once native plants have been thus ranked, there are at least four opportunities for them to be managed and exploited in refugia. These are:

1. The use of selective herbicides in field margins to remove potentially invasive weeds to leave the desirable species. New experimental herbicides have potential to improve refugia quality. A matrix of differential selectivity of herbicides (Boatman 1989) represents a management tool for field boundary maintenance. Quinmerac (BASF: BAS 518H) and fluoroxypyr (Dow Agriculture) are relatively specific control agents for cleavers (*Galium aparine*) (Boatman 1989).

2. The reduction of agrochemical inputs in the outermost crop. This encourages the development in the headland of weeds and associated insect fauna—the so-called "Conservation Headland" approach (Sotherton 1991). This method was developed in an attempt to enhance levels of the natural insect supply for chicks of grey partridge (*Perdix perdix* L.); the weeds in the Conservation Headland supported an insect fauna that led to up to a threefold increase in chick survival rate. Table 2 summarises the protocol for the creation of these headlands. The flowering weeds in the headlands are also a useful resource for hoverflies; more syrphids and syrphid eggs were laid in these areas than in adjacent replicated control plots (Cowgill et al. 1993b), and the difference in the behaviour of the flies confirmed that they were indeed foraging more in the conservation headland treatment. Cowgill et al. (1993b) demonstrated two possible disadvantages associated with hoverfly use of headlands. The increase in syrphid densities occurred only in the headland, and eggs were laid mainly on the aphid-infested weeds in the crop. If crops had sustained higher aphid densities, the egg distribution pattern may have been different.

Chiverton and Sotherton (1991) trapped arthropods in conservation headlands and compared their densities and diet (via dissection) with those from fully-sprayed plots. Unsprayed headlands had more weed species, higher weed densities, greater weed biomass, higher percentage weed cover and higher non-pest densities, especially the species important in the diet of insect-eating gamebird chicks. They also had more predatory arthropods, especially polyphagous species and their alternative prey. Within 6 x 10m plywood enclosures, which were designed to prevent

Table 2. Guidelines for the establishment of conservation headlands (from Sotherton 1991).

	Autumn-sown cereals	Spring-sown cereals
Insecticides	Yes (avoiding drift into hedgerows) but only until 15 March	No
Fungicides	Yes (not pyrazophos after 15 March)	Yes (except pyrazophos)
Growth regulators	Yes	Yes
Herbicides Grass weeds	Yes (but avoid broad-spectrum residual products—use only those compounds approved for use*)	
Broad-leaved weeds	No (except those compounds approved for use against specific problem weeds, e.g., cleaver control with fluoroxypyr). If broad-leaves weeds are a problem in the headland, consult the Field Officer.	

*Tri-allate, dichlofop-methyl, difenzoquat, flamprop-m-isopropyl, fenoxaprop-ethyl.

beetle movement, pitfall traps were set up and emptied twice weekly in June and July, two months after the herbicide was applied to half the plots. There were no significant between-treatment differences in the total pitfall trap catch of the two most numerous carabid beetle species, *Pterostichus melanarius* (Illiger) and *Agonum dorsale* (Pontoppidan). However, a significantly greater proportion of female *A. dorsale* were caught in the enclosures in treated plots than in untreated plots. The guts of gravid and non-gravid female and male *P. melanarius* and gravid female and male *A. dorsale* from untreated plots contained more solid arthropod food remains than did those from sprayed headlands. Both carabids had also taken significantly more meals and preyed upon a wider variety of arthropod food in untreated plots. As a consequence fewer pest cereal aphids were consumed by beetles in unsprayed plots. Higher numbers of eggs per female were found for both species of carabid in untreated plots. This work was carried out in June and July, a period when predator species overwintering in the field boundary would be well dispersed in the crop. In summer, therefore, refugia of this type could improve recruitment to the predator population via improved individual fitness. In spring, when species such as *A. dorsale* pass through the headland area (Coombes & Sotherton 1986), these refugia could have a positive or negative effect on further emigration by such species. Mols (1979) showed that a carabid's searching rate was affected by gut extension, suggesting these refugia could delay predator emigration to the field beyond the headland.

Other groups of insects with aesthetic if not predatory value on farmland are the butterflies. Although the Norwegian work referred to earlier showed that boundaries could be an impediment to butterfly movement, conservation headlands appear to have a positive effect (Sotherton 1991), and may affect the long-term population ecology of this group through improved fitness and survival (c.f. Carabidae in the work of Chiverton & Sotherton [1991] above). Sotherton (1991) showed that eleven non-pest and two pest butterfly species occurred in significantly higher numbers in unsprayed headlands.

3. The drilling of native non-crop species in the field margin strip (see Figure 1). This approach has been pioneered by Marshall and Nowakowski (1991) and by Harwood et al. (1992). A range of native species are drilled in the margin strip (2–3 metres wide in this case) and various cutting and herbicide regimes are used to manage competitive interactions between the aggressive and less aggressive species. This is a powerful manipulative technique that can help the establishment of drilled or unsown beneficial broad-leaf species. Use of the phenoxyproprionic acid grass herbicide fluazifop-P-butyl (sold by ICI as "Fusilade 5" in the U.K.) leads to good stands of the umbellifer wild carrot (*Daucus carota*), for instance. The sulfonanilide plant growth regulator mefluidide (sold by ICI as "Echo" in the U.K.) had more mixed, and less useful effects (Table 3). These plots have several aesthetic and biocontrol advantages:

1. They are attractive and could be an aesthetically acceptable way of removing land from commercial cultivation.

2. They form a biological barrier between the field margin and the crop, minimizing the colonization of the crop by invasive weeds such as *Galium aparine*.

3. The nectar and pollen resources would be beneficial to syrphids and parasitoids, contributing to biocontrol in the adjacent crop. The seed mixtures on trial cost around £200/ha but as the drilled margin strips are only 2–3 metres wide, this represents an area of ca. 0.2–0.3ha in a square 20ha field for which such seed costs would be £40–£60; herbicide costs should, and labour costs

Table 3. Frequencies (%) in 1990 of sown and unsown plant species in field margin plots treated in 1989 with herbicides of differing selectivity (from Marshall and Nowakowski, 1991).

	Untreated	Fluazifop-P-butyl	Mefluidide Low Rate	High Rate	Quinmerac	SED
SOWN SPECIES						
Daucus carota	33	87	64	36	30	11.8
Leucanthemum vulgare	7	37	5	5	13	8.6
Prunella vulgaris	13	70	27	17	37	10.8
Rumex acetosa	10	43	4	9	33	11.8
UNSOWN SPECIES						
Agrostis spp.	77	3	84	56	80	12.4
Lolium multiflorum	100	27	69	81	57	18.4
Plantago major	3	67	11	T	10	11.3
Heracleum sphondylium	17	3	24	6	7	7.1

could be added to these figures. The take-up of these wild flower margins by farmers in the U.K. is increasing, and they represent a useful multi-function field margin treatment, with high potential as biocontrol refugia.

4. Drilling single-species stands of pollen/nectar plants across or around fields. This approach has the advantage that particular plant species can be positioned as refugia where and when they are needed. In some current work in Europe, seeds of the species used are available in commercial quantities and are relatively cheap. An active recent project in the south of the U.K. concerns the drilling of the U.S. native annual plant *Phacelia tanacetifolia* around and across wheat and maize fields as a pollen source for Syrphidae. *Phacelia* has been used in this context before (see p. 380), but, given the mobility of adult Syrphidae, experiments of this type really need to be at the landscape scale, ideally with fields as replicates. *Phacelia tanacetifolia* was chosen for the U.K. work because it will grow readily in almost any soil, is reasonably resistant to frost (withstanding up to $-8°C$), and the flowering period is very long. *Phacelia tanacetifolia* drilled in N. Hampshire (southern U.K.) in mid-March started to flower at the end of May and some plants were still in flower at the end of July (Hickman & Wratten 1996). Although not a U.K. native plant, *P.*

tanacetifolia seed is inexpensive and widely obtainable, as it is one of a number being used in "Nitrogen Sensitive Areas" where leaching is a problem (D. Christian personal communication). It is also grown by beekeepers and shows potential as a trap crop for cyst nematodes (*Heterodera schachtii* Schmidt) in sugar beet (Cooke 1985). In California, mirids (*Calocoris norvegicus* (Hemiptera: Miridae) are very abundant on *P. tanacetifolia* (R.L. Bugg, personal communication), but they are not pests of arable crops in the U.K. They are an important food item for partridge chicks, however (Sotherton 1991).

The labia of hoverflies are not long enough to remove nectar from the corollae of *P. tanacetifolia* flowers, so for this group the plant is a pollen resource only. In view of the great importance of pollen for female hoverflies in particular, and the fact that honeydew exudate from aphids within the crop could provide an alternative source of energy for the flies, the absence of available nectar was not felt to militate against the use of *Phacelia*.

In the U.K., *Phacelia tanacetifolia* cv 'Lisette' was drilled at a rate of 12kg/ha along two sides of each of three maize fields. Two drills 30cm apart were sown on three dates to provide flower cover over the period when the maize crop in the adjoining field would be in flower. The control fields were chosen primarily as having crops at the same state of development as the experimental fields.

Water traps painted fluorescent yellow, highly attractive to hoverflies (Finch 1992), were located in transects out from the *P. tanacetifolia* borders into the adjacent crop. Hoverfly numbers entering the crop were monitored via these traps, giving five categories of information: date, trap position, fly species and sex together with information on pollen types in the flies' guts. (*P. tanacetifolia* pollen is easily identified as it differs from that of the native U.K. farmland flora.) This study was repeated in New Zealand giving two seasons' data in one year (Lövei et al. 1992). As well as fly counts in traps, syrphid eggs were counted in colonies of the oat aphid (*Rhopalosiphum padi* L., Homoptera: Aphididae) on the maize plants. Aphid populations were also assessed in control and *P. tanacetifolia*–bordered fields. Results were encouraging, with higher trap catches of flies in the experimental fields (Figure 8), together with a higher syrphid egg/aphid ratio and lower aphid numbers/plant. Unfortunately, the farmer sprayed the control fields with insecticide! However, the data of Hickman and Wratten (1996) confirm these results.

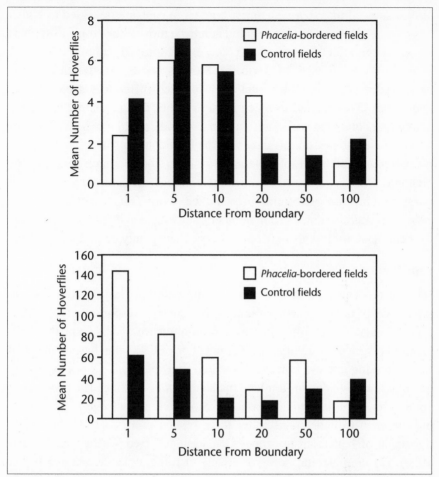

Figure 8. Mean daily number of aphidophagous syrphids caught per trap at different distances from field boundaries: a) 1–8 August 1991; b) 9–11 August 1991. Hickman unpublished.)

In the New Zealand work, vacuum suction sampling was carried out on the strips and large numbers of adult and larval Coccinellidae and lacewings (Neuroptera) were found. This is an exciting result as it suggests that this non-native plant may harbour non-pest aphids, which would be a useful early-season alternative food/host for predators and parasitoids. Current U.K./New Zealand work is evaluating coriander (*Coriandrum sativum*) and buckwheat (*Fagopyrum esculentum*) as sources of pollen and, in these cases, nectar too—the nectaries are readily available to short-tongued insects feeding on the shallow flowers in the umbels. In New Zealand, White et al.

(1995) showed that *P. tanacetifolia* borders to cabbage fields led to higher hoverfly populations and lower aphid numbers; most of the crop was sprayed so only edge effects, where the crop was not sprayed, could be detected. Work by C. M. France, M. Irvin, M. Stevens and S. Wratten at Lincoln, New Zealand, has involved buckwheat planted in replicated plots under apple trees. This has led to much higher catches of a parasitoid above the buckwheat compared with controls, and a virtual doubling of parasitism rate in the trees. The pest in this case was the lepidopteran leaf roller *Epiphyas postvittana* (Walker) (Lipidoptera:Tortricidae) and the parasitoid was the braconid *Dolichogenidea tasmanica* (Cameron). In some experiments, a parasitoid of the beneficial brown lace–wing (*Micromus tasmaniae* [Walker]) was also enhanced in numbers, showing that habitat manipulation for beneficial arthropods may not always enhance only beneficial species.

Conclusion

Most of this chapter concerns work begun within the last decade. The implied level of interest is encouraging and is surely a response to the "greening" of agriculture following over-production of cereals via the EEC Common Agricultural Policy. Government or EEC funding has not matched this increasing interest, however. Potential for expansion of work exploiting ecological knowledge in farmland management is exciting. The prospect of combining empirical studies based on ecological knowledge is promising, but we need to understand better three related processes: (1) the spatial dynamics of beneficial arthropods on farmland (see Wratten & Thomas 1990); (2) the potential negative effects of refugia (see above and Fry & Main 1992); (3) the mechanisms involved in the functioning of refugia (see Thomas et al. 1991). Without this knowledge, the true implications and potential of farmland refugia will not be understood.

References

Anonymous. 1986. Monitoring landscape change, volume 1: main report. Huntings Surveys and Consultants Ltd. Dept. of the Environ. & Countryside Commission, London.
Anonymous. 1991. Practical conservation: boundary habitats. J. Tait [ed.], p. 52. Open University. Hodder and Stoughton, East Kilbride, U.K.
Boatman, N. D. 1989. Selective weed control in field margins. Brighton Crop Protection Conference—Weeds 2:785-795.
Bonkowska, T. 1970. The effect of shelter belts on the distribution of Carabidae. Ekologia Polska 28:561-568.

Brown, V. K. & T. R. E. Southwood. 1987. Secondary succession: patterns & strategies, pp 315-337. *In* A. J. Gray, M. J. Crawley & P. J. Edwards [eds.], Colonisation, Succession & Stability. Blackwell, Oxford.

Cameron, R. A., D. K. Down & D. J. Pannett. 1980. Historical and environmental influences in hedgerow snail faunas. Biol. J. Linnean Soc. 13:75-87.

Chambers, R. J., K. D. Sunderland, P. L. Stacey & I. L. Wyatt. 1982. A survey of cereal aphids and their natural enemies in winter wheat in 1980. Ann. Appl. Biol. 101:175-178.

Chiverton, P. A. 1989. The creation of within-field overwintering sites for natural enemies of cereal aphids. Brighton Crop Protection Conference—Weeds. 3:1093-1096.

Chiverton, P. A. & N. W. Sotherton. 1991. The effects on beneficial arthropods of the exclusion of herbicides from cereal crop edges. J. Appl. Ecol. 28:1027-1039.

Cilgi, T. & P. C. Jepson. 1995. The risks posed by delta methrian drift to hedgerow butterflies. Environ. Pollut. 87:1-9.

Cooke, D. A. 1985. The effect of resistant cultivars of catch crops on the hatching of *Heterodera schachtii*. Ann. Appl. Biol. 106:111-120.

Coombes, D. S. & N. W. Sotherton. 1986. The dispersal and distribution of polyphagous predatory Coleoptera in cereals. Ann. Appl. Biol. 108:461-474.

Cowgill, S. E., S. D. Wratten & N. W. Sotherton. 1993a. The selective use of floral resources by the hoverfly *Episyrphus balteatus* (Diptera:Syrphidae) on farmland. Ann. Appl. Biol. 122:223-231

1993b. The effects of weeds on the number of hoverfly adults and the distribution and composition of their eggs in winter wheat. Ann. Appl. Biol. 123:499-515

Davies, E. T. & W. J. Dunford. 1962. Some physical and economic considerations of field enlargement. Univ. of Exeter, Dept. of Agric. Econ. Publ. No. 133.

Dennis, P. & S. D. Wratten. 1991. Field manipulation of populations of individual staphylinid species in cereals and their impact on aphid populations. Ecol. Entomol. 16:17-24.

Desender, K. 1982. Ecological and faunal studies on Coleoptera in agricultural land II. Hibernation of Carabidae in agro-ecosystems. Pedobiologia 23:295-303.

Desender, K. & M. D'Hulster. 1982. Ecological and faunal studies on Coleoptera in agricultural land. III. Seasonal abundance and hibernation of Staphylinidae in the grassy edge of a pasture. Pedobiologia 23:403-414.

D'Hulster, M. & K. Desender. 1983. Ecological and faunal studies on Coleoptera in agricultural land. IV. Hibernation of Staphylinidae in agro-ecosystems. Pedobiologia 26:65-73.

Deveaux, D. 1976. Répartition et diversité des peuplements en carabiques en zone bocagère et arassé, pp 377-384. *In* Les bocages: histoire, ecologie, economie. Univ. of Rennes, Rennes, France.

Duffield, S. J. & N. Aebischer. 1994. The effect of spatial scale of treatment with dimethoate on invertebrate population recovery in winter wheat. J. Appl. Ecol. 31:263-281.

Edwards, A. J. 1970. Field size and machinery efficiency, pp. 28-31. *In* M. D. Hooper & M. W. Holdgate [eds.], Hedges and hedgerow trees. Monks Wood Symposium No. 4. The Nature Conservancy.

Edwards, C. A., K. D. Sunderland & K. S. George. 1979. Studies on polyphagous predators of cereal aphids. J. Appl. Ecol. 16:811-823.

Evans, R. 1985. Soil erosion—the disappearing trick. *In* Better Soil Management for Cereals and Oilseed Rape. Proceedings of a Conference, November 1985, Natl. Agric. Cent., Stonleigh.

Evans, R. & S. Cooke. 1987. Soil erosion in Britain, pp 28-59. *In* C. P. Burnham & J. I. Pitman [eds.], Soil Erosion. SEESOIL 3, J. South East Soils Discussion Group.

Felki, G. 1988. First investigations on the abundance of epigeal arthropods, cereal aphids and stenophagous aphid predators in herbicide-free border strips of winter-wheat fields in Besse. Ges. Pflanzen. 40:483-490.

Finch, S. 1992. Improving the selectivity of water traps for monitoring populations of the cabbage root fly. Ann. Appl. Biol. 120:1-7.

Forman, R. T. T. & J. Baudry. 1984. Hedgerows and hedgerow networks in landscape ecology. Environ. Manage. 8:495-510.

Fougeroux, A., C. Bouchet, J. N. Reboulet & M. Tisseur. 1988. Importance of microhymenoptera for aphid population regulation in French cereal crops, pp 61-68. *In* R. Cavalloro & K. D. Sunderland [eds.], Integrated crop protection in cereals. Proc. of a meeting of the EC Experts' Group, Littlehampton, 1986. A. A. Balkema, Rotterdam.

Fry, G. L. A. & A. R. Main. 1993. Restoring seemingly natural communities on agricultural land, pp 224-241. *In* D. A. Saunders, R. J. Hobbs & P. R. Ehrlich [eds.], Nature conservation 3:reconstruction of fragmental ecosystems. Global & Regional Perspectives. Surrey Beatty & Sons, Chipping Norton, U.K.

Fry, G.L.A. & W.J. Robson. 1994. The effects of field margins on butterfly movememt, pp 111-116. *In* N. Boatman [ed.], Field margins: integrating agriculture and conservation. British Crop Protection Council Monogr. No. 58. Croyton, United Kingdom.

Gorny, M. 1970. The problem of ecological role played by shelterbelts in the light of complex pest control methods. Sywlan. 114:27-32.

Greaves, M. P. & E. J. P. Marshall. 1987. Field margins: definitions & statistics, pp. 3-11. *In* J. M. Way & P. W. Greig-Smith [eds.], Field Margins. British Crop Protection Council Monograph No. 35.

Gurr, G., S. D. Wratten & H. F. van Emden. (In press). Habitat manipulation and natural enemy efficiency: implications for the control of pests. *In* Barbosa, P. [ed.], Conservation Biocontrol. Academic Press.

Harwood, R. W. J., S. D. Wratten, & M. Nowakowski. 1992. The effect of managed field margins on hoverfly (Diptera: Syrphidae) distribution and within-field abundance, pp 1033-1037. Proc. Brighton Crop Protection Conf.—Pests & Diseases. British Crop Protection Council, Croydon, U.K.

Harwood, R. W. J., J. M. Hickman, A. Macleod, T. N. Sherratt & S. D. Wratten. 1994. Managing field margins for hoverflies, pp. 147-152. *In* N. Boatman [ed.],

Field margins: integrating agriculture and conservation. British Crop Protection Council Monogr. No. 58. Croyton, United Kingdom.

Haslett, J. R. 1989. Interpreting patterns of resource utilization: randomness and selectivity in pollen feeding by adult hoverflies. Oecologia 78:433-442.

Helliwell, D. R. 1975. The distribution of woodland plant species in Shropshire hedgerows. Biol. Conservation. 7:61-72.

Hickman, J. M. & S. D. Wratten. 1996. Use of *Phacelia taracetifolia* strips to enhance biocontrol of aphids by hover fly larvae in cereal fields. J. Econ. Entomol. 89:832-840.

Kühner, C. H. 1988. Investigations in Hesse (FRG) on effects and impact of unsprayed crop edges 2: population growth of cereal aphids and their specific antagonists, pp. 43-55. Proceedings of the symposium "Crop edges-positive effects for agriculture."

Le Clezio, P. 1976. Les ambiguités de la notion de maille optimal, pp. 551-554. *In* J. Missonnier [ed.], Les bocages: histoire, ecologie, economie. University of Rennes, Rennes, France.

Levins 1970. Extinction. *In* M. Gerstenhaber [ed.], Some mathematical questions in biology. American Mathematical Society, Providence, R.I., U.S.A.

Lövei, G. L., D. MacDoughall, G. Bramley, D. J. Hodgson & S. D. Wratten. 1992. Floral resources for natural enemies: the effect of *Phacelia tanacetifolia* (Hydro-phyllaceae) on within-field distribution of hoverflies (Diptera: Syrphidae), pp. 60-61. Proc. 45th New Zealand Plant Protection Conference.

Luff, M. L. 1966. The abundance and diversity of the beetle fauna of grass tussocks. Ecology. 35:189-208.

Marshall, E. J. P. & M. Nowakowski. 1991. The use of herbicides in the creation of an herb-rich field margin, pp. 655-660. Proc. Brighton Crop Protection Conference—Weeds. 655-660.

Mauremootoo, J. R., S. D. Wratten, S. P. Worner & G. L. A. Fry. 1995. Permeability of hedgerows to predatory carabid beetles. Agric. Ecosyst. Environ. 52:141-148.

May, R. M. 1973. Stability and complexity in model ecosystems. Princeton University Press.

Mols, P. J. M. 1979. Motivation and walking behaviour of the carabid beetle *Pterostichus coerulescens* L. at different densities and distributions of the prey, pp. 185-198. *In* P. J. den Boer, H. U. Thiele & F. Weber [eds.], On the evolution of behaviour in carabid beetles. Miscellaneous Papers Landbouwhogeschool, Wageningen, Veenman, No. 18.

Molthan J. & V. Ruppert. 1988. Sur Bedentung blühender Wildkräuter in Feldrainen und Ackern für Blüten besuchende Nutzinsekten. Mitteilungen aus der Biologischen Bundesanstalt für Land und Forstwirtschaft 247:85-99.

Nentwig, W. 1988. Augmentation of beneficial arthropods by strip management I. Succession of predacious arthropods and long-term change in the ratio of phytophagous and predacious arthropods in a meadow. Oecologia 76:597-606.

Opdam, P. 1990. Dispersal in fragmented populations: the key to survival, pp. 3-17. *In* R. G. H. Bunce & D. C. Howard [eds.], Species dispersal in agricultural habitats. Belhaven Press, London.

Perrin, R. M. 1977. Pest management in multiple cropping systems. Agro-Ecosystems 3:93-118.

Pihan, J. 1976. Bocage et erosion hydrique des soils en Bretagne, pp. 185-192. *In* Les bocages: histoire, ecologie, economie. University of Rennes, Rennes, France.

Pollard, E. 1971. Hedges. VI. Habitat diversity and crop pests. A study of *Brevicoryne brassicae* and its syrphid predators. J. Appl. Ecol. 8:751-780.

Pollard, E., M. D. Hooper & N. W. Moore. 1974. Hedges. Collins & Sons, London.

Potts, G. R. 1991. The environmental and ecological importance of cereal fields, pp. 3-21. *In* L. G. Firbank, N. Carter, J. F. Darbyshire & G. R. Potts [eds.], The ecology of temperate cereal fields. Blackwell, Oxford, U.K.

Sengonça, C. & B. Frings. 1988. Einfluss von *Phacelia tanacetifolia* auf Schädlings und Nutzlingspopulationen in Zuckerrübenfeldern. Pedobiologia 32:311-316.

Sotherton, N. W. 1984. The distribution and abundance of predatory arthropods overwintering on farmland. Ann. Appl. Biol. 105:423-429.

1985. The distribution and abundance of predatory Coleoptera overwintering in field boundaries. Ann. Appl. Biol. 106:17-21.

1991. Conservation headlands: a practical combination of intensive cereal farming and conservation, pp. 373-397. *In* L.G. Firbank, N. Carter, J. F. Darbyshire & G. R. Potts [eds.], The ecology of temperate cereal fields. Blackwell, Oxford.

Sotherton, N. W., S. J. Moreby & M. G. Langley. 1987. The effects of the foliar fungicide pyrazophos on beneficial arthropods in barley fields. Ann. Appl. Biol. 111:75-87.

Sunderland, K. D. & G. P. Vickerman. 1980. Aphid feeding by some polyphagous predators in relation to aphid density in cereal fields. J. Appl. Ecol. 17:389-396.

Thomas, M. B. 1991. Manipulation of overwintering habitats for invertebrate predators on farmland. Ph.D. thesis, University of Southampton, U.K.

Thomas, M. B., S. D. Wratten & N. W. Sotherton. 1991. Creation of "island" habitats in farmland to manipulate populations of beneficial arthropods: predator densities and emigration. J. Appl. Ecol. 28:906-917.

van Emden, H. F. 1965. The effect of uncultivated land on the distribution of the cabbage aphid *(Brevicoryne brassicae)* on adjacent crop. J. Appl. Ecol. 2:171-196.

1988. The potential for managing indigenous natural enemies of aphids on field crops. Philos. Trans. R. Soc. London 318:183-201.

van Emden, H. F. & S. D. Wratten. 1991. Tri-trophic interactions between host plant, aphids and predators, pp 29-43. *In* J. A. Webster & D. C. Peters [eds.], Aphid-plant interactions: populations to molecules. Oklahoma University Press, Stillwater.

von Klinger, K. 1987. Effects of margin-strips along a winter wheat field on predatory arthropods and the infestation by cereal aphids. J. Appl. Entomol. 104:47-58.

Varley, G. C. & G. R. Gradwell. 1960. Key factors in population studies. J. Animal Ecol. 29:399-401.

Vorley, W. T. & S. D. Wratten. 1985. A simulation model of the role of parasitoids in the population development of *Sitobion avenae* (Hemiptera: Aphididae) on cereals. J. Appl. Ecol. 22:813-823.

Vorley, W. T. & S. D. Wratten. 1987. Migration of parasitoids (Hymenoptera: Braconidae) of cereal aphids (Hemiptera: Aphididae) between grassland, early-sown cereals and late-sown cereals in southern England. Bull. Entomol. Res. 77:555-568.

Wallin, H. 1985. Spatial and temporal distribution of some abundant carabid beetles (Coleoptera: Carabidae) in cereal fields and adjacent habitats. Pedobiologia. 28:19-34.

Wallin, H. 1986. Habitat choice of some field inhabitating carabid beetles (Coleoptera: Carabidae) studied by recapture of marked individuals. Ecol. Entomol. 11:457-466.

Way, M. J. 1977. Pest and disease status in mixed stands vs. mono-cultures: the relevance of ecosystem stability, pp 127-138. *In* J. M. Cherrett & G. R. Sagar [eds.], Origins of pest, parasite, disease and weed problems. Oxford, Blackwell.

White, A. J., S. D. Wratten, N. A. Berry & U. Weigmann. 1995. Habitat manipulation to enhance biological control of *Brassica* pests by hoverflies (Diptera: Syrphidae). J. Econ. Entomol. 88:1171-1176.

Williamson, K. 1969. Habitat preferences of the wren on English farmland. Bird Study. 17:30-96.

Wratten, S. D. 1988. The role of field boundaries as reservoirs of beneficial insects, pp. 144-150. *In* Environmental management in agriculture: European perspectives. EEC/Pinter Publishers Ltd, London.

Wratten, S. D., K. Bryan, D. Coombes & P, Sopp. 1984. Evaluation of polyphagous predators of aphids in arable crops, pp 261-270. Proc. 1984 British Crop Protection Conf.: Pests & Diseases. Lavenham Press, Suffolk.

Wratten, S. D., M. Mead-Briggs, G. P. Vickerman & P. C. Jepson. 1988. Effects of the fungicide pyrazophos on predatory insects in winter barley, pp. 327-334. British Crop Protection Council Monograph No. 40. Environ. Effects of Pesticides. BCPC, Croydon, U.K.

Wratten, S. D. & C. F. G. Thomas. 1990. Farm-scale spatial dynamics of predators and parasitoids in agricultural landscapes, pp. 219-237. *In* R. G. H. Bunce & D. C. Howard [eds.], Species dispersal in agricultural habitats. Bellhaven Press, London.

Wratten, S. D., A. D. Watt, N. Carter & J. C. Entwistle. 1991. Economic consequences of pesticide use for grain aphid control in winter wheat in 1984 in England. Crop Protection 9:73-77.

Wratten, S. D. & W. Powell. 1991. Cereal aphids and their natural enemies, pp. 233-257. *In* L. G. Firbank, N. Carter, J. F. Darbyshire & G. R. Potts [eds.], The ecology of temperate cereal fields. Blackwell, Oxford.

Index

Barylypa fulvescens 365
Barylypa prolata 365
bastard clover 52
bean 91, 95, 220
bean aphid 55, 362, 363
bean-maize 105
bean-tomato 105
Beauveria brongniartii 190
Bechsteinia terminalis 244
beetle 185, 225
beets 180, 194, 358
bell bean 76, 78
Bembidion 130, 142, 145, 147, 148
Bembidion gilvipes 146
Bembidion guttula 145, 146
Bembidion lampros 130, 145, 146, 378
Bembidion properans 145, 146
Bembidion quadrimaculatum 145, 146
Berberis 62
Bessa harveyi 91
Bethylidae 241
Bethyloidea 99
big-leaf yarrow 76
bigeyed bug 30, 242, 248, 273, 279
Biolysia tristis 368
bird cherry–oat aphid 144, 183, 363
bird rape 181
bird's eyes 76
bishop's flower 76
biting midge 30
black-veined white butterfly 30
blackberry 196, 298
blackberry leafhopper 298
blackbird, Brewer's 342
blackbird, red-winged 342
blue elderberry 361, 369
blue gum 340
blue knapweed 52, 53, 54, 55, 62
blue wildrye 243, 348, 352
bluebird, mountain 342
bluebird, western 342

boll weevil 127, 247, 358
bollworm 127
borage 52, 53, 57
Boraginaceae 58
Borago officinalis 52, 53, 56, 57, 61, 63
Brachyantherum 350
Brachycaudus helichrysi 362, 363
Brachycaudus lychnidis 54, 55
Braconidae 53, 102, 367, 368
braconids 95, 106
Brassica 42, 55, 194, 361
Brassica napus 50, 53, 54, 55, 61, 63
Brassica napus var. *oleifera* 181
Brassica oleracea 239
Brassica oleracea var. *capitata* 220
Brassica oleracea var. *gemmifera* 220
Brassica rapa 53, 56, 63, 181
Brassicaceae 53, 54, 55, 58, 122, 180
Brevicoryne 54
Brevicoryne brassicae 54, 104
broad bean 52, 260
broad bean weevil 257
broccoli 223
brome grass 194
Bromus arvensis 194
Bromus carinatus 348
Bromus sterilis 186
brown-headed cowbird 342
brussels sprouts 27, 220, 222, 239
buckwheat 50, 76, 78, 221, 397
burclover 362
burrowing owl 342
bush-tit 342
butterfly 386, 387, 388, 394

C

C7 267
cabbage 220, 222
cabbage aphid 55
cabbage looper 131
cabbage root fly 30, 42, 97, 122, 129

G

Gaeumannomyces graminis 179

Galeopsis tetrahit 53, 55, 56, 57, 61, 62, 63

Galinsoga ciliata 50, 55, 56, 63

Galinsoga parviflora 186

Galium 194

Galium aparine 57, 383, 392, 394

Galium mollugo 60, 61

gall midges 196

gallant soldier 50, 55

Gamasidae 179, 182

garden lupine 57

Gelechiidae 256, 360

Gemmifera 240

Geocoris 9, 27, 75, 79, 131, 132, 140, 248, 280, 283, 284, 285

Geocoris atricolor 248, 367

Geocoris bullatus 242

Geocoris pallens 9, 131, 248, 367, 368

Geocoris punctipes 10, 131, 248, 279, 284, 367, 368

Geranium pyrenaicum 61

Geocoris uliginosus 131

Geum urbanum 61

Gilia 78

Gilia, tricolor 76

Globodera 180

Glycyrrhiza lepidota 350

Glypta 365

Glypta simplicipes 288

Gnaphosidae 226, 244

goldfinch, lesser 342

Goniozus legneri 241

Goodling's willow 369

Gossypium hirsutum 10, 248, 259

gout fly 183

grain aphid 389, 390

grain moth 256

grape 181, 240, 297, 300, 304, 312, 333, 360

grape, wild 360

grape berry moth 183

grape leafhopper 192, 240, 297, 299, 300, 302, 303, 305, 306, 307, 360

grape moth 190, 191

grapevine 318, 325

grass 183, 185, 194, 196, 228, 232, 315, 352, 359, 380, 381, 383, 384, 388, 390, 391

grasshopper 265, 352

great hedge bedstraw 60

green muscardine fungus 190

green peach aphid 74, 242, 262, 362, 363

green rice leafhopper 316

Grevillea robusta 358

grey partridge 392

Grindelia camporum 350

grosbeak, black-headed 342

ground beetle 28, 33, 127, 130, 179, 186, 230, 273, 359, 389

Gryllus pennsylvanicus 215

Gymnosoma fuligonsum 368

H

Hahniidae 226

hairy vetch 9

Halictidae 367

halticine chrysomelids 54

Haplodiplosis equestris 180

Harpalus 130

Harpalus rufipes 378

harrier, northern 342

hawk, red-tailed 342

hawk, rough-legged 342

hawk, Swainson's 342

hawthorn 243, 378

hazelnut 340

Heleocharis 350

Helianthus annus 188

Helicoverpa 140, 142

Helicoverpa zea 92, 104, 132, 312

Heliothis 6, 99

N

O

Compositors:	The Publications Department
Text:	11/14 Times
Display:	Times
Printer and Binder:	Braun-Brumfield, Inc.